61 Springer Series in Solid-State Sciences

Edited by Peter Fulde

Springer Series in Solid-State Sciences

Editors: M. Cardona P. Fulde H.-J. Queisser

40 **Semiconductor Physics**
An Introduction By K. Seeger

41 **The LMTO Method**
Muffin-Tin Orbitals and Electronic Structure
By H. L. Skriver

42 **Crystal Optics with Spatial Dispersion, and Excitons**
By V. M. Agranovich and V. L. Ginzburg

43 **Resonant Nonlinear Interactions of Light with Matter**
By V. S. Butylkin, A. E. Kaplan, Yu. G. Khronopulo, and E. I. Yakubovich

44 **Elastic Media with Microstructure II**
Three-Dimensional Models
By I. A. Kunin

45 **Electronic Properties of Doped Semiconductors**
By B. I. Shklovskii and A. L. Efros

46 **Topological Disorder in Condensed Matter**
Editors: F. Yonezawa and T. Ninomiya

47 **Statics and Dynamics of Nonlinear Systems**
Editors: G. Benedek, H. Bilz, and R. Zeyher

48 **Magnetic Phase Transitions**
Editors: M. Ausloos and R. J. Elliott

49 **Organic Molecular Aggregates,** Electronic Excitation and Interaction Processes
Editors: P. Reineker, H. Haken, and H. C. Wolf

50 **Multiple Diffraction of X-Rays in Crystals**
By Shih-Lin Chang

51 **Phonon Scattering in Condensed Matter**
Editors: W. Eisenmenger, K. Laßmann, and S. Döttinger

52 **Superconductivity in Magnetic and Exotic Materials**
Editors: T. Matsubara and A. Kotani

53 **Two-Dimensional Systems, Heterostructures, and Superlattices**
Editors: G. Bauer, F. Kuchar, and H. Heinrich

54 **Magnetic Excitations and Fluctuations**
Editors: S. Lovesey, U. Balucani, F. Borsa, and V. Tognetti

55 **The Theory of Magnetism II**
Thermodynamics and Statistical Mechanics
By D. C. Mattis

56 **Spin Fluctuations in Itinerant Electron Magnetism** By T. Moriya

57 **Polycrystalline Semiconductors,**
Physical Properties and Applications
Editor: G. Harbeke

58 **The Recursion Method and Its Applications**
Editors: D. Pettifor and D. Weaire

59 **Dynamical Processes and Ordering on Solid Surfaces**
Editors: A. Yoshimori and M. Tsukada

61 **Localization, Interaction, and Transport Phenomena**
Editors: B. Kramer, G. Bergmann, and Y. Bruynseraede

Volume 1-39 are listed on the back inside cover

Localization, Interaction, and Transport Phenomena

Proceedings of the International Conference, August 23–28, 1984
Braunschweig, Fed. Rep. of Germany

Editors: B. Kramer, G. Bergmann, and Y. Bruynseraede

With 125 Figures

Springer-Verlag
Berlin Heidelberg New York Tokyo

Professor Dr. Bernhard Kramer
Physikalisch-Technische Bundesanstalt Braunschweig, Postfach 3345
D-3300 Braunschweig, Fed. Rep. of Germany

Professor Dr. Gerd Bergmann
Institut für Festkörperforschung, Kernforschungsanlage Jülich, Postfach 1913
D-5170 Jülich, Fed. Rep. of Germany

Professor Dr. Yvan Bruynseraede
Laboratorium voor Vaste Stof-Fysika en Magnetisme, Celestijnenlaan 200 D
B-3030 Leuven, Belgium

Organizing Committee
B. Kramer *(Chairman)* (Physikalisch-Technische Bundesanstalt Braunschweig, F.R.G.)
G. Bergmann (Kernforschungsanlage Jülich, F.R.G.)
Y. Bruynseraede (Universiteit Leuven, Belgium)
L. Van Gerven (Universiteit Leuven, Belgium)
W. Gey (Technische Universität Braunschweig, F.R.G.)

Advisory Committee

R. Huguenin, Switzerland	L. P. Gor'kov, USSR	W. Weller, G.D.R.
M. Pepper, U.K.	A. A. Abrikosov, USSR	M. Ya. Azbel', Israel
P. Chaudhari, USA	A. G. Aronov, USSR	G. Deutscher, Israel
P. W. Anderson, USA	A. A. Rogachev, USSR	S. Kobayashi, Japan
E. Abrahams, USA	R. Taylor, Canada	S. Hikami, Japan
A. W. Overhauser, USA	W. Götze, F.R.G.	P. Wyder, The Netherlands

Secretary Elisabeth Meyer (Physikalisch-Technische Bundesanstalt Braunschweig, F.R.G.)

Sponsors
The conference was organized under the auspices of the Deutsche Physikalische Gesellschaft

Financial support was obtained from:
Physikalisch-Technische Bundesanstalt Braunschweig
Stadt Braunschweig
Volksbank Braunschweig
Firma Richard Borek Braunschweig
IBM Deutschland

ISBN-13:978-3-642-82518-7 e-ISBN-13:978-3-642-82516-3
DOI: 10.1007/978-3-642-82516-3

Softcover reprint of the hardcover 1st edition 1985

2153/3130-543210

Preface

LITPIM

When we first had the idea of organizing the International Conference on Localization, Interaction, and Transport Phenomena in Impure Metals we expected to bring together at most a hundred physicists. The fact that more than a hundred and fifty participated clearly shows that the topic of the meeting was of great interest to an important fraction of the solid state physics community. In fact, remembering that the localization problem is already a quarter of a century old, it is quite amazing to see how, during the last five years, new and very successful theoretical models emerged which were confirmed by sometimes ingenious experiments. The number of groups involved in the study of localization or related problems in the transport properties of matter even seems to be increasing.

The main purpose of this conference was to review the present status of activities in the localization field and hopefully to stimulate new ideas. A study of the Conference Proceedings ascertains that we were successful in reaching these two goals. Moreover, the presence of the authors of the about ninety contributed papers published in the supplement volume assured the very lively atmosphere which characterizes successful conferences. We think that this was the most important ingredient for achieving the second goal in particular.

We thank our sponsors for their support, which was given unreluctantly and generously. Especially, we gratefully acknowledge the hospitality of the PTB and the city of Braunschweig during the time of the meeting. Without this help the organization of the conference would hardly have been possible in the form which we all have experienced. We also thank the numerous people who have contributed by helping during the preparation of the conference and in the daily affairs, especially Wolfgang Wöger, Ludwig Schweitzer, and Robert Johnston. In particular, we would also like to thank Mr. Wolfgang Pogrzeba, who was responsible for the technical arrangements, and our conference secretary, Mrs. Elisabeth Meyer. She was extremely efficient in organizing all the details of LITPIM, preparing the Conference Program, the Abstract Booklet, and the Proceedings. It would have been very hard, if not impossible, to run the meeting without her continuous and extremely reliable work.

Braunschweig,
November1984

B. Kramer
G. Bergmann
Y. Bruynseraede

Contents

Part II **Titles of Contributed Papers**

Localization, Interaction, and Transport Phenomena in Impure Metals. An Introduction

B. Kramer
Physikalisch-Technische Bundesanstalt, Bundesallee 100
D-3300 Braunschweig, Fed. Rep. of Germany

G. Bergmann
Institut für Festkörperforschung der KFA Jülich, Postfach 1913
D-5170 Jülich, Fed. Rep. of Germany

Y. Bruynseraede
Laboratorium voor Vaste Stof-Fysika en Magnetisme, Universiteit Leuven
Celestijnenlaan 200D, B-3030 Leuven, Belgium

1. Overview

Transport properties of solids are determined by disorder due to the presence of impurities or imperfections of the lattice, by various interaction effects such as electron-electron, electron-phonon interaction and by spin-dependent processes due to spin-orbit and spin-spin coupling. At room temperature the details of these interactions are unimportant. They are usually incorporated into one of several mean free paths. As the temperature is lowered these effects become however more and more important. Thus the study of the transport properties of solids is a typical field of low-temperature physics.

A number of "classical" low-temperature effects in metals are a signature of the presence of well defined microscopic processes [1]. The *residual resistance* is due to impurity scattering, the *Kondo effect* to spin scattering, and the "classical" temperature-dependence of the resistance, T^5, is within Bloch's theory described as an electron-phonon effect. There are "classical" rules, how to combine various scattering processes, such as Matthiesen's rule. Many of these effects and "laws" can be derived neglecting quantum mechanical interference effects within the frame work of the relaxation-time approximation. One of the main subjects of the research during the past thirty years has been to include quantum mechanics into a microscopic transport theory. This goal is not yet reached. However, one very important step has become more and more transparent during the past five years: The problem of localization due to disorder has been formulated in a way allowing systematic theoretical, and what is perhaps more important, experimental studies. In addition, one may expect fruitful interaction with the modern technology of very large-scale integration (VLSI).

It is now little more than 25 years ago that P.W. Anderson published "The Absence of Diffusion in Certain Random Lattices" [2]. He simultaneously formulated the problem, made the link between localization and transport, and gave the first quantitative estimate for the critical disorder for the transition between the diffusive and the non-diffusive regimes. Localization may be viewed as one of the most fundamental quantum mechanical phenomena related to real condensed matter. The most simple model to discuss is that of a spinless, noninteracting particle moving in a random potential. As it forms the basis for the far more complicated theory including spin and interactions, we want to discuss its qualitative aspects in some detail in the following sections. More details will be presented by the various authors in this volume.

2. Definition of Localization

In this section we consider a system at zero temperature. For a classical particle in a random potential it is comparatively simple to decide whether its motion is restricted to finite portions of the space (Fig. 1).

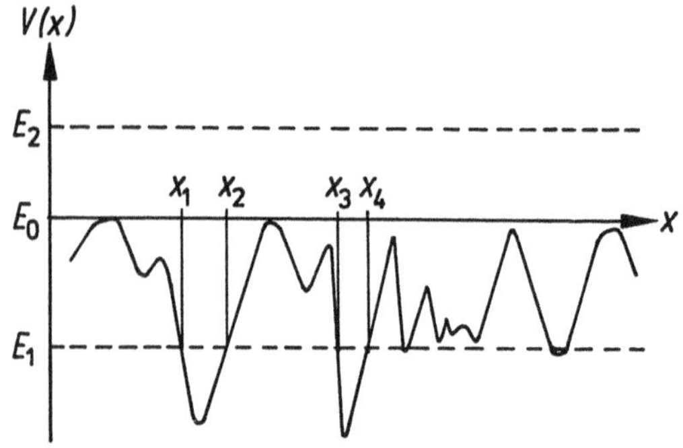

Fig. 1: Random potential

If the energy of a particle is higher than E_0 it can move through the whole space. On the other hand, if the energy is smaller than E_0, say E_1, the motion is restricted to finite intervals, (x_1, x_2), (x_3, x_4), etc. In this case, one would assign to E_0 a *mobility edge*. For the corresponding quantum mechanical problem the situation is more complicated for two reasons: Firstly, there is the tunnel effect, which allows the particle to tunnel through the potential wells, thus delocalizing the particle. Secondly, there might be interference effects within the wave function scattered by the random potential. These may lead to *localization* of the particle via destructive superposition for energies above E_0. Thus, quantum mechanically, the conditions for localization are not as simple as in classical mechanics.

An example for the first effect, delocalization due to tunneling, are the Bloch states in perfect crystals, which can be found in every text book of solid state physics. An example for localization via interference is the quantum mechanical behaviour of a particle in a one-dimensional random potential. In this case only localized states occur, independent of the magnitude of the disorder and/or the energy of the particle [3].

We have deliberately used the words "localization", "delocalization", and "disorder" without further details. Let us now specify what we mean by localization. One of the interesting aspects of the problem of localization is the fact that it depends strongly on the dimensionality of the system. In a hypercube of volume L^d all states are d-dimensional plane waves, either standing or propagating, depending upon the boundary conditions. These states are extended since the probability to find an electron at site x in a volume d^dx is constant and equal to d^dx/L^d. A trivial case of exponential localization is obtained if a charged nucleus is placed somewhere within the hypercube. A bound electron will be restricted to a volume around the nucleus. The corresponding probability density is exponentially decreasing far away from the nucleus. If we introduce however a *random potential* v(x) into the hypercube the situation is more complicated.

First of all we have to specify the potential. This is usually done by replacing the system by a statistical ensemble, which is characterized by a probability distribution. In the case of a Gaussian white noise potential, v(x), one has

$$\bar{v} = 0 \tag{1}$$

$$\overline{v(x)v(x')} = v^2 \, \delta(x-x') \tag{2}$$

$\overline{....}$ means the ensemble average. The degree of the disorder is given by v^2. In general it is more complicated to specify the disorder although v^2 is often quite useful.

The states in such a system will vary more or less randomly in space (Fig. 2).

Fig. 2: Random electron state

In addition to the random phase the amplitude may be decreasing or increasing as a function of x. A state may be defined as exponentially localized if for $|x-x'| \to \infty$

$$\overline{|\psi(x)|^2 \; |\psi(x')|^2} \approx \exp(-|x-x'|/\xi) \tag{3}$$

ξ is the localization length (Fig. 3).

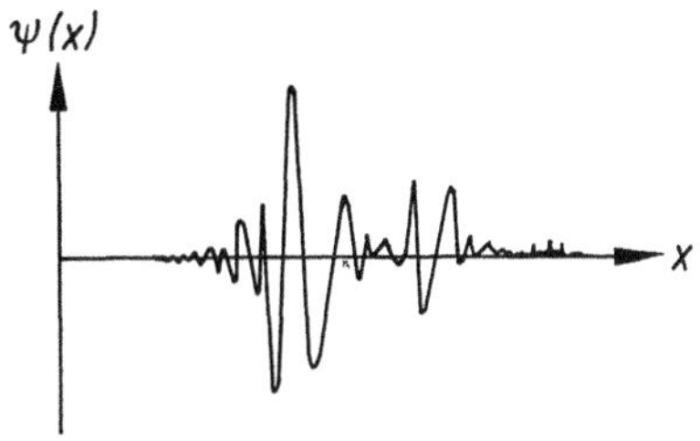

Fig. 3: Localized state

As a consequence of the disorder, the electron will have a finite *mean free path* l, which determines in a Drude-like theory the conductivity or resistivity of the sample at high temperature. l can be defined in terms of the wave function ψ.

$$\overline{\psi(x)\psi(x')} \approx \exp(-|x-x'|/2l) \tag{4}$$

This mean free path of the electron describes an *averaged* phase correlation of the electronic wave function $\psi(x)$. It is obvious that always $l \le \xi$.

However, it is important to have in mind that for a given wave function one has a well defined local phase relation between two positions x and x' as long as their distance is not (much) larger than the localization length.

Only when integrated over x the correlation decays exponentially with the free electronic path. Of course, in a conduction process one hardly probes the single wave function $\psi(x)$ but integrates over a small but finite energy range. For example an ac-measurement with the frequency ω averages over the energy ω. *But the phase correlation of the wave function ψ changes only slightly with energy.* Therefore, the local phase coherence is of great importance for the transport properties, as will be discussed in section 5.

Summarizing, at T = 0 we may identify two characteristic length in a d-dimensional disorder system: i) the localization length ξ, and ii) the mean free path l. In the following sections we shall discuss the interplay between these characteristic lengths and their effect on the electronic transport. However, we will see that finite temperature, magnetic field, and frequency, for instance, introduce an additional class of length scales.

3. Localization and Transport

As long as the disorder is very small, the localization length will be very large if not infinite. Classical transport theory (Drude theory) assumes that in the regime where $l \gg 2\pi/k_F$ the particle moves quasiclassically between the collisions. However, as will be discussed below, there are corrections to the Drude-like behaviour due to *quantum mechanical interference*. In particular, in one- and two-dimensions quantum interference is not merely a perturbation but changes the character of the eigenstates qualitatively. *The wave functions become exponentially localized.* For finite system size (or, equivalently, finite frequency or temperature) the conductance can be calculated perturbationally. In three dimensions, the quasiclassical theory gives essentially the correct result in this regime. When the scattering events become so frequent that the mean free path is of the order of $2\pi/k_F$, transport has to be described quantum mechanically.

For an infinite system, the conductivity is given by the Kubo formula

$$\sigma = \lim_{L\to\infty} 2\pi e^2/(\hbar m^2 L^d) \sum_{\psi\psi'} \overline{|\langle\psi|p|\psi'\rangle|^2}$$

$$\overline{\delta(E_\psi - E_{\psi'} - \hbar\omega)\,(f(E_\psi) - f(E_{\psi'}))/\hbar\omega} \quad . \tag{5}$$

ψ, ψ' are the eigenstates of the Hamiltonian of the system corresponding to the eigenvalues E_ψ, $E_{\psi'}$. f(E) is the Fermi distribution, and L the length of the system.

A little manipulation shows that this may be connected with the definition of the localization length in (3). In particular, if there are only localized states near the Fermi level it turns out that

$$\lim_{T\to 0} \lim_{\omega\to 0} \sigma(\omega,T) = 0 \tag{6}$$

(except, perhaps, in certain pathological cases).

Let us now discuss a little more in detail the relation to transport. In classical transport theory current density and electrical field are related linearly by Ohm's law

$$\vec{j} = \sigma\cdot\vec{E} \quad . \tag{7}$$

The conductivity is here a material constant, which does not contain any boundary effects. Einstein's relation gives the connection with diffusion.

$$\sigma = De^2 \, n(E_F) \tag{8}$$

where D is the diffusion constant, and $n(E_F)$ the density of states at the Fermi level. In the localized energy regime the conductivity vanishes. Therefore, σ is no longer useful for the description of the relation between current and voltage for a *finite* sample. One has to use a length-dependent conductivity $\sigma(L)$. For a hypercube of length L we obtain for the conductance G

$$G = \sigma(L) \, L^{(d-2)} \; . \tag{9}$$

4. Size Scaling

In the classical limit $\sigma(L) = \sigma = \text{const.}$ Eq. (9) can then be considered as the scaling law for the classical conductance. When quantum corrections become important for the transport (weak localization, quantum interference), the scaling properties of G are no longer given by this simple relation. In the case of exponential localization [4] we expect that for $\xi \ll L$ the conductance decreases exponentially with increasing size of the system.

$$G = G_0 \exp(-L/\xi) \; . \tag{10}$$

These two scaling laws (9) and (10) have been connected using a very simple arguement by ABRAHAMS et al [5]. They define

$$\beta = d\ln G/d\ln L \; . \tag{11}$$

From (9) and (10) one readily obtains asymptotically

$$\beta = \begin{cases} d-2 & \text{for } L/\xi \to 0 \\ \ln G & \text{for } L/\xi \to \infty \; . \end{cases} \tag{12}$$

One may calculate corrections to this asymptotic behaviour. Of particular importance is the regime of large conductance. Here one can show that taking into account the "maximally crossed diagrams" yields a negative correction to (12). If now $\beta(v,L) = \beta(G)$, and is monotonically increasing with G then $\beta(G)$ should behave qualitatively as shown in Fig. 4.

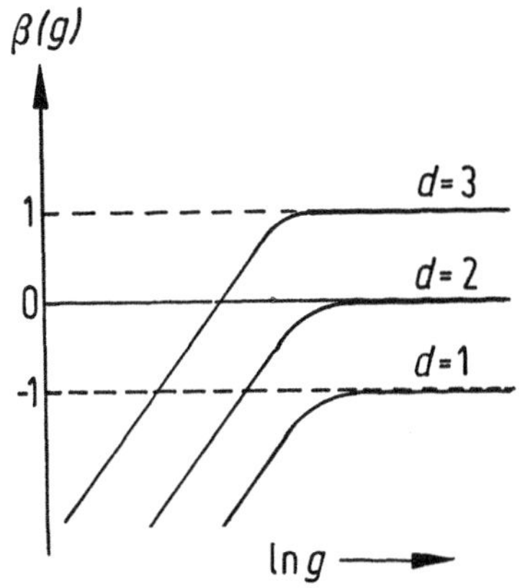

Fig. 4: Scaling of conductance

The β-function describes the geometrical scaling of the conductance with the size of the system. Consider the curve for d = 1. For a given disorder and length of the system we have some value for the conductance G and a corresponding value for β. Increasing the length of the system will decrease the conductance,since β is negative. As a consequence, β becomes more negative. We move on the curve for d = 1 in Fig. 4 towards the regime were $\beta = \ln G$, i. e. towards the localized state. The same happens for d = 2. Only for d = 3 there is a different behaviour possible. If we start with a conductance corresponding to $\beta > 0$ the system scales to the classical regime, i. e. the extended state. On the other hand, for $\beta < 0$ the system scales as in one and two dimensions to the localized state. The behaviour near the fix-point $G_c = 0$ describes the transition from extended to localized states. An Anderson transition would then only be possible for d = 3. In d = 1 and 2 the system is always localized in the thermodynamic limit.

An important question connected with the latter statement is about ergodicity. As localization results in nonergodic behaviour of the system one would expect strong fluctuations in the conductance which cannot be described within the central limit theorem. Whether the resistance fluctuations observed experimentally by various groups in quasi one-dimensional systems (see papers by Azbel, Prober, Landauer and Webb) are of this nature,or are due to the discreteness of the spectrum in a *small* system,remains to be seen.

It is clear that these conclusions were rather spectacular and needed further support from theory as well as from experiment. The theoretical situation became clear very soon. WEGNER [6] had already reformulated Thouless' scaling arguements into a renormalization group for the localization problem, and had paved the way for a complete field theory of localization. Numerical evidence was provided only two years later by using recursive techniques [7]. A review of the numerical work is given by A. MacKinnon in this volume.

How the scaling theory has to be modified in cases where one-parameter scaling is not valid is presently not clear. This concerns in particular the problem of a disordered system placed into a magnetic field and leads directly to consideration of the recent experiments on the transport properties of inversion layers in high magnetic fields which show the quantum Hall effect. The quantum Hall effect might be thought of as an ideal tool for studying localization and interaction effects in two-dimensional disordered systems in a magnetic field, although the theory is certainly not yet complete. The paper by Ebert later in this volume will concentrate on this.

In order to compare the results of the localization theory with experiment we need to include one additional ingredient, namely the influence of a finite temperature. This will be done explicitly for the regime of "weak localization" in the next section.

5. The Quantum Interference Regime

The regime of quasi-classical propagation of the electrons (limit of large G in Fig. 4) with quantum interference corrections (often called weak localization) allows a good insight into the interplay between disorder and interference. The basic idea of the interference can be already extracted when one considers the classical diffusion of an electron in a system with random scattering centers. Depending on the dimensionality of the system,the probability to find the electron at a given distance from the starting point after the time t is given by

$$p(r,t) = (4\pi Dt)^{-d/2} \exp(-r^2/(4Dt)) \ . \tag{13}$$

There is a finite probability $p(r=0,t)$ which is proportional to $t^{-d/2}$ that the electron returns to the origin. The corresponding diffusion path is a closed loop. If one considers now the electron as a quantum mechanical probability density wave, then the two partial waves propagating on the closed loop in opposite direction interfere (at the origin at the time t). This gives twice the probability to return to the origin as compared to the classical diffusion, i.e. one obtains a quantum peak in the diffusion profile which can be considered as an echo. The tendency of the electron to return to its starting point is called weak localization because it is thought of as a precursor of localization.

This quantum peak reduces the effective diffusion of the electron and causes a reduction of the conductance. In other words that part of the electron that returns to the starting point does not contribute to the conductance and the correction is proportional to the time integral of the echo which yields

$$\Delta\sigma \ \alpha \int_{\tau_0}^{\tau_\phi} \frac{1}{t^{d/2}} \, dt \tag{14}$$

In three dimensions the upper limit of the integral does not yield a divergency for τ_ϕ approaching infinity. However, in two and one dimensions one obtains a divergency of the integral, and even this interference description which corresponds to a perturbation approach shows the break down of the quasi-classical treatment and the occurrence of the localization phenomenon. Here the interference picture can only be used when the upper time limit of the integral has a finite cut-off so that the quantum correction to the conductance is small in comparison to the quasi-classical conductance.

At finite temperature the phase coherence between the two partial waves (on the closed loop) is only maintained for a finite time τ_ϕ, the phase coherence time, which is often replaced by the inelastic lifetime of the electron. There are several other mechanisms which yield a finite upper time limit: i) an ac-field with the frequency $\omega(\tau_\phi \approx 1/\omega)$, ii) a finite magnetic field $(\tau_\phi \approx \hbar/(4eDH))$, iii) magnetic impurities $(\tau_\phi \approx \tau_s)$.

As a consequence of the interference, one obtains for the resistance of two-dimensional disordered systems anomalies as a function of temperature and magnetic field.

6. Experiments on Quantum Interference and the Role of Interactions

Quantum interference exists in one, two, and three dimensions as well [8]. For an experimental investigation the 2D case is most favorable. Here the corrections to the resistance are of the order of 10^{-2} to 10^{-3}. They can easily be measured with an accuracy of 1%. In addition, one can apply a magnetic field perpendicular to the plane of the electron system, which is very useful, as will be discussed below, in distinguishing localization corrections from the corrections due to interaction effects.

One example of quasi two-dimensional systems are thin metallic films. The condition for a film for being two-dimensional is that the film thickness is smaller than the inelastic diffusion length. Since one can prepare thin films of every metal and most alloys, one can study many materials with quite different properties, such as simple metals, transition metals, superconductors nearly magnetic metals etc. The paper by S. Kobayashi will concentrate on these systems.

Electron inversion layers in MOSFETS and heterostructures are examples for nearly ideal two-dimensional systems, and their transport properties have been studied thoroughly, as will be reported by D.J. Bishop.

It was very stimulating for the developement of the field of "weak localization" that logarithmic corrections to the classical resistance have been measured in alloys [9] and pure metals [10]. Although it was not in every case proved that they were due to quantum interference at defects, the discovery as such had great impact on both experiment and theory. The difficulty was that in 2D logarithmic corrections to the resistance could also be caused by screened Coulomb interaction between the electrons, as was calculated by ALTSHULER et al [11]. Since the Coulomb interaction is retarded in a disordered metal, the conduction electrons "need the diffusion time to screen an electric charge". The screening is less perfect than in a pure metal. The question was, how to distinguish between the two mechanisms. These problems will be discussed by H. Fukuyama and A. Schmid.

In this connection, a most useful discovery was that quantum interference could be influenced systematically by applying a magnetic field or introducing spin-orbit scattering. A magnetic field destroys the quantum interference. HIKAMI et al [12] showed that in the presence of strong spin-orbit scattering, the quantum interference is such that the conductance is increased instead of decreased. This seemed to contradict the picture of incipient localization. As mentioned above, we have to interpret weak localization as an interference effect of the conduction electrons at defects. This interference is generally constructive in the backward direction (yielding an echo). However, in the presence of spin-orbit scattering, the spin wave function of the electrons leads to destructive interference into the backward direction.

Quantum interference became experimentally important because it corresponds to a time-of-flight experiment with the conduction electrons. It allows the measurement of characteristic electronic times, particularly the inelastic scattering time, the spin-orbit scattering time, and the magnetic scattering time. The inelastic scattering time is the life time of an electron in an energy eigenstate. Its measurement was rather difficult before the discovery of weak localization (see paper by J. E. Mooij). The investigation of systems with magnetic impurities has only started, but the measurement of the magnetic scattering time will certainly be very useful for a better microscopic understanding of those systems.

A particularly interesting application of quantum interference is the periodic oscillation of the resistance of cylindrical films in a magnetic field which is parallel to the axis of the cylinder. The period of the oscillation is given by the (superconducting) magnetic flux quantum h/2e. The effect has been predicted by ALTSHULER and ARONOV [13] and the first experimental demonstration was by SHARVIN and SHARVIN [14]. Meanwhile, as one can see from the Symposion of Flux Quantization in Normal Rings, a number of experiments confirmed the effect [15].

Particularly interesting, but because of the interaction between the electrons rather complex, systems are superconducting films above their transition temperature. Besides weak localization and Coulomb anomaly they show also superconducting fluctuations. These result in additional temperature-dependent anomalies in the resistance and the magneto resistance. Reviews about these effects are given by G. Deutscher and S. Maekawa.

7. The Metal Insulator Transition

Perturbation calculations, although very useful in providing an overview of possible phenomena in the limit of "weak localization", fail to describe the behaviour of a disordered system if the disorder potential is increased. It is partucularly difficult to treat the transition between the conducting and the insulating regime - the metal insulator transition.

As we have seen in section 2, disorder induces a metal insulator transition in a 3D system, if one parameter scaling is fulfilled, i. e. when there is only one relevant parameter necessary for the description of the transport - the localization length. This condition excludes additional scattering processes which may introduce additional parameters. It excludes also a magnetic field, which introduces the cyclotron length as an additional parameter. For the one electron system it is most important to have information about the behaviour of the conductivity (at the metallic side of the transition) and the localization length (at the insulating side) as a function of the disorder (for a given energy) or as function of the energy (for a fixed disorder).

Mott was one of the first to think about the conductivity near the "mobility edge" [16]. His suggestion was that the conductivity should jump discontinuously to zero when going from the conducting to the insulating regime. On the basis of the "random phase model" he gave an estimate for the "minimum metallic conductivity", which seems to explain many experiments. Other experiments do not show any finite step in the conductivity, but rather suggest a continuous decrease at the transition, in agreement with the conclusions from the one parameter scaling theory. Hence, there is considerable uncertainty in our knowledge about transport near the metal insulator transition. This remains so, even if we would accept that the nature of the transition might depend on the nature of the electronic system under consideration, aud that there are samples which show the minimum metallic conductivity and others do not.

Let us assume that one parameter scaling is right and that the metal insulator transition may be described as some kind of a phase-transition [17]. Then, the critical behaviour of the relevant quantities should be given by power laws, namely

$$\xi = \text{const}\cdot(E_c - E)^{-\nu} \qquad E < E_c \qquad \text{and} \tag{15}$$

$$\sigma = \text{const}\cdot(E - E_c)^{s} \qquad E > E_c \tag{16}$$

where E_c is the "mobility edge", and ν,s are the critical exponents. It is not difficult to see, using Wegner's renormalization group analysis [6], for instance, that s and ν should be equal. Many of the existing theories agree on this point.

However, the actual numbers for the critical exponents are quite difficult to evaluate. This holds for theory as well as for experiment. The reason is that the power law behaviour can only be expected "sufficiently close" to the "phase-transition". Most of the existing analytical theories are approximative, and only valid in the asymptotic regimes, i. e. in the limit of weak scattering and/or strong scattering. They are not necessarily correct in the regime of the transition. These yield usually $s = \nu = 1$. The numerical solution of the problem yields $s = \nu = 1.5 \pm 0.02$, for instance. Any numerical estimate is, however, confronted with the difficulty that either the accuracy of the data used is not sufficient or the data are taken not close enough to the transition. This is, in principle, also the problem one is faced with in experimental work. As a consequence, experiments on the metal insulator transition provide numbers between 0.5 and 1.7 for the critical exponent of the conductivity. The paper by G. A. Thomas concentrates on these questions.

Interaction effects and/or external fields complicate the critical behaviour quite seriously. One parameter scaling is certainly violated.

The transport properties near the metal insulator transition is one of the most exciting problems in the field of disordered materials. The evaluation of these is a basic problem of quantum mechanics, even in one-electron approximation. Its solution will certainly require additional approximations, as is discussed by W. Götze. It may eventually not rule out the analogy with phase-transition,but certainly will lead to a deeper understanding of "macroscopic quantum transport phenomena".

References

1 See, for instance, N.W. Ashcroft, N.D. Mermin: Solid State Physics (Holt, Rinehart and Winston, 1976) and J.M. Ziman: Principles of the Theory of Solids (Cambridge, University Press, 1972)
2 P.W. Anderson: Phys. Rev. 109, 1492 (1958)
3 N.F. Mott, W.D. Twose: Adv. Phys. 10, 107 (1961)
4 D.C. Licciardello, D.J. Thouless: Phys. Rev. Letters 35, 1475 (1975)
5 E. Abrahams, P.W. Anderson, D.C. Licciardello, T.V. Ramakrishnan: Phys. Rev. Letters 42, 673 (1979)
6 F.J. Wegner: Z. Phys. 25, 327 (1976)
7 A. MacKinnon, B. Kramer: Phys. Rev. Letters 47, 1546 (1981)
8 G. Bergmann: Phys. Reports 107, 3 (1984)
9 G.J. Dolan and D.D. Osheroff: Phys. Rev. Letters 43, 721 (1979)
10 L. Van den Dries, C. Van Haesendonck, Y. Bruynseraede, and G. Deutscher: Phys. Rev. Letters 46, 565 (1981)
11 B.L. Altshuler, A.G. Aronov, P.A. Lee: Phys. Rev. Letters 44, 1288 (1980)
12 S. Hikami, A.I. Larkin, Y. Nagaoka: Progr. Theor. Phys. 63, 707 (1980)
13 B.L. Altshuler, A.G. Aronov, B.Z. Zpivak: JETP Letters 33, 94 (1981); Pis'ma Eksp. Teor. Fiz. 33, 101 (1981)
14 D.Y. Sharvin, Y.V. Sharvin: JETP Letters 34, 272 (1981); Pis'ma Eksp. Teor. Fiz. 34, 285 (1981)
15 See Supplement of the Proc. of the International Conference LITPIM (PTB-PG-1, 1984)
16 N.F. Mott, E.A. Davis: Electronic Processes in Non-Crystalline Solids (Oxford University Press, London 1971)
17 For further details, see the papers in Anderson Localization, Y. Nagaoka and H. Fukuyama (Eds.), Series in Sol. St. Sciences 39 (Springer, 1982) and Disordered Systems and Localization, C. Castellani, C. DiCastro, and L. Peliti (Eds.), Lecture Notes in Physics (Springer, 1981)

Part I

Invited Papers

Some Unresolved Questions in the Theory of Localization

P.W. Anderson
AT&T Bell Laboratories, 600 Mountain Avenue, Murray Hill, NJ 07974, USA, and
Princeton University*, Princeton, NJ 08544, USA

Extended Abstract

History

(a) Localization as an unrecognizable monster: e.g., "singular continuous" spectra in almost periodic potentials (Science Magazine, August 14, 1984: "Singular continuous spectra have no physical applications"). ∿ 10 years of real history - 1974 begins "era of belief". (Note: 1975, Göttingen Acad. Sci., first recognition.)

(b) Localization as a paradigm: Study of dirt effects, e.g., tunneling centers, aggregation, percolation, Kohlrausch relaxation in glass.

Localization presaged the study of coherence and "reproducible noise" or chaos. Source of motivation in 1957: Microscopic nature of irreversibility. Could coherence prevent quantum transport? (Note Azbel's emphasis on reproducible noise in σ of small samples.)

This is the essence of many beautiful experiments on weak localization - coherence can weaken or even enhance quantum transport. Spin orbit scattering reverses phases (Fukuyama, Bergmann, Dynes, and Bishop). Oscillatory phenomena: Sharvin and Sharvin, etc. In sum: "We can no longer doubt the fact of interference modulation of transport processes no matter how much we may dispute the details of the metal insulator transition".

(c) A neglected area: Non-electrical analogues, e.g., propagation of light in fogs + pigments, sound in random media, etc. When diffusive, reflective, transmitting? Role of conservation laws. Some ill-conceived work on excitons and on phonons in random media, but no conclusive work. One might name this field "dirty bosons": It is relevant, for instance, to many phase-transition problems.

*Work done at Princeton University supported in part by NSF Grant DMR 8020263

Unresolved questions

But, for the substantive part of my talk, I return to two as yet open problems in electron transport theory: Quantum Hall Effect, and Si:P and the $(n-n_c)^{1/2}$ vs. $(n-n_c)^1$ phenomenon. Classify QHE under four questions:

(1) How does conventional 2d theory fit with QHE? Conjectural answer by Khmel'nitskii, "Schlumbergerons" (Pruisken, Libby, Levine): "Levitation" of the extended states. According to their field theory, scaling maps one in the σ_{xy}, σ_{xx} space in such a way that states which start at $\sigma_{xy}^0 = \frac{e^2}{h}(i+½)$ become extended. But,

$$\sigma_{xy}^0 = \frac{ne^2}{m}\tau \frac{\omega_c\tau}{1+\omega_c^2\tau^2} \tag{1}$$

and, if $\omega_c\tau$ is small, this is very different from the appropriate quantized value. Things change over precisely when

$$\frac{ne^2\tau^2\omega_c}{m} \simeq \frac{e^2}{2h} , \tag{2}$$

which turns out to be

$$\frac{r_H{}^2}{\ell^2} \simeq 1 , \tag{3}$$

when the magnetic length becomes the "microscopic" length. Everything seems to fit smoothly, more or less.

(2) What is the nature of the scaling diagram? For strong scattering the field theory gives, as I mentioned, a diagram as shown in Fig. 1.

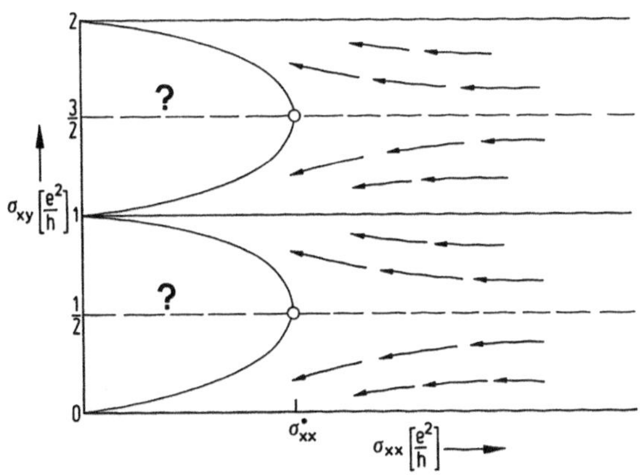

Fig. 1

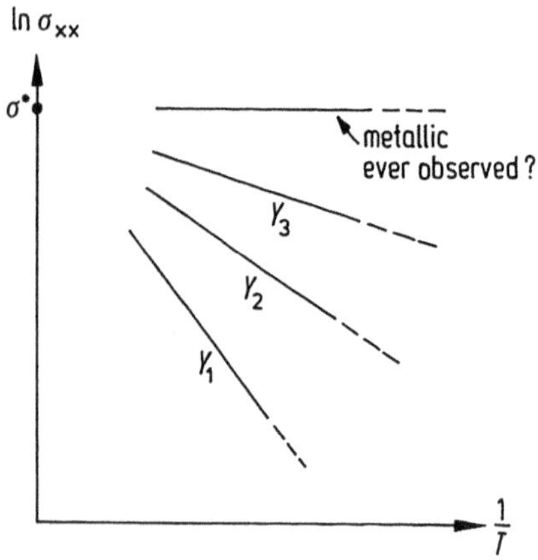

Fig. 2

The left-hand region is purely conjectural except for qualitative arguments and MacKinnon's and Aoki's calculations, which seem to agree with the "Schlumbergerons". Such a diagram explains the observations. But what happens at the left?
Also see (4). Query: Is it ever a metal? Is σ^*_{xx} significant? Note the Mott-type diagrams of conductivity σ_{xx} for different densities (Fig. 2). These are an outstanding question!

(3) Why or if we can neglect interactions? What about the fractional QHE?
It is, as usual, not at all obvious why the basic phenomenon is independent of interactions except for the very important role of the exchange splitting between bands. But for the FQHE, a fascinating conjecture has been made by a group at Aspen involving Pruisken, Kosterlitz, Laughlin, Cohen and others: That the diagram with interactions has two sets of fixed points, "boxes" and "circles", the former representing the condensation phenomenon, the latter localization, presumably not much affected by interactions. These people have proposed the following complicated diagram (Fig. 3), which has very attractive features, not least that at each level the fractional entities localize. Does this occur at a σ_{xx} which scales with $\frac{(e^*)^2}{h} = \frac{(\nu e)^2}{h}$: The fractional charge?

(4) This, to me, is the big question: How can scaling and a field theory work at all? Are we sure that this is not a case where, unlike the standard one (see Shapiro), the crossover is from localization to percolation and not vice versa?

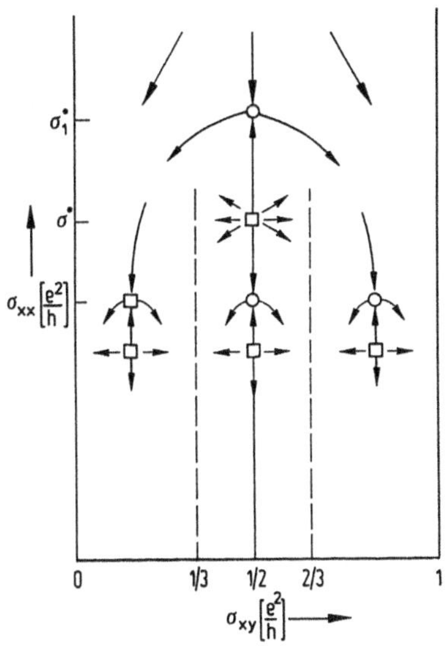

Fig. 3

This is suggested by the semiclassical picture of Trugman, Luryi, Littlewood and others (including myself). Here one imagines a smooth random potential with correlation length large compared to r_H, and semiclassical orbitals circling the peaks and lakes in opposite directions, with localized states being classical orbits closed on themselves. There is one percolating energy dividing the peaks from the lakes. Scattering by point-like perturbations does not change this much - it fuzzes out the orbits a bit, but it is <u>all forward scattering</u> and can lead to no coherent back-scattering. Thus, the coherence is all rather long range and caused by orbits closing on themselves. It is hard to see how a field theory can relate to this picture; yet the conclusions are not vitally different as, e.g., that everything scales to $\sigma_{xy} = \frac{e^2}{h}\, i$.

The second question to discuss in some detail is localization and interactions in three dimensions. There are two serious questions:

(a) Why does $\upsilon = 1$ work so well in most cases?

(b) Why does $\upsilon = ½$ work so well for Si:P and where does it come from?

A lot of people are working toward what seems to be a similar resolution of these questions: With me, B.G. Kotliar, E. Abrahams; Finkelstein in the USSR; Lee, di Castro et al; and A. Ruckenstein. A very primitive answer to (b) which is unmotivated but right was given by Ting.

In the talk I discuss four points:

(1) The data, especially on Si:P - this will appear in other talks and need not be repeated, but it is unequivocal.

(2) That the standard "McMillan" theory is almost inescapable at T = 0. At T = 0, interactions can do nothing but screen the existing random potential; imaginary parts play no role, since all the action is right at the Fermi surface. I was convinced by Abrahams that there is an exchange term in the screening coming from the wave function correlations enforced by the diffusion kernel which mimics the standard localization $\frac{1}{g}$ term, but this still leads to $\upsilon = 1$, and is even more universal and unavoidable. As was first emphasized by McMillan, conductivity at T = 0 is a purely static, geometrical effect and the energy scale, i.e., dynamics, is irrelevant. It is a common theme of both the FQHE and Si:P problems that we have to postulate a <u>qualitative change in the excitation spectrum</u> to get anything new. So there are two problems: Where does $-\frac{1}{g}$ go? and where does $\upsilon = \frac{1}{2}$ come from?

(3) Our hypothesis is that Si:P differs by having spin as well as charge conservation, and hence a localizing <u>spin</u> diffusion constant D_s as well as charge Diffusion D. This makes contact with the spin-dependent potential case of localization which has a β-function $1 - \frac{g_c^2}{g^2}$, which leads via scaling to:

$$\sigma \sim \sqrt{\sigma_0^2 - \sigma_c^2} \tag{4}$$

which fits the experiment beautifully.

(4) Why should this occur? We postulate that spin fluctuations are strong so that $D_s^0 \ll D^0$ in the microscopic theory. Here we appeal to (a) mean field spin fluctuation theory which says that, if

$$\chi_0(q\omega) = \frac{\chi_0 D q^2}{i\omega + Dq^2}, \qquad \text{then} \tag{5}$$

$$\chi(q) = \frac{\chi_0(q)}{1 - \bar{I}\chi_0} \rightarrow D_s = (1 - \bar{J}\chi_0)D\ , \tag{6}$$

while $D_Q = D(1 + \bar{I}\chi_0)$, and (b) Brinkman-Rice theory of strongly repulsive Fermi liquids, which emphasizes the strong frequency-dependence of the self-energies, leading to a diverging wave-function renormalization Z which can both remove the conventional term and, via spin fluctuation scattering, lead to strong spin-dependent scattering, so long as spin fluctuations are dominant leading to $D_s << D$. The large Z also accounts for the pseudogap in G_1.

Conclusion

You see that I present a scenario rather that a theory; so far it is better supported by experiment than by theory in any true sense. But, the key idea is, if I may repeat: We must go smoothly from the metallic state to an insulating state where we know that all of the low-energy excitations are spins. It must make sense to assume that the situation is dominated by these spins on the one side of the mobility edge; why not on the other?

Localization and Interaction in Quasi-Two-Dimensional Metallic Films

Shun-ichi Kobayashi

Faculty of Science, University of Tokyo, 7-3-1 Hongo, Bunkyo-ku
Tokyo, Japan 113

1. Introduction

It was only five years ago when the first lnT dependence of resistivity at low temperatures in PdAu films was observed and attributed to the localization of electrons in dirty films by DOLAN and OSHEROFF [29]. Since then, the experimental works in metallic films, together with theoretical ones, have made remarkable progress and have contributed to a large extent to the understanding of the localization and interaction effects in dirty metals.

Among one-, two- and three-dimensions, the two-dimension has been most extensively studied, mainly because of easiness of sample preparation and of richness of the physics. The peculiar prediction of the scaling theory [54], that the two-dimensional metal cannot be a metal but an insulator in its ground state, also stimulated the study.

Generally speaking, the understanding of localization and electron-electron interaction in weakly localized regime has already been well established in normal metal films both theoretically and experimentally. However, there remain many problems in the interplay between localization and other cooperative phenomena such as superconductivity and magnetism. The strongly localized regime is still a new world.

Besides these studies of localization itself, there appeared a new field where the localization phenomena are used as a tool to measure physical properties which have not been able to be measured before. This direction seems very promising, and will contribute to many other fields of solid state physics.

In the following, we present a short review of the experimental works, and try to clarify what has been done and what has not. The works related to superconductivity are excluded.

2. Early experiments

Since the scaling [54] and the microscopic [55] theories of localization predicted lnT dependence of conductivity, several tens of lnT have been reported till now. In most of the papers appeared in 1979-1981, the authors were interested in observing lnT dependence and in fitting the coefficients to the localization theory. The first was the work in PdAu [29] and it was followed by Cu [1], Pt [30], AuPd [31], Cu [2] and many others, as listed in the references.

In the meanwhile, the more detailed localization theory was presented, which enabled the analyses of the magnetic field dependence of the conductivity [56]. By this theory, the material parameters as inelastic scattering time, spin-orbit scattering time and spin-flip scattering time became possible to be obtained by the fitting. Therefore, shortly after this theory, a number of experiments on magnetoconductance appeared. The theory predicted lnH dependence of magnetoconductance in high field with positive and negative sign, depending on whether the material has weak or strong spin-orbit interaction. The experiments in high field [3,9,15,23,24] well agreed with the theory, showing a clear discrimination for light and heavy metals.

The localizaion theory [56] said that in the field high enough to observe lnH dependence of magnetoconductance, the lnT dependence should vanish. However, the experiments showed that even in high magnetic field the coefficient of lnT changed little [3,9,49]. Based on this fact, the importance of electron-electron interaction [60,61] in metallic films for lnT dependence was pointed out. A possibility of interaction effect was also proposed in the experiments of dc conductivity in the presence of microwave electric field [4]. The magnetoconductance was found to be strongly suppressed in magnetic films [19,26,49]. This is consistent with the theory [56] ; the localization is easily broken by perturbations which violate the time-reversal symmetry.

By measuring magnetoconductance at many temperature points, the temperature-dependence of inelastic scattering time can be obtained. These measurements [7,42,17] revealed that in many cases such as noble and transition metals, the increase of the inelastic scattering time with decreasing temperature reaches to the spin-flip scattering time (the origin of which is not well known) and below that temperature the coefficient of lnT which arises from the localization effect vanishes. This finding suggested that lnT dependence in these films are due mostly to the interaction effect which does not vanish even in high field. The first observed lnT [29] should now be attributed to the interaction but to the localization.

At present, the temperature and magnetic field dependence of conductivity is, of course, analyzed by using full expression of localization and interaction effects, and the results are quite satisfactory as will be shown in <u>5</u>.

3. Samples and Measurements

In this section, we quickly review the materials and the methods to fabricate thin metallic films used in the experiments. The methods of measurements are also briefly summarized.

The materials which have been studied are Cu, Au, Ag, Mg, Pt, Pd, Bi, InO_x etc. as listed in the references. Some superconductive meterials as Al and Nb are also studied in normal phase [50,51,52].

The most common method of making films is to deposit metals by simple vacuum evaporation on room temperature substrates. While this is the easiest way, it is difficult to make very thin film by this method, because the evaporated material forms small islands on the substrate instead of a continuous film. Therefore the samples by this method inevitably become relatively thick and very conductive, unless resistive metals or alloys are used. The rf and dc sputtering is also used in many cases. Glass, sapphire, mica and crystalline Si are used as the substrates. No systematic difference depending on deposition methods and substrates are reported [35]. To prepare InO_x films, material was evaporated in low-pressure oxygen gas [44].

The quench condensing method [23 and other papers by the same author] ,i.e., vacuum deposition onto cold (helium temperature) substrate, is useful to make very thin continuous films with high resistivity. The films can be annealed if necessary.

A novel method was developed by the group of the present author [1,3,9,10,13]. The film made by this method is an ensemble of very small metallic particles with diameters smaller than 30A which are slightly oxidized and packed two-dimensionally on substrate. The oxide layer serves as the scattering centers and by controlling its thickness, the resistivity can be varied from several tens Ω to infinity. Further, by stacking the number of layers, three-dimensional system which is resistive enough to measure the changes in conductivity can be achieved.

To vary the physical properties such as spin-orbit interaction and spin-flip scattering rate selectively, a method to deposit a subatomic layer of foreign material on the main film was employed [24,25]. These parameters were also varied by changing the alloy concentration [10], or by replacing one layer in multilayered structure with a different material [9,13,14].

Through these studies with a variety of sample preparation methods, it became clear that, as far as the systems are in weakly localized regime, the physics is insensitive to detailed microscopic structures which depend on the methods. This is because the phenomena have long characteristic length which are usually 100-1000A and the detailed structures are smeared out. To study the strong localization regime, on the other hand, these structures will play essential roles.

In most of the works, dc or very low frequency ac conductance is measured with conventional method. The electric field is kept as low as possible to avoid the heating except in some measurements to obtain nonlinear conductivity or electron phonon relaxation time [4,16]. The temperature range for the measurements extends from room temperature to 10mK [53]. The magnetic field as high as 14T is used [36].

4. Theories

The theories of the localization and interaction are constructed by using the most modern tools of physics; renormalization group theory and quantum field theory. Therefore, it is sometimes very difficult to get physical pictures behind the theories, especially for experimentalists who are not familiar with these tools.

In the following,we summarize the current theoretical results for the temperature and magnetic field dependence of conductivity in two-dimensional dirty metal in weakly localized regime [54,55,56,57,58,59].

The important parameters which appear in the theory are,

D: the diffusion constant,
τ_0: the elastic scattering time,
τ_ε: the inelastic scattering time,

τ_{so}: the spin-orbit scattering time,
τ_s: the spin-flip scattering time,
F: the parameter for the screening,
and p: in $\tau_\varepsilon \propto T^{-p}$.

The films can be considered two-dimensional when the thickness t is smaller than $\sqrt{D\tau_\varepsilon}$ for the localization effect and $\sqrt{hD/k_BT}$ for the interaction effect. The correction to the sheet conductivity (conductance of square-shaped sample) due to the localization effect, including spin-orbit and spin-flip scattering and the Zeeman splitting as well as the finiteness of the thickness, are given as

$$\Delta\sigma(H,T) = -\frac{e^2}{2\pi^2\hbar}\left[\psi\left(\frac{1}{2}+\frac{1}{a\tau_0}\right) - \left(\frac{1}{2}+\frac{1}{a\tau_1}\right) - \frac{1}{2\sqrt{1-\gamma}}\left\{\psi\left(\frac{1}{2}+\frac{1}{a\tau_+}\right) - \psi\left(\frac{1}{2}+\frac{1}{a\tau_-}\right)\right\}\right] \quad (1)$$

in the field perpendicular to the film, and

$$\Delta\sigma(H, T) = \frac{e^2}{4\pi^2\hbar}\left[2\ln\left(\frac{\tau_0}{\tau_1}+\frac{D\tau_0 t^2}{3\ell_H^4}\right) - \frac{1}{\sqrt{1-\gamma}}\left\{\ln\left(\frac{\tau_0}{\tau_+}+\frac{D\tau_0 t^2}{3\ell_H^4}\right) - \ln\left(\frac{\tau_0}{\tau_-}+\frac{D\tau_0 t^2}{3\ell_H^4}\right)\right\}\right] \quad (2)$$

in the field parallel to the film. Here,

$$\tau_1^{-1} = \tau_\varepsilon^{-1} + 2\tau_s^{-1} + 4\tau_{so}^{-1},$$

$$\tau_\pm^{-1} = \tau_\varepsilon^{-1} + 6\tau_s^{-1} + 2(\tau_{so}^{-1} - \tau_s^{-1})(1 \pm \sqrt{1-\gamma}),$$

$$\gamma = \{g\mu_B H/2\hbar(\tau_s^{-1} - \tau_{so}^{-1})\}^2, \quad \ell_H = \sqrt{\hbar c/eH}, \quad (3)$$

and ψ is the di-gamma function which has the limiting forms as

$$\psi(x) = \begin{cases} \ln(x) & x \gg 1 \\ -1.9635 & x = 0.5. \end{cases} \quad (4)$$

The correction due to the interaction effect has more complicated form [60,61,62,63]. Therefore,we show the results for some limiting cases. In the following expressions, the localization effect is also included.

a) $\tau_{so}, \tau_s \gg \tau_\varepsilon$

$$\Delta\sigma(T) = [P + g_1 + g_2 - 2(g_3 + g_4)]\frac{e^2}{2\pi^2\hbar}\ln T \quad (5)$$

in zero magnetic field

$$\Delta\sigma(T) = (g_1 - g_3)\frac{e^2}{2\pi^2\hbar}\ln T \quad (6)$$

in perpendicular magnetic field higher than $\frac{\hbar c}{4De\tau_\varepsilon}$ and $\frac{4\pi k_B T}{g\mu_B}$

$$\Delta\sigma(T) = (g_1 + g_2 - g_3 - g_4) \frac{e^2}{2\pi^2\hbar} \ln T \tag{7}$$

in parallel field much higher than $\frac{4\pi k_B T}{g\mu_B}$

b) $\tau_s \gg \tau_\varepsilon \gg \tau_{so}$

$$\Delta\sigma(T) = [-\frac{P}{2} + g_1 - \frac{1}{2}(g_2 + g_3 + g_4)] \frac{e^2}{2\pi^2\hbar} \ln T \tag{8}$$

in zero magnetic field

$$\Delta\sigma(T) = (g_1 - \frac{1}{2}g_3) \frac{e^2}{2\pi^2\hbar} \ln T \tag{9}$$

in the field much higher than $\frac{\hbar c}{4De\tau_\varepsilon}$ and $\frac{4\pi k_B T}{g\mu_B}$

c) $\tau_{so}, \tau_\varepsilon \gg \tau_s$

$$\Delta\sigma(T) = (g_1 - g_3) \frac{e^2}{2\pi^2\hbar} \ln T \tag{10}$$

Here, g_i's are the effective interaction constants ; g_1 and g_2 are the sum of the exchange type corrections with particle-hole and particle particle diffusion processes and g_3 and g_4 are sum of the corresponding Hartree type corrections. When the interaction is of screened Coulomb type, $g_1=1$ and $g_2=g_3=g_4=F/2$. F is unity when the screening is complete and zero for bare interaction.

The magnetic field dependence of the correction due to the interaction effect in higher magnetic field than $hc/(4De\tau_\varepsilon)$ and $4\pi k_B T/(g\mu_B)$ is logarithmic and given as

a) τ_{so} , $\tau_s \gg \tau_\varepsilon$

$$\Delta\sigma(H) = (1 + g_2 - g_3 - 2g_4) \frac{e^2}{2\pi^2\hbar} \ln H$$

in perpendicular field, and

$$\Delta\sigma(H) = (2 - g_3 - g_4) \frac{e^2}{2\pi^2\hbar} \ln H$$

in parallel field.

b) $\tau_s \gg \tau_\varepsilon \gg \tau_{so}$

$$\Delta\sigma(H) = - (1 + g_3 + g_4) \frac{e^2}{4\pi^2\hbar} \ln H$$

in perpendicular field, and

$$\Delta\sigma(H) = - (2 + g_3 + g_4) \frac{e^2}{4\pi^2\hbar} \ln H$$

in parallel field.

It may be worthwhile to mention the localization length in two dimensional sample which exhibits lnT dependence. The lnT term arises from lnL dependence of conductivity, where L is the sample size, by replacing L with $T^{-p/2}$ which is valid when the inelastic scattering time is the power function of T. If p=1 as often observed at low temperatures, when the temperature is decreased by one decade the change of conductivity is $\Delta\sigma = -[e^2/(2\pi^2\hbar)]\ln 10 = -2.8 \cdot 10^{-5}$ mho. The temperature-independent term of conductivity is typically $\sigma_0 = 10^{-3}$ mho. Therefore, to make $\Delta\sigma$ comparable to σ_0, the temperature should decrease in several tens decades. On the other hand, the value of L $(=\sqrt{D\tau_\varepsilon})$ is usually of the order of 10^{-5} cm at 1K. Again converting the temperature to the length, this means that the localization length, which is of the order of the length within which $\Delta\sigma \sim \ln L$ at T=0 holds, is about $10^{-5} \cdot 10^{\sigma_0/2\Delta\sigma} \sim 10^{12}$ cm. This value is of course very much larger than the sample size of several mm. Therefore, in the usual condition where lnT dependence is observed, it is impossible to get into strong localization regime even if the temperature can be lowered as much as one wants, and this is the reason why lnT is observed in so wide range of temperature. In Fig.1 we show an example of lnT dependence which holds in three decades of temperature [53]. (Strictly speaking, this lnT is due mostly to the interaction effect for which similar arguments of scaling may hold [68].)

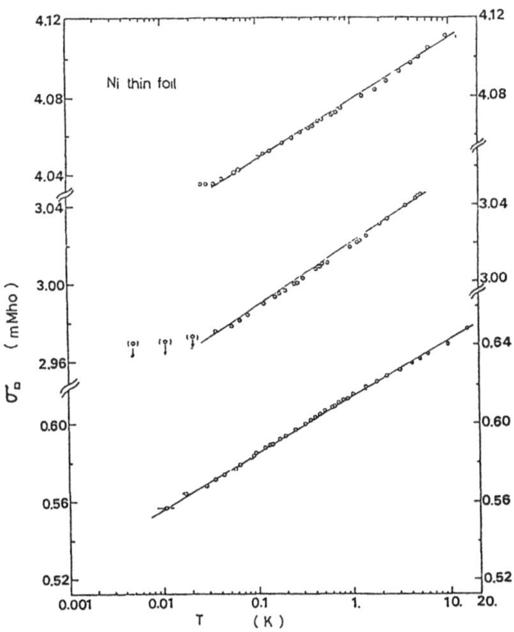

Fig.1: Temperature dependence of conductivity of Ni films [53]

5. Typical Experiments and Discussions

In this chapter, we discuss the typical experimental results in the light of current theories of localization and interaction [54-67].

(1) Localization effects

The effect of localization can be seen more clearly in magnetic field dependence of conductivity than in temperature dependence. Further, the magnetoconductance is very sensitive to spin-orbit and spin-flip scattering times. In Fig.2, the magnetoconductance measured in quench-condensed Mg film is shown [23]. The magnetic field in abscissa is normalized so that the data at different temperatures fall on a common curve. The solid curve in the figure is the fitting to eq. 1 by adjusting τ_ε. The second term in eq. and the interaction effects are neglected because the used field of 0.5T is low enough. The agreement with experimental points are excellent. The normalization factor for each temperature gives the temperature-dependence of τ_ε. τ_ε thus obtained is proportional to T^{-2} and has atrend to saturate at lower temperatures. According to the author, this suggests that τ_{so} is comparable to τ_ε in Mg.

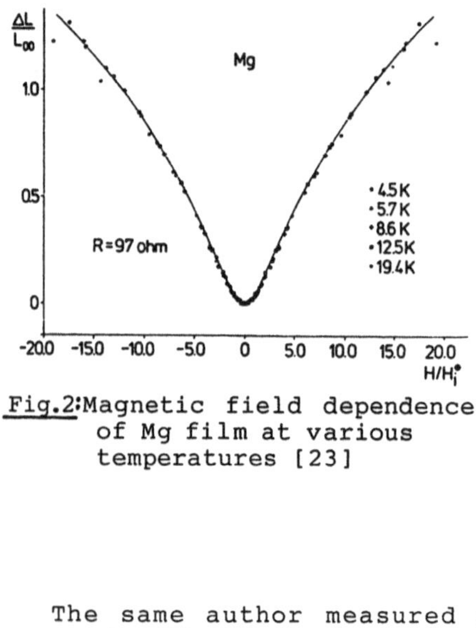

Fig.2: Magnetic field dependence of Mg film at various temperatures [23]

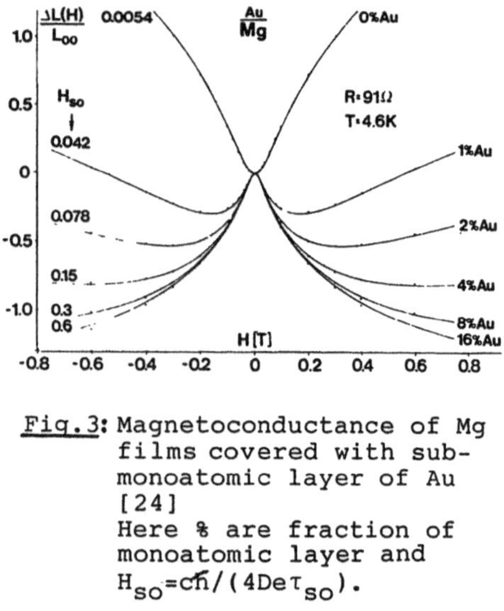

Fig.3: Magnetoconductance of Mg films covered with sub-monoatomic layer of Au [24]
Here % are fraction of monoatomic layer and $H_{so}=c\hbar/(4De\tau_{so})$.

The same author measured magnetoconductance in Mg films covered with subatomicmonolayer of Au which modifies the spin-orbit interaction selectively [24]. The results are shown in Fig.3. Because Au is very thin, only τ_{so} is changed. As shown in the figure, all the samples with different Au coverage can be fitted to (1) by adjusting only τ_{so}.

Similar results were obtained in Cu particle films in which a part of Cu was replaced with Au and Ag [9]. The agreement between experiment and theory was again excellent. The magnetoconductance in magnetic field parallel to the film surface also agreed with the theories.

Besides these two systematic studies, magnetoconductance was measured in many independent materials [28,27,8,12,5,7,36,44,47,17,39,41,43]. Agreements with theories were obtained except in a few cases [51,34].

Another important parameter τ_s was studied in CuMn alloy films with various Mn concentration [10]. The results are shown in Fig.4, where the fitted theoretical curves (1) are also shown. The value of τ_s given in the figure are reasonable for these Mn concentrations. It is clearly seen that the magnetic impurities break the localization. The magnetoconductance completely vanished in pure Mn film while the temperature-dependence is $[e^2/(2\pi^2\hbar)]\ln T$. Although some anomalies which arise from the Kondo effect are expected, there appeared no evidence. The effect of spin-flip scattering is also studied in Mg film covered with a very small amount of Fe [25].

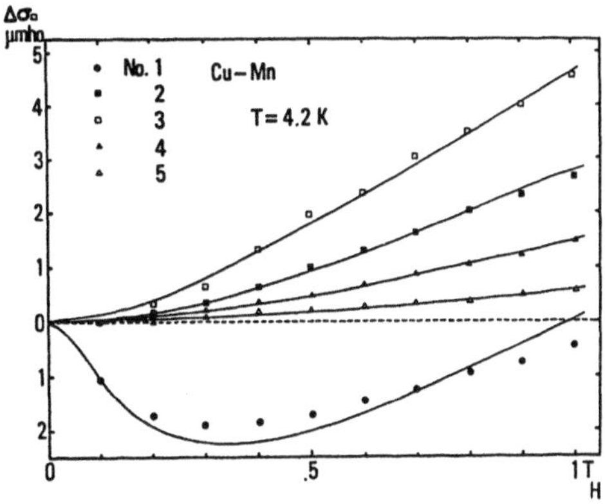

Fig.4: Magnetoconductance of CuMn alloy films [10]

Mn concentration and fitted values of τ_s are

No.1:	10^{-3}	60
No.2:	0.3	4.8
No.3:	0.9	2.4
No.4:	3.3	1.9
No.5:	12 atm.%	$0.3 \cdot 10^{-12}$

The temperature-dependence of energy relaxation time has been measured in a number of papers [24,26,28,27,12,7,47, 42,48,17,15,41]. The data are log-log plotted to obtain p by assuming $\tau_\varepsilon \propto T^{-p}$. The value of p commonly observed in higher temperatures than 1-10K is 2, with some exceptions such as p=1.65 in Cu, Au, and Ag films [15] and p=1.3 in $InSnO_x$ were reported [47], while p=2 were also observed in Cu [12]. At lower temperatures, all the data showed smaller p than 2, in many cases p=1. Fig.5 shows an example. Further decrease of p at lower temperatures was observed in some transition and noble metals [9,42,17]. This saturation is probably due to finite spin-flip scattering time caused by magnetic impurities in films. As can be seen in (1) , it is hard to determine τ_ε and τ_s individually by fitting when these two have nearly the same value. As p becomes smaller, the lnT dependence due to the localization vanishes and becomes insensitive to magnetic field.

Theoretically, the electron-electron interaction results in $\tau_\varepsilon^{-1} \propto T\ln T$ when $kT < \hbar/\tau_0$ and otherwise $\tau_\varepsilon^{-1} \propto T^2$ [64]. These two temperature-dependences are qualitatively in agreement with the experimental results which are proportional to T^2 and to T at high and low temperatures respectively. But the theoretical temperature which devides high and low temperature regions is too high by many orders of magnitude. Therefore, the observed T^2 dependence is attributed to the electron-phonon scattering [15,20]. However, there are still difficulties to explain the absolute values of τ_ε. The discussion on the equivalence of τ_ε and τ_ϕ is not settled, the former being essential in the localization [7]. Further theoretical elaboration is desirable.

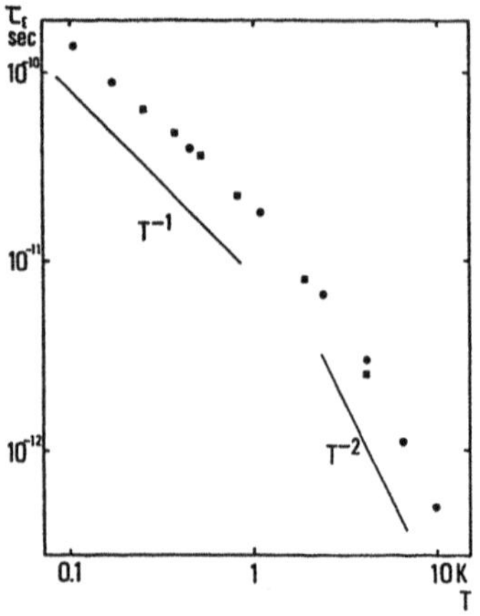

Fig.5: Energy relaxation time in Bi film [42]

The effect of the electric field on the localization has been theoretically discussed soon after the first lnT experiment [65]. The simplest model is that of hot (or warm) electron; a high electric field causes a higher temperature of electrons than that of lattice. The other theories [66,67] argue whether the vector potential which causes the electric field destroys the interference of electrons or not. Experimentally, the temperature rise in electron and lattice is unavoidable when high electric field is applied, and thus the thermometers for both systems are needed to check the theories [29,18,34]. Such an experiment was performed in Au and Ag films [16] and the results supported the theory [66] which predicts no electric field dependence of conductivity.

(2) Antilocalization effect

A striking prediction of the theory of localization is that when the spin-orbit scattering time is much shorter than the inelastic scattering and spin-flip scattering times, the conductivity increases as lnT with decreasing temperature [56]. This is often called the antilocalization effect. This feature arises from the symmetry of the hamiltonian of the system which are also important in determining the energy levels in heavy nuclei and small metallic particles. It is interesting to ask whether the ground state of the system is a perfect conductor or an insulator, but the theory cannot answer because it is a perturbational calculation in the weakly localized regime.

Experimentally, the antilocalization effect was clearly observed in Bi [39,40,41], in which the temperature-dependence of τ_ε does not saturate at low temperatures, so that the temperature dependence of the localization effect can be switched on and off by magnetic field. Fig.6 shows the conductivity versus temperature of a Bi film [40]. The coefficient of lnT which is about a half of $e^2/(2\pi^2\hbar)$ in zero magnetic field recovers its full value in the field of 1T. This increase is caused by the suppression of negative contribution of localization (antilocalization) by the field, in good agreement with eq. 1 .

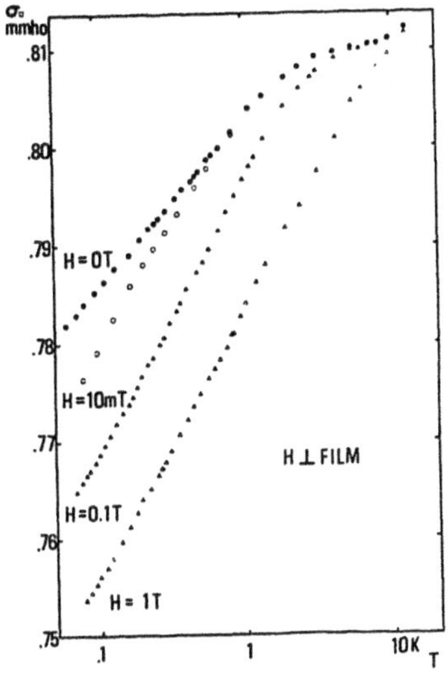

Fig.6: Temperature-dependence of conductivity in Bi with and without magnetic field [40]

What is left in 1T is the interaction effect,which is less sensitive to the field. As the experiments show, even if the localization effect is negative, the total conductivity always decreases because of the interaction effect (except in the case of superconductivity).

(3) Interaction effect

By the parameter fitting of magnetoconductance to the theories, we obtain F as well as τ_ε, τ_{so} and τ_s. Most of the data in resistive metallic films show that F is as small as 0.3-0.1 or smaller [26,28,7,36,17,22,39]. The temperature-dependence of conductivity are consistently explained with these values of F for individual materials and the temperature-dependence of τ_ε.

The parameter F measures the degree of screening,which is equal to unity when the screening is complete and zero when it is absent [60]. By using the values of usual bulk metals, F is calculated to be close to 0.5. Therefore,the observed values are considerably smaller. The reason for this disagreement is not known, but the effect of disorder to the screening seems important,because there is a tendency that the more conductive the film is, the larger F becomes [8].

The interaction effect modifies the density of state at the Fermi level, as a precursor of the Coulomb gap. This can be directly measured by tunneling experiment. The data obtained in InO_x [46] clearly showed anomalous decrease of tunneling conductance which is proportional to lnV, with voltage V. The density of state anomaly can be measured also by the Hall effect. The results are somewhat controverial ; in Bi [41] and Au [22], the lnT dependence of the Hall coefficients were observed giving F=0.3, while in InO_x, which showed tunneling anomaly, did not show the Hall anomaly [44]. The precise Hall measurement is generally very difficult because of high electron density in metallic films.

(4) Crossover to three-dimension

The conditions for two-dimensionality, i.e. the thickness is smaller than $\sqrt{D\tau_\varepsilon}$ for localization and than $\sqrt{D\hbar/(k_B T)}$ for interaction, can be violated by simply increasing the film thickness or by increasing the temperature. In both cases the system becomes three-dimensional and the temperature and magnetic field dependence of conductivity becomes $T^{1/2}$ and $H^{1/2}$. These crossovers were observed in several papers [45,47,43,13,14]. The effect of spin-orbit interaction was also investigated [13].

References

[Cu]

1 S. Kobayashi, F. Komori, Y. Ootuka and W. Sasaki, J. Phys. Soc. Jpn. **49**, 1635 (1980).

2 L. Van den dries, C. Van Haesendonck, Y. Bruynseraede and G. Deutscher, Physica **107B**,7 (1981) and Phys. Rev. Letters **46**, 565 (1981).

3 F. Komori, S. Kobayashi, Y. Ootuka and W. Sasaki, J. Phys. Soc. Jpn. **50**, 1051 (1981).

4 M. E. Gershenzon and V. N. Gubankov, JETP Letters **34**, 32 (1981).

5 M. E. Gershenzon and V. N. Gubankov, Solid State Commun. **41**, 33 (1982).

6 F. Komori, S. Kobayashi and W. Sasaki, Surface Science **113**, 540 (1982).

7 M. E. Gershenzon, B. N. Gubankov and Yu. E. Zhuraviev, Sov. Phys. JETP **56**, 1362 (1982).

8 C. Van Haesendonck, L. Van den dries, Y. Bruynseraede and G. Deutscher,Phys. Rev. B **25** 5090 (1982).

9 F. Komori, S. Kobayashi and W. Sasaki, J. Phys. Soc. Jpn. **51** 3136 (1982).

10 F. Komori, S. Kobayashi and W. Sasaki, J. Mag. Mag. Mater. **35**, 74 (1983).

11 D. Abraham and R. Rosenbaum, Phys. Rev. B **26**, 1413 (1983).

12 D. Abraham and R. Rosenbaum, Phys. Rev. B **27**, 1409 (1983).

13 F. Komori, S. Okuma and S. Kobayashi, J. Phys. Soc. Jpn. **53**, 1606 (1984).

14 F. Komori, S. Okuma and S. Kobayashi, in this volume.

[Ag, Au]

15 G. Bergmann, Z. Phys. B **48**, 5 (1982).

16 G. Bergmann, Z. Phys. B **49**, 133 (1982).

17 M. E. Gershenzon, B. N. Gubankov and Yu. E. Zhuraviev, JETP Letters **35**, 576 (1982).

18 S. I. Dorozhkin and V. T. Dolgopolov, JETP Letters **36**, 15 (1982).

19 T. Kawaguti and Y. Fujimori, J. Phys. Soc. Jpn. **51**, 703 (1982).

20 G. Bergmann, Solid State Commun. **46**, 347 (1983).

21 T. Kawaguti and Y. Fujimori, J. Phys. Soc. Jpn. **52**, 722 (1983).

22 G. Bergmann, Solid State Commun. **49**, 775 (1984).

[Mg]

23 G. Bergmann, Phys. Rev. B **25**, 2937 (1982).

24 G. Bergmann, Phys. Rev. Letters **48**, 1046 (1982).

25 G. Bergmann, Phys. Rev. Letters **49**, 162 (1982).
26 G. Bergmann, Phys. Rev. B **28**, 515 (1983).
27 A. E. White, R. C. Dynes and J. P. Garno, Phys. Rev. B **29** (1984).
28 G. Deutscher, M. Gijs, C. Van Haesendonck and Y. Bruynseraede, preprint.

[Pd, Pt]
29 G. J. Dolan and D. D. Osheroff, Phys. Rev. Letters **43**, 721 (1979).
30 R. S. Markiewicz and L. A. Harris, Phys. Rev. Letters **46**, 1149 (1981).
31 J. T. Masden and N. Giordano, Physica **107B**, 1 (1981).
32 W. C. McGinnis, M. J. Burns, R. W. Simon, G. Deutscher and P.
33 M. Chaikin, Physica **107B**, 5 (1981).
34 H. Hoffmann, F. Hoffmann and W. Schoepe, Phys. Rev. B **25**, 5563 (1982).
35 G. Dumpich, H. Kristen and E. F. Wassermann, Z. Phys. B **51**, 251 (1983).
36 R. S. Markiewicz and C. J. Rollins, Phys. Rev. B **29**, 735 (1984).

[Bi]
37 Yu. F. Komnik, E. I. Baukhshtab and Yu. V. Nikitin, Sov. J. Low Temp. Phys. **7**, 656 (1981).
38 A. K. Savchenko, V. N. Lutskii and A. S. Rylik, JETP Letters **34**, 349 (1982).
39 Yu. F. Komnik, E. I. Bukhshtab, A. V. Butenko and V. V. Andrievsky, Solid State Commun. **44** 865 (1982).
40 F. Komori, S. Kobayashi and W. Sasaki, J. Phys. Soc. Jpn.52, 368 (1983).
41 P. H. Woerlee, G. C. Verkade and A. G. M. Jansen, J. Phys. C **16**, 3011 (1983).
42 F. Komori, S. Kobayashi and W. Sasaki, J. Phys. Soc. Jpn. **52**, 4306 (1983).
43 D. S. McLachlan, Phys. Rev. B **28**, 6821 (1983).

[InO_x]
44 Z. Ovadyahu and Y. Imry, Phys. Rev. B **24**, 7439 (1981).
45 Y. Imry and Z. Ovadyahu, J. Phys. C **15**, L327 (1982).
46 Y. Imry and Z. Ovadyahu, Phys. Rev. Letters **49**, 841 (1982).
47 T. Ohyama, M. Okamoto and E. Otsuka, J. Phys. Soc. Jpn. **52**, 3571 (1983).
48 Z. Ovadyahu, Phys. Rev. Letters **52**, 569 (1984).

[other metals]
49 S. Kobayashi, Y. Ootuka, F. Komori and W. Sasaki, J. Phys. Soc. Jpn. **51**, 689 (1982).
50 G. Bergmann, Phys. Rev. B **29**, 6114 (1984).
51 S. Moehlecke and Z. Ovadyahu, Phys. Rev. B **29**, 6203 (1984).
52 P. Santhanan and D. E. Prober, Phys. Rev. B **29**, 3733 (1984).
53 N. Nishida, F. Komori and S. Kobayashi, unpublished.

[theories]
54 E. Abrahams, P. W. Anderson, D. C. Licciardello and T. V. Ramakrishnan, Phys. Rev. Letters **42**, 673 (1979).
55 L. P. Gorkov, A. I. Larkin and D. E. Khmelnitzkii, JETP Letters, **30**, 228 (1979).

56 S. Hikami, A. I. Larkin and Y. Nagaoka, Prog. Theor. Phys. **63**, 707 (1980).
57 S. Maekawa and H. Fukuyama, J. Phys. Soc. Jpn. **50**, 2516 (1981).
58 B. L. Altshuler and A. G. Aronov, JETP Letters **33**, 499 (1981).
59 B. L. Altshuler, A. G. Aronov, D. E. Khmelnitzkii and P. A. Lee, Phys. Rev. B **22**, 5142 (1980).
60 B. L. Altshuler, A. G. Aronov and P. A. Lee, Phys. Rev. Letters **44**, 1288 (1980).
61 H. Fukuyama, J. Phys. Soc. Jpn. **48**, 2169 (1980), ibid. **50**, 3407 (1981) and ibid. **51**, (1982).
62 B. L. Altshuler, A. G. Aronov, A. I. Larkin and D. E. Khmelnitzkii, Sov. Phys. JETP **54**, 411 (1981).
63 A. Kawabata, J. Phys. Soc. Jpn. **50**, 2461 (1981).
64 E. Abrahams, P. W. Anderson, P. A. Lee and T. V. Ramakrishnan, Phys. Rev. B **24**, 6783 (1981).
65 P. W. Anderson, E. Abrahams and T. V. Ramakrishnan, Phys. Rev. Letters **43**, 718 (1979).
66 B. L. Altshuler and A. G. Aranov, JETP Letters **30**, 482 (1979).
67 T. Tsuzuki, Physica **107B**, 679 (1981).
68 B. L. Altshuler and G. Aronov, JETP Letters **37**, 410 (1983).

Weak Localization in Silicon Mosfets: Isotropic and Anisotropic Two-Dimensional Electron Gases

D.J. Bishop, D.C. Tsui[+], and R.C. Dynes
AT&T Bell Laboratories, 600 Mountain Avenue, Murray Hill, NJ 07974, USA

Abstract

In this paper we review measurements in the weak localization regime of silicon mosfets. We first present data on isotropic (100 and 111) devices and discuss the magnetoresistance and temperature-dependent resistivity which provide support for the scaling theory of localization in two dimensions. We then discuss more recent measurements on anisotropic devices (110) which provide the first parameter-free test for the scaling theory in two dimensions.

Following ideas of THOULESS, ABRAHAMS, ANDERSON, LICCIARDELLO and RAMAKRISHNAN [1] (AALR) developed a single-parameter scaling theory of transport which showed that there exists no true metallic behavior in two dimensions. These authors found that in the high conductivity limit, the scaling length dependence of the conductivity of a two-dimensional electron gas is given by:

$$\sigma(L) = \sigma(L_0) - \frac{\alpha e^2}{\pi^2 \hbar} \ln (L/L_0) \tag{1}$$

where α is a constant expected to be of order 1. It was also shown that the temperature-dependence of the conductivity can be derived from (1) via the temperature-dependence of the inelastic scattering rate

$$\frac{1}{\tau_{in}} \sim T^P \tag{2}$$

where P depends on the dominant inelastic scattering process. Using the idea that the length scale sampled at a finite temperature is given by the inelastic diffusion length $L = \sqrt{\frac{1}{2}D^2\tau_{in}}$ (1) and (2) produce a conductance given by:

$$\sigma(T) = a(T_0) + \left(\frac{\alpha P}{2}\right) \frac{e^2}{\pi^2 \hbar} \ln \left[\frac{T}{T_0}\right] \tag{3}$$

Here D is the electron diffusivity.

Physically, what happens is that as the temperature is increased, inelastic scattering shortens the length scale over which the system is probed and as is shown by (1) the conductivity is increased.

Motivated by the experiments of DOLAN and OSHEROFF [2] who had apparently seen these effects, we examined the conductance of silicon mosfets at low temperatures [3] and observed the logarithmic temperature-dependences shown

+ Present Address: Princeton University, Princeton, New Jersey, USA

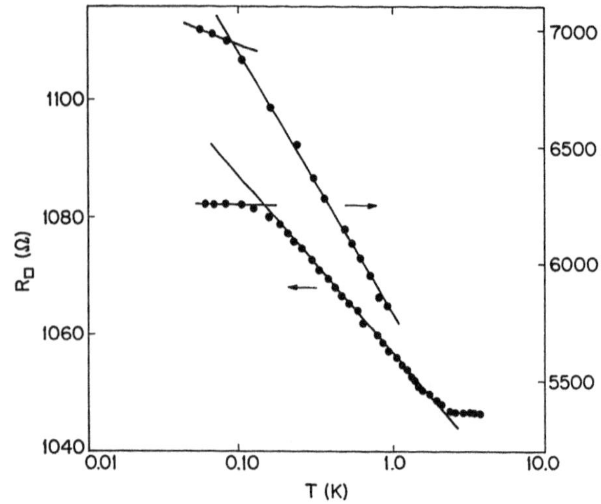

Fig. 1: The resistance as a function of temperature is shown for a (111) silicon mosfet with electron densities of $2.03 \times 10^{12} cm^{-2}$ (right) and $5.64 \times 10^{12} cm^{-2}$ (left)

in Fig. 1. In retrospect, we now know that the observations of logarithmic dependences by Dolan and Osheroff were Coulomb interaction effects, and that the data shown in Fig. 1 represent the <u>first</u> observation of weak localization in a two-dimensional system.

In Fig. 2 we have plotted the strength of these logarithmic dependences as a function of $R_\square$. Using the definitions implicit in (1) to (3) our data from Fig. 2 indicates that $\alpha P = 1.04 \pm 0.1$

FUKUYAMA [4], HIKAMI, LARKIN and NAGAOKA [5] and LEE and RAMAKRISHNAN [6] have studied the effects of a magnetic field on transport in the weakly lo-

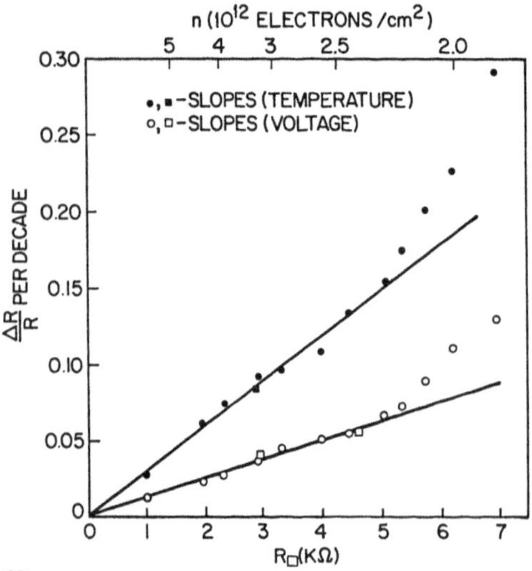

Fig. 2: The logarithmic slopes vs temperature and voltage are shown as a function of $R_\square$ (T=1K) and electron density

calized regime. For a perpendicular magnetic field these workers have shown that the localization effects are diminished when the size of the first Landau orbit is comparable to the inelastic diffusion length $\sqrt{\frac{1}{2} l_{in} l_{el}}$. Thus, in the weakly localized regime, one finds a negative magnetoresistance which depends on the magnitude of the perpendicular component of field. Lee and Ramakrishnan have shown that the change in conductivity at a fixed temperature is given by:

$$\Delta\sigma(H,T) = \frac{\alpha e^2}{2\pi^2\hbar}\left[\Psi\left(\frac{1}{2} + \frac{\hbar c}{2eH l_{in} l_{el}}\right) - \Psi\left(\frac{1}{2} + \frac{\hbar c}{2eH l_{el} l_{el}}\right) + \ln\left(\frac{l_{in}}{l_{el}}\right)\right] \quad (4)$$

Where l_{el} and l_{in} are the elastic and inelastic scattering lengths and Ψ is the digamma function.

Several groups have reported measurements in the low magnetic field regime for silicon mosfets [7], [8], [9]. Shown in Fig. 3 are our results on a Si (111) device at a density of 4.52×10^{12} electrons/cm^2. Also shown in Fig. 3 are fits to the data using (4). From negative magnetoresistance data such as this one obtains a value of α and $\tau_{inelastic}$ at a given temperature. By varying temperature one can then obtain α and τ_{in} as a function of temperature. Our results indicate that α is independent of temperature, and as is shown in Fig. 3 the best fits are obtained for $\alpha = 1.0 \pm 0.1$.

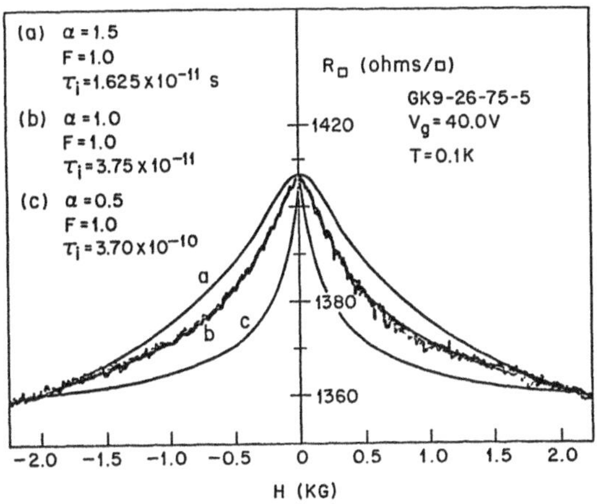

Fig. 3: The negative magnetoresistance is shown for a density of 4.52×10^{12}cm^{-2}. Also shown are fits for various values of α and τ_{in}

Shown in Fig. 4 are the values of $\tau_{inelastic}$ obtained as a function of temperature and electron density. In agreement with the earlier measurements we find an inelastic time which varies as:

$$\tau_{inelastic} \sim T^{-P} \quad \text{where } P \sim 1 \qquad (5)$$

Therefore we have found from the temperature-dependent resistivity measurements that $\alpha P \sim 1$ and independently from the magnetoresistance that $\alpha = 1$ and $P \sim 1$.

The data shown in Fig. 4 suggests that the relevant inelastic scattering mechanism is electron-electron scattering in the dirty limit. ABRAHAMS,

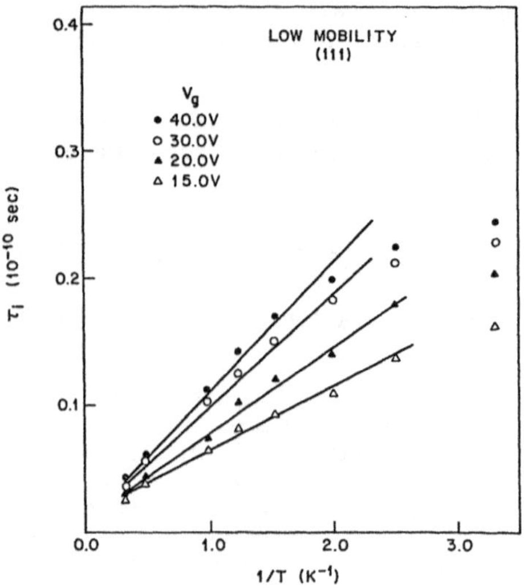

Fig. 4: $\tau_{inelastic}$ obtained from the fitted magnetoresistance of Fig. 3 as a function of 1/T

ANDERSON, LEE and RAMAKRISHNAN [10] have shown that in this limit the electron-electron scattering rate is given by:

$$\frac{1}{\tau_{ee}} = \frac{e^2}{\varepsilon \hbar^2 DK} kT(\ln \frac{T}{T_1}) \qquad (6)$$

where $kT_1 = \frac{\hbar^3 D^3 K^4 \varepsilon^2}{e^4}$

ε is the dielectric constant and K the screening constant. This correctly explains the temperature-dependence that we find as well as the dependence on diffusity that we obtain. Shown in Fig. 5 is our data for τ_{in} vs diffusivity as well as the predictions of (6). Clearly we obtain good agreement with the predictions of (6). In addition, more recent measurements by WHITE [11], et

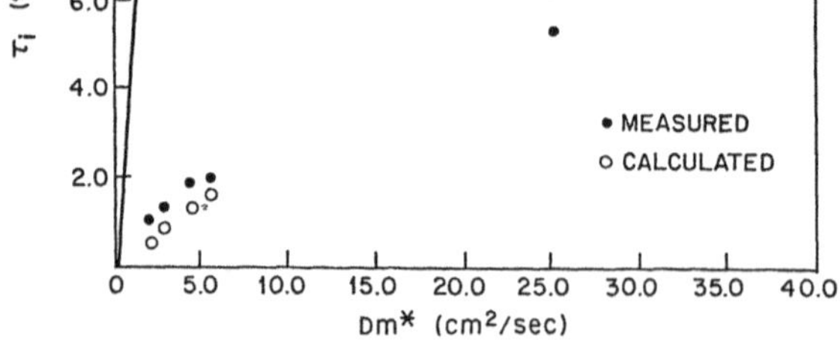

Fig. 5: $\tau_{inelastic}$ vs Dm at 0.5K for two different mosfets. The open circles are calculated using (6) and the solid line is calculated ignoring the $\ln(T/T_1)$ enhancement.

al on quench condensed Mg films give strong support for the mechanism suggested by Abrahams, et al.

Summarizing our results for isotropic devices (100 and 111) we find $\alpha P \sim 1$, $\alpha \sim 1$ and $P \sim 1$. We now turn to measurements on anisotropic (110) devices. Our samples were N channel mosfets fabricated on a (110) surface of P type silicon with a mobility of $\sim$1000 cm²/V at 4.2 K. As shown in Fig. 6 our samples had a Van de Pauw geometry with the current and voltage leads oriented along the 110 and 100 directions. The resistances were measured using a four-terminal bridge and the anisotropic resistances per square calculated using the algorithm of MONTGOMERY and LOGAN, RICE and WICK [12].

At low temperatures the samples had an anisotropic conductivity ratio less than the ratio of effective masses. For example, for the data shown in Fig. 6 $R_\square(1\bar{1}0) = 1903\ \Omega/\square$ and $R_\square(100) = 1041\ \Omega/\square$. Because $m(1\bar{1}0) = .585$ and $m(100) = .190$ we conclude that on the 110 surfaces there exists anisotropic elastic scattering as has been shown by previous workers [13]. In Fig. 6 we show typical low-field perpendicular magnetoresistance curves for the two orthogonal directions ([100] and [110]) of the silicon mosfet.

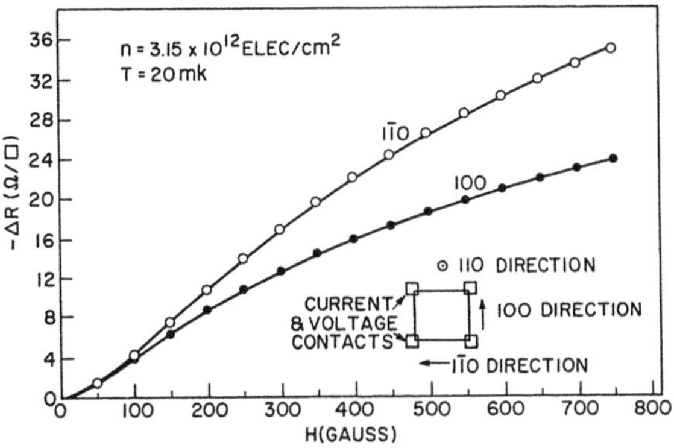

Fig. 6: Shown is the change in resistance with field for the two orientations on the (110) surface of a Si mosfet. The solid line is a fit to the data.

In our analysis l_{el} is calculated from the value of the conductivity at 4.2K and the $l_{inelastic}$ and α are obtained from a fit to the data using (4). Such a fit is shown in Fig. 6. For the data shown we obtain $\alpha[1\bar{1}0] = .693$ and $\alpha[100] = 1.256$. In Fig. 7 we show the values of α for the two directions as a function of temperature and in Fig. 8 as a function of electron density. As for the isotropic case α is independent of both temperature and density. In Fig. 9 we show the inelastic scattering lengths as a function of temperature. From the data in Fig. 9 we see that $P[100] = P[1\bar{1}0] = 1$. We have also measured the temperature-dependence of the zero field conductivity and obtain $(\Delta R/R)[100] = (\Delta R/R)[110] = 4.5\%$/decade. Therefore, as for the isotropic case αP as measured from the temperature-dependence of the resistivity is consistent with α and P as determined from the magnetoresistance.

WÖLFLE and BHATT [14] have considered the case of an anisotropic two-dimensional electron gas with both scattering time and effective mass anisotropy. In their results they obtain a logarithmic divergence similar to (1) indicating the localization of all states at T=0 in two dimensions. However, they find that the anisotropy changes the universal coefficient $e^2/2\pi^2\hbar$ in (1). They obtain:

$$\sigma(T) = \sigma(T_0) - \frac{\sigma_{\mu\mu}}{\bar{\sigma}} (\alpha P) \frac{e^2}{2\pi^2\hbar} \ln(T_0/T) \tag{7}$$

where $\bar{\sigma} = (\sigma_{xx}\ \sigma_{yy})^{\frac{1}{2}}$

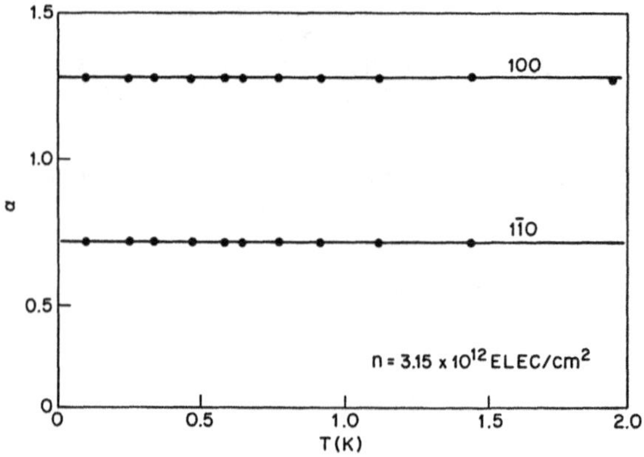

Fig. 7: The alpha parameters for the [1$\bar{1}$0] and [100] directions are shown as a function of temperature

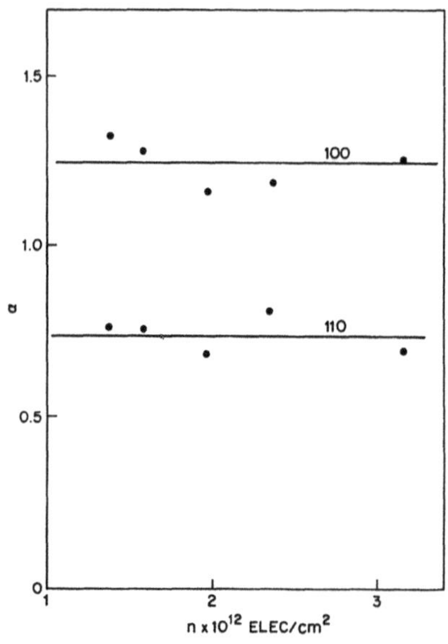

Fig. 8: The alpha parameters are shown as a function of electron density

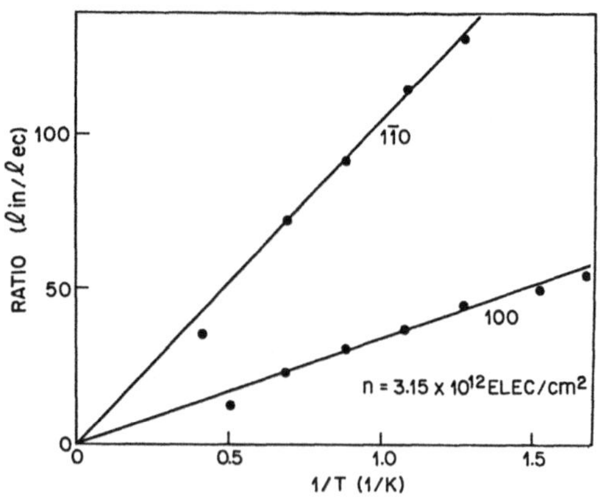

Fig. 9: The ratio l_{in}/l_{el} is shown as a function of temperature

Therefore they find that the effects of anisotropy can be completely absorbed into an anisotropic diffusion coefficient. This result can be compared to our experimental observations. By taking the ratio $(\alpha(1\bar{1}0)/\alpha(100))_{exp}$

and comparing it with $(\alpha(1\bar{1}0)/\alpha(100))$ theory, we can obtain a parameter-free test of the scaling theory used to calculate (1) and (7).

As an example for the data shown in Fig. 7 we obtained $\alpha(100)$= 1.256 and $\alpha(110)$ = .693. Using $R_{\square}(100)$= 1041$\Omega/_{\square}$ and $R_{\square}(110)$ = 1903 $\Omega/_{\square}$ we obtain $(\alpha(100)/\alpha(1\bar{1}0))$ theory = 1.828 in excellent agreement with the experimental value of 1.812. Our measurement therefore represents the first quantitative parameter-free test of the scaling theory of localization and we find excellent numerical agreement.

In conclusion we have measured the conductivity as a function of temperature and magnetic field for isotropic and anisotropic silicon mosfets. We find excellent numerical agreement with the scaling theory of localization and its extensions to finite magnetic fields. The inelastic lengths that we obtain agree well with the theory of Abrahams, et al. Finally, our measurements in anisotropic devices provide the first parameter-free test of the scaling theory of localization and we find excellent numerical agreement.

References

1. E. Abrahams, P. W. Anderson, D. C. Licciardello and T. V. Ramakrishnan: Phys. Rev. Lett. 42, 673 (1979).
2. G. J. Dolan and D. D. Osheroff: Phys. Rev. Lett. 43, 721 (1979).
3. D. J. Bishop, D. C. Tsui and R. C. Dynes: Phys. Rev. Lett. 44, 1153 (1980); Phys. Rev. B26, 773 (1982).
4. H. Fukuyama: Surf. Sci. 113, 489 (1982).
5. S. Hikami, A. I. Larkin and Y. Nagaoka: Prog. Theor. Phys. 63, 707 (1980).
6. P. A. Lee and T. V. Ramakrishnan: Phys. Rev. B26, 4009 (1982).
7. Y. Kawaguchi and S. Kawaji: J. Phys. Soc. JPN. 48, 699 (1980).
8. R. Wheeler: Phys. Rev. B24, 4645 (1981).
9. R. A. Davies, M. J. Uren and M. Pepper: J. Phys. C14, L531 (1981).
10. E. Abrahams, P. W. Anderson, P. A. Lee and T. V. Ramakrishnan: Phys. Rev. B24, 6783 (1981).
11. A. E. White, R. C. Dynes and J. P. Garno: Phys. Rev. B29, 3694 (1984).
12. H. C. Montgomery: J. Appl. Phys. 42, 2971 (1971); B. F. Logan, S. O. Rice and R. F. Wick: J. Appl. Phys. 42, 2975 (1971).
13. H. Sakaki and T. Sugano: JPN. J. Appl. Phys. Suppl. 41, 141 (1972).
14. P. Wölfle and R. N. Bhatt: Phys. Rev. B30 (1984)

Transport as a Consequence of Incident Carrier Flux

Rolf Landauer

IBM Thomas J. Watson Research Center, P.O. Box 218, Yorktown Heights, New York 10598, U.S.A.

1. Introduction

The electrical resistance of a one-dimensional metal, with no inelastic scattering, is [1,2]

$$\mathscr{R}_{el} = (\hbar\pi/e^2)R/(1-R). \tag{1.1}$$

where R is the reflection probability for a carrier entering the specimen, at the Fermi level. Equation (1.1) allows for a two-fold spin degeneracy. This relationship received widespread attention in the localization literature, following the work of ANDERSON, et al. [3]. As an example of the many further extensions, we cite only [4]. The opportunities for generalizations of (1.1), as well as its limitations, are, perhaps, still inadequately appreciated. In that connection, I will stress the utility of a viewpoint taken over from circuit theory. It was originally invoked [5] in a paper treating the metallic residual resistance problem from a viewpoint in which the current flow into the specimen is taken as the causal agent, and the electric field is built up as the consequence of a continued flow of charges against scattering centers. The duality of currents and fields as source agents is, of course, very familiar in electrical engineering, but has not become a common viewpoint in transport theory.

The principal purpose of [5] was to investigate the spatial variation of electric fields and currents in the neighborhood of a point defect scatterer, in a metal. Reference [5] asked questions which were not fashionable, at that time, and also suffered from a nontrivial level of mathematical complexity. As a result [5] received little attention when it appeared, and a simplified discussion [6] was given subsequently. Reference [5] showed that the additional transport field introduced into a metal, for a fixed current flow, upon the introduction of an additional localized lattice defect scatterer, is a highly localized dipole field. Reference [5] also treated the spatial nonuniformities in the current flow. The highly localized "residual resistivity dipole" fields, associated with transport past a localized scatterer, have received some measure of acceptance in the electromigration literature. Electromigration is the motion of lattice defects in metals in the presence of electronic current flow. The driving force, for motion of a defect, depends on the transport field at the defect site, and on the momentum delivered to the defect by the carriers actually incident on the defect. The defect motion can, thus, serve as a probe for spatial nonuniformities. Electromigration is a subject of great technological importance, and also is a subject in which a number of noted theoreticians (e.g. Friedel, Nozières, Peierls, Sham) have arrived at differing viewpoints, and which has been characterized by considerable controversy. It is remarkable that despite all that, the field has not attracted the attention of the broader transport theory community. We will not discuss the subject in further detail, and only cite a few references [7] to lead to the citation trail.

We would like to emphasize that, in the derivation of (1.1), and its generalizations, we can proceed in two ways. We can, as in [5], follow the carriers incident on an obstacle, and follow their subsequent motion, at first ignoring the resulting pile-up of charges. We can then calculate the charge pile-up in different parts of space, and let this be screened self-consistently, by the electron gas, to find a spatial potential distribution. Alternatively, if we only want to know the electrical resistance, we can follow the purely diffusive motion, of supposedly neutral carriers, and then appeal to the Einstein relation to obtain an electrical resistance. This was done in [8], in the derivation of (1.1). The point we want to emphasize: The two procedures are not really different. In the first case, e.g. as in [5], we are simply rederiving the Einstein relation along the way, without appealing to it explicitly, as in [8].

2. One-dimensional Systems

Reference [5] also discussed plane scatterers, exemplified by grain boundaries, or by tunnel junctions. The latter did not yet exist, in any clear form, at the time of [5], but had been analyzed, theoretically, long before then [9]. The results for the plane scattering surface were similar to those given in (1.1), but not identical. Reference [5], after all, invoked an incident three-dimensional velocity distribution, whereas (1.1) is a strictly one-dimensional result. Reference [5] stressed a point which we re-emphasize, and which arises whenever we go beyond the simplest one-dimensional models. The electrical resistance of a system, characterized by its scattering action of incident carriers, depends on the particular velocity distribution incident on the scatterer. One can, in general, expect this to depend on the kinetics in the "leads" which bring the carriers up to the system of interest.

Equation (1.1) can be derived by considering the unperturbed wave functions of the specimen, illustrated in Fig. 1, and then evaluating both the current flow and density gradient associated with these. Wave functions representing carriers incident from the reservoir on the right are also taken into account, but not shown in Fig. 1. The complex conjugates of the two wave functions that we have mentioned are not invoked, because the conjugates require *coherent* incident streams from the two reservoirs. The reservoirs are assumed to cause phase randomization, and cannot produce emerging streams with well-defined phases. It is this elimination of the time-reversed wave functions which yields a positive value of resistance in (1.1) [8]. Thus, the existence of the reservoirs is essential to that result. We can eliminate the reservoirs by closing the specimen on itself, making it into a closed loop. The loop must then be driven through a changing magnetic flux. How will this loop behave? This question is discussed in detail in another contribution at this conference [10], and in closely related work [11]. We stress, however, that in the absence of inelastic scattering the resistance of (1.1) does not appear.

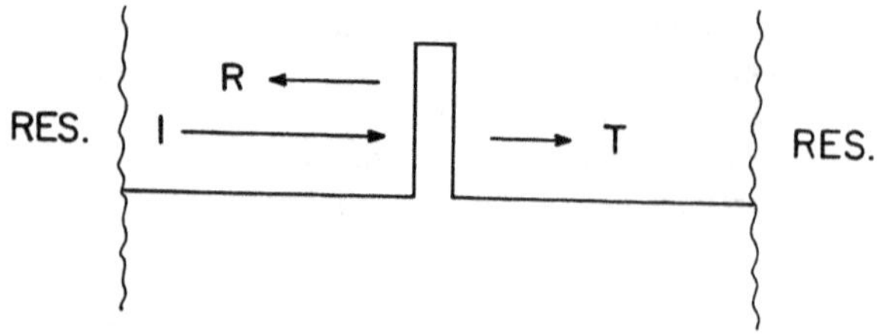

Fig. 1. Wave incident from left produces transmitted and reflected waves. The complex conjugate wave function would have two *incident* streams, with a well-defined phase relationship, and a single wave of unit amplitude leaving the obstacle.

One-dimensional systems, with degenerate Fermi statistics, have a particular simplicity. The current flow into a specimen defines the velocity distribution. We must, in general, expect greater complexity. Nevertheless, it is apparent that Fig. 1 represents a general prescription. The transmitted streams, perhaps weighted with suitable velocity factors, define the current flow. The reflected portions add to the incident streams, to define a density on the incident side. The smaller transmitted portions define a lesser density on the other side. Thus, after summing over all incident velocity classes, a density gradient or driving force is defined. It is clear that in this way generalizations of (1.1) can be found, which go beyond the one-dimensional case. They can allow for inelastic scattering, and probably also for sophisticated many-body interactions, in the sample under consideration. A discussion for the case of interacting electrons has, in fact, been provided by APEL [12]. Section 4 will discuss the generalization to many dimensions, following the work of IMRY, et al. [13]. A generalization for the case of inelastic scattering has been given by GEFEN and SCHÖN [14].

The treatment of a reflecting obstacle, embedded in a region with a spatially more uniform source of scattering, e.g. lattice scattering, has been given in [15], using semi-classical approximations. Thus, for example, [15] ignores the Friedel oscillations produced by the interference of incident and reflected waves, near the localized obstacle. Figure 2, adapted from [15] summarizes the results, using quasi-Fermi levels. The quasi-Fermi level ψ, for a particular class of electrons is the value of the Fermi-level needed to give the actual electron population in that class.

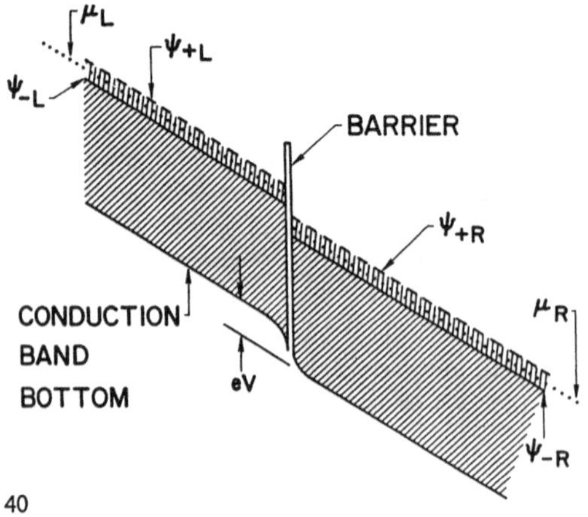

Fig. 2. Localized barrier embedded in a conductor with spatially uniform scattering. The quasi-Fermi levels, denoted by ψ with various subscripts, are discontinuous at the barrier. The potential drop, however, reaches exponentially (with the range of the screening length) into the regions on each side of the barrier. Symbols explained in detail in text.

The potential drop eV, associated with the localized barrier, extends over a screening length on each side of the obstacle. (In the case of an extended obstacle, not as localized as the one in Fig. 2, the location of the space charges associated with the voltage drop leads to a more complex discussion, which we avoid here.) The uniform slope of the bands, far from the obstacle, represents the field required for transport in the presence of the spatially uniform background scattering. ψ_{-L} and ψ_{-R} represent the levels to which electron states moving to the left are occupied, with the letter subscripts denoting the side of the barrier. Similarly the ψ_{+} denote the levels to which right moving electrons are present; the difference between ψ_{+} and ψ_{-} produces the current. μ_L is half-way between ψ_{-L} and ψ_{+L} and represents the total number of electrons present; similarly for μ_R. These μ's, as shown in [15], are needed to discuss the self-consistent screening that is present. They are also, as explained by ENGQUIST and ANDERSON [16], the potentials measured by an ideal potentiometer, connected to a particular point in space. The potentiometer tends to come into equilibrium equally with both directions of electron motion, and therefore averages between them.

To keep Fig. 2 from becoming too cluttered, one important aspect is omitted. Consider the electrons to the left of the barrier, moving to the right, in the energy range between ψ_{-L} and ψ_{+L}. These electrons, which represent the excess of right-moving electrons over left-moving electrons, will be transmitted elastically through the barrier. Thus, the transmitted electrons will arrive on the right hand side of the barrier, with an excess of kinetic energy, above that indicated by ψ_{+R}. This excess kinetic energy will be dissipated, to the right of the barrier, by whatever inelastic collisions are available for such thermal relaxation. Similarly the "holes", incident on the barrier from the right hand side, will emerge on the left hand side with an energy below that shown in Fig. 2. The thermal relaxation processes we have just invoked can only result from the fact that different energy levels, at the same position in real space, are out of equilibrium. Such a "vertical" flow of current results in an externally measured potential drop, just as much as the more normally considered "horizontal" flow. Note, however, that the distances, in energy, between the various transport related potentials in Fig. 2, are themselves proportional to the current flow. We are, therefore, moving current vertically through a distance which is proportional to the current. If the rate of equilibration between different energy levels increases as the levels come closer, then the extra potential drop, caused by the vertical flow, will vary more rapidly than the current flow, and represents a source of nonlinearity. It will not contribute to the linear resistance. We will have occasion to return to "vertical" current flow, in a very different connection, in Sec. 3.

We have emphasized the obvious fact that the energy dissipation required by current flow past a reflecting obstacle, characterized by (1.1), does not occur at the obstacle, but elsewhere. We have also emphasized that the derivation of (1.1) requires a phase randomization, presumably through inelastic processes, in the reservoirs connecting to the obstacle. These two inelastic processes can, of course, be caused by the same physical event. If there really are no inelastic processes available in the circuit, then, as indicated in [11], we cannot expect elastic scattering to give rise to a resistance. Reference [11], in its specific detailed results, is tied to a highly idealized one-dimensional model. Its principal aspect, however, the relationship of electrical resistance to inelastic processes, is likely to be more general. In that connection we comment on the recent paper by CARINI, et al. [17]. Reference [17] disagrees with [11] in its details, but also in connection with the point we have just emphasized. Reference [11] emphasizes that a small one-dimensional loop, with no inelastic scattering, and no leads, does not exhibit dissipative effects, but acts remarkably like a superconducting loop with a Josephson junction. In contrast, [17] asserts that the behavior of the eigenstates in such a loop can explain experimental observations on the resistance of a cylinder long compared to the inelastic scattering length, and with attached leads. A simple and appealing physical interpretation of the resistance oscillation in these cylinders has been given by BERGMANN [18].

Equation (1.1) gives a conductance proportional to the transmission coefficient only in the limit of small transmission probability. This was a point of some temporary dispute, as discussed in [2], but seems to be settled now. Outside of the one-dimensional localization field, however, the point is still not widely appreciated, as evidenced, for example, by [19].

3. Interface Resistances and Azbel Resonances

When two different conducting media are placed in series, we can, in general, expect an interface resistance arising from the fact that the distribution of carriers in real space, or in momentum space, in one medium, does not match that of the other. Thus a thin wire, connected to the end of a conductor with a larger cross-section, gives rise to the well known spreading resistance. The current flow in the thicker conductor; near its end, cannot be parallel to the conductor walls, but must converge toward the thin conductor. Similarly with an n^+-n semiconductor junction. The current on the n^+ side flows at lower energy levels than on the n side. Consider electron flow incident on the junction from the n^+ side. At the junction, carriers must flow up in energy to the higher energy levels, on the less heavily doped side. This *vertical* flow, as we have called it in Sec. 2, must be associated with a disequilibration between energy levels. This deviation from equilibrium will show up as part of the total voltage drop in the circuit, just as disequilibration along the spatial coordinates does. The upward flow in energy, requiring heat absorption from the surrounding lattice, is well know as a thermoelectric effect. The associated voltage drop, however, is not generally discussed. Note that such a vertical flow requires inelastic processes, and *takes place more easily if these are frequent.* That is contrary to the common expectation that inelastic collisions are an obstacle to current flow. This same observation applies to the escape rate from a metastable potential well, if motion in the well is severely underdamped [20]. In the case of escape from the potential well, this point was discussed perceptively, decades ago, by KRAMERS [21]. In the case of one-dimensional localization it has, of course, been understood for some time, that at very low temperatures inelastic collisions are needed to permit current flow as a result of jumps between the localized states.

We have given examples of interface resistances. As a result of their existence, we cannot, in general, assign a value of resistance to a sample in such a way that the resistance is independent of the kinetics in the connecting leads. This point was already stressed in [5], in connection with the resistance of a plane

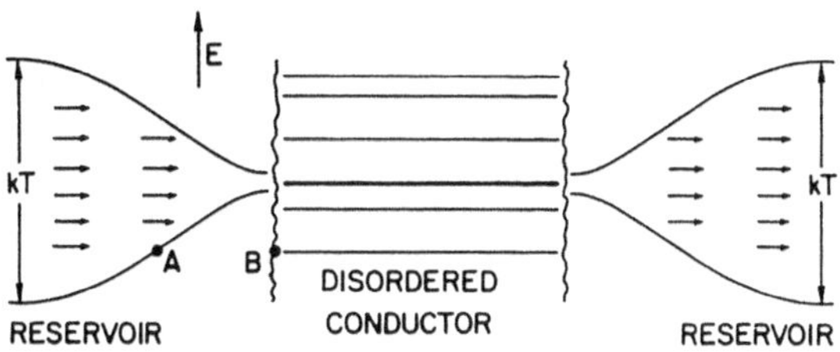

Fig. 3. Disordered conductor in middle, with its localized states, giving pronounced transmission resonances. Darkest horizontal line represents level with best transmission. Current flow in reservoirs is distributed over normal energy range and has to be funneled into good level.

barrier. We illustrate this, in further detail, by application to a one-dimensional sample which has a drastic dependence of transmission probability, on energy, as will be characteristic of a highly disordered sample [22]. Equation (1.1) assumes that the transmission coefficient is independent of energy, over the relevant range of ~kT, near the Fermi surface. Azbel pointed out that for a long and highly disordered sample this is not true; the transmission probability can vary drastically. It peaks at the localized eigenstates, but the height of these peaks can also vary greatly from one energy level to the next. Figure 3 illustrates the situation, and will be invoked in our analysis. First, however, we point out that the situation sketched in Fig. 3 is by no means the only possibility. We could, instead, imagine our sample as one in a long linear sequence of identical disordered conductors. If we allow no inelastic scattering, then the resistances put in series do not add. After all, the wave functions can be subject to multiple coherent scattering events back and forth between adjacent sections, and the phase relaxation required for (1.1) will be missing. We can, however, assume enough inelastic scattering between adjacent samples to give the required phase randomization, but still let most of the scattering and resistance come from the elastic disorder in the samples. Imagine, for example, that the adjacent identical samples are separated by regions which produce inelastic scattering, but inelastic scattering causing only small energy and momentum changes. Then the current can clearly concentrate in the "band" defined by the energy of the most highly transmitting state and by the width of its resonance.

One can easily derive a generalization of (1.1), leading to a conductance for energy dependent transmission [23].

$$G = \frac{e^2}{\pi\hbar}\left(\int_{-\infty}^{+\infty} dE(-\partial f/\partial E)T(E)\right)\Big/\int_{-\infty}^{+\infty} dE(-\partial f/\partial E)(1 - T(E)). \qquad (3.1)$$

Equation (3.1) can be derived by assuming that the excess of electrons incident from one side, over those incident from the other, is given by a simple shift in Fermi-levels. Equation (3.1) tells us that in the presence of large variations in T(E), the contributions to G come primarily from the resonances, and very likely, from the best resonance, within the range in which $\partial f/\partial E$ is nonvanishing. This leads to the situation shown in Fig. 3. That figure makes it clear that we have the equivalent of a spreading resistance, due to the need for "vertical" flow toward the transmitting energy level. Figure 3 also reminds us of another point, of lesser importance. Little or no current flows into the energy levels with poor transmission. Points A and B in Fig. 3 are essentially in equilibrium. Their respective population is given by the same quasi-Fermi level, equivalent to an absence of a voltage drop between A and B, if it were a simple spreading resistance problem. Therefore, the poorly conducting energy regions see a somewhat greater driving force than the highly transmitting level. This is ignored in (3.1), but can be presumed to be a minor source of error. Figure 3 also assumes that inelastic scattering is effective in the reservoirs, but not in the sample itself. An Appendix gives the detailed derivation of the extra spreading resistance implied by Fig. 3.

The discussion given in the Appendix is casual and illustrative, and not intended to be a definitive and final answer to the problem. The result derived in the Appendix is

$$R = \frac{1}{e\pi}(kT/\hbar\omega)\sqrt{\tau_i(\partial E/\partial n)/\sigma}\,\log\frac{kT}{\Delta E}. \tag{3.2}$$

$\hbar\omega$ is the energy exchange in the typical inelastic event. ΔE is the width of the resonance. M. Azbel has asked: What does (3.2) yield, if we assume that the connecting reservoirs in Fig. 3 only exhibit inelastic scattering, and no elastic scattering? In that case, σ varies as τ_i, and the result in (3.2) is independent of τ_i. Using $\sigma = (4\tau_i e^2/\pi^2 h^2)\partial E/dn$, (3.2) becomes

$$R = (\hbar/2e^2)(kT/\hbar\omega)\log(kT/\Delta E). \tag{3.3}$$

The $(\hbar/e^2)$ in (3.3) is familiar enough, and the logarithmetic term is typical of two-dimensional spreading resistances. The $kT/\hbar\omega$ factor measures the effectiveness of each scattering event, in producing the required change in energy.

We could elaborate on (3.2), invoking models for τ_i, and compare it to the actual result for the resistance of the disordered section [24], derived from (3.1). Our purpose, however, was to demonstrate the nature of contact resistances, through an example. There is a limited purpose in further elaboration on playful one-dimensional models, and we only add a few observations. Our result, (3.2), varies as log ΔE, whereas ΔE tends to occur as a simple multiplier in (3.1). ΔE, in turn, decreases exponentially with sample length. This suggests, as pointed out to us by BÜTTIKER [25] and STONE [26], that for long samples the spreading resistance becomes unimportant. The range of applicability of the *linear* resistance of (3.1) is, however, likely to be very small. First of all, as stressed by FOWLER [27], we are depending on the conductance of a *single* level, partially occupied by two electrons. These electrons, as a result of the resonance, cannot traverse the sample with a typical Fermi velocity, but bounce back and forth for a much longer time. Additionally, the applied field will perturb the delicate phase relationships which cause the resonance to occur at a particular energy [28]. Another point worth mentioning: We have not alluded to the coherence length of the wave packets emitted by our "leads." Coherence length is usually considered in optical interference experiments and in the case of electron beams utilized in an Aharanov-Bohm experiment. Our omission, here, is customary of papers in this field. That doesn't imply that it is a correct procedure, or that it is even immediately apparent what the value of the coherence length is.

4. Multichannel Conductance

We now generalize (1.1) to the case where we have many channels coming into the specimen and leaving it. In that case, as already stated, we lose the simplicity leading to (1.1); the total current no longer defines the distribution of current among the channels. Reference [5] already provided many-channel formulations, in connection with the point-scatterer, as well as in the case of a partially reflecting barrier. Both here, and in [5], we assume a distribution for the flux of carriers incident on the scatterer. *The transmitted and reflected fluxes are then determined by the kinetics of the specimen.* We can exercise some choice and taste in the assumed distribution of the incident flux; therefore, there can be many results. Unless we are given a totally defined system, we cannot say that one of these results is right, and that the others are wrong. To the best of this author's knowledge, a treatment of a complete system, including the power source and the (presumably necessary) inelastic effects, is not yet in existence.

Let us first review the multi-channel analysis of LANGRETH AND ABRAHAMS [29]. They assume that the chemical potential for all channels on the left, as determined by the average between both left and right moving carriers, is the same. Similarly the chemical potential for all channels on the right is the same. The chemical potential on the left is, of course, different from that on the right. At first sight, this seems plausible and reasonable. It corresponds to our common sense notion that there should be a fixed driving force across a sample. Furthermore, the fact that all the channels on the left have the same value of chemical potential means that they are in equilibrium with each other. That may seem appropriate: They are in contact with the same source. On closer inspection, however, the assumption of [29] becomes more questionable. It is a most remarkable mechanism which keeps the population between different channels perfectly in equilibrium, but permits the left and right moving components in the same channel to be out of equilibrium. Furthermore, in [29], the way in which the current is distributed among the channels is a function only of the scattering behavior of the conductor, and is unrelated to the scattering mechanism in the leads, which produces the presumed equilibration between the channels. In fact the left moving electrons, on the left side, are in part electrons which have just come out of the specimen, after transmission from the other side. Why should these newly transmitted electrons be part of an equilibrium maintained by the left hand side channels?

There is, however, a way to give the LANGRETH and ABRAHAMS [29] approach a proper physical basis. Let us assume that we have a long series of identical samples, as already discussed in Sec. 3, separated by phase randomizing agents. Then we must have spatial periodicity, and this requires the well defined chemical potential drop invoked in [29]. In a periodic array it becomes natural that the current distribution, among the channels, depend only on the scatterer. Such a periodic array of scatterers was already invoked in [5], in the treatment of partially reflecting planes. The solution, there, was obtained by requiring that the distribution of the outgoing flux, between channels, say on the right side, and in turn incident on the next scatterer, match the flux distribution incident on the left side of the original scatterer.

Alternatively, we could try to emphasize the actual scattering kinetics in the leads, emulating the multi-channel treatment of the point scatterer of [5]. Let us assume inelastic scattering in the leads, as invoked in the one-dimensional analysis of Fig. 2. Then we would, presumably, be carrying current with emphasis on those channels which have the largest velocity component parallel to the accelerating field. This would be unlikely to match the channel preferences of the scatterer, and we are left with a messy interface resistance problem.

Let us present one more possibility, in some detail. We assume that the *incident* flux from all of the channels on the left is characterized by one value of the quasi-Fermi level, that from the right by another. That is the approach which yields (3.1). Following ENGQUIST and ANDERSON [16] we note that the chemical potential in each channel is determined by averaging the quasi-Fermi level between the two directions within that channel. For all channels on the left, the incident (or right moving) carriers have the same quasi-Fermi level. But the number of transmitted or reflected electrons coming into the channel, out of the specimen, will, as already mentioned, depend on the scatterer kinetics. Thus, the chemical potential of the various channels, as determined by their total population including both directions of motion, will differ. They are not in equilibrium with each other. This approach is taken from IMRY, et al. [13]. Reference [13] re-establishes an earlier result of AZBEL [30], which seems to have received very little attention.

As already specified, we assume that the particle density in channels incident from the left is characterized by a quasi-Fermi level $\Delta\mu$ in excess over that of the channels arriving from the right. For simplicity, we assume a relatively energy independent transmission coefficient within the range kT and $\Delta\mu$. Further generalizations, however, of the results of this section, in the same way that (3.1) generalizes (1.1), are obviously possible. Let T_{ij} be the *probability* that the current incident in channel j is transmitted to channel i. Similarly, R_{ij} is the *probability* for reflection from channel j into channel i on the same incident side. For our present purposes, it will be adequate to limit ourselves to incident channels on the left. We assume that different channels represent orthogonal transverse wave functions and their interference, after integration over the transverse cross-section of the input leads, vanishes. We also assume that different incident channels are incoherent with each other. Therefore, *probabilities* due to different incident channels can be added. Finally, the interference effects between an incident wave and reflection back into the same channel are unavoidable, and must occur. But, as in the derivations of (1.1) we average over a number of wavelengths, to make these standing wave effects (or Friedel oscillations) disappear. Let $\Sigma_j T_{ij} = T_i$. This measures the total transmission probability into channel j. Similarly, let $\Sigma_j R_{ij} = R_i$. The excess incident carriers from the left in channel i carry a current $v_i\ \Delta\mu\ dn_i/d\mu$. v_i is the velocity component parallel to the direction of incidence on the sample, and $dn_i/d\mu$ a density of states. (To achieve generalizations of the form of (3.1), $dn_i/d\mu$ must be interpreted as a change in occupation density, as the Fermi-level is moved.) The excess incident current, from the left, in channel j is $ev_j\Delta\mu(dn_j/d\mu)$. Using $v_j = (1/\hbar)d\mu/dk_j$ this becomes $(e/2\pi\hbar)\Delta\mu$, assuming that different spins are treated as different channels. The portion of this current transmitted into channel i is $T_{ij}(e/2\pi\hbar)\Delta\mu$. The total current then becomes

$$j = \Sigma_{ij} T_{ij} (e/2\pi\hbar)\Delta\mu = \Sigma_i T_i (e/2\pi\hbar)\Delta\mu. \tag{4.1}$$

Now we must ask about the accompanying potential difference. Let us, at first, just follow the incident carriers, without regard to space charge neutrality, and see how their density accumulates. In channel i, on the right, the particle density is j_i/ev_i, or $v_i^{-1}\Delta\mu/2\pi\hbar$. Thus the total density on the right is

$$\Delta n_R = \Sigma_i v_i^{-1} T_i \Delta\mu/2\pi\hbar. \tag{4.2}$$

On the left we have the incident stream, as well as the reflected stream, giving

$$\Delta n_L = \Sigma_i v_i^{-1}(1 + R_i)\Delta\mu/2\pi\hbar. \tag{4.3}$$

The density difference between Eqs. (4.3) and (4.2) can then be written

$$\Delta n = \Sigma_i v_i^{-1}(1 + R_i - T_i)\Delta\mu/2\pi\hbar. \tag{4.4}$$

We can now go to a potential difference, ΔV, from (4.4), by invoking either the Einstein relation, or self-consistent screening. This gives

$$e\Delta V dn/d\mu = \Delta n. \tag{4.5}$$

$dn/d\mu$ is now the total density of states, including both left and right moving carriers in each channel. Using

$$dn/d\mu = \Sigma_i (1/\pi\hbar v_i), \tag{4.6}$$

Eqs. (4.4) - (4.6) yield

$$\Delta V = \frac{1}{2e}\left(\Sigma_i v_i^{-1}(1 + R_i - T_i)\Delta\mu\right)\Big/\Sigma_i v_i^{-1}. \tag{4.7}$$

Equation (4.7) together with (4.1) defines the conductance

$$G = \frac{e^2}{\pi\hbar}\frac{(\Sigma_i T_i)\left(\Sigma v_i^{-1}\right)}{\Sigma_i v_i^{-1}(1 + R_i - T_i)}, \tag{4.8}$$

in agreement with [13] and [30]. This is, of course, all within the single particle approximation; generalizations beyond that are likely to be possible. We have also assumed that the two leads attached to our sample have the same physical structure, permitting a one-to-one identification of left hand side channels with right hand side channels. Generalizations beyond that are in existence [13].

We add a final comment intended primarily for prospective authors in this field. We have stressed that the occurrence of a dissipative resistance requires inelastic processes. Elastic scattering, by itself, is not enough, as shown in [10] and [11]. In view of the need for inelastic scattering we must, in general, allow for the possibility that electrons can gain energy and move *vertically,* as in Fig. 3.

Therefore, the highest electron energy found in the sample, or in perfectly conducting leads, need not necessarily be the exact energy of some electron reservoir, or the energy which electrons are given by a battery.

ACKNOWLEDGMENTS

P. Marcus helped in the exposition of Section 4, with a critical question. M. Büttiker has been a constant discussion and debating partner. Y. Imry, as the principal author and architect of [13], provided much of the motivation for this paper.

APPENDIX

We take a semi-classical viewpoint under which we assume that there is a quasi-Fermi level ψ, giving the population density for each position (E, x) in the two-dimensional space of Fig. 3. We then assume that the current flow is determined by

$$j_x = \frac{\sigma_x}{e}\frac{\partial\psi}{\partial x}, \quad j_E = \frac{\sigma_E}{e}\frac{\partial\psi}{\partial E}. \tag{A.1}$$

The factor (1/e) arises because ψ is an energy, not a voltage. Setting div j=0 yields

$$\sigma_x \partial^2\psi/\partial x^2 + \sigma_E \partial^2\psi/\partial E^2 = 0. \tag{A.2}$$

Let us re-scale, using $E' = (\sigma_x/\sigma_E)^{1/2}$ E, $x' = x$. In the new coordinates the conductivities become $\sigma'_x = \sigma'_E = \sqrt{\sigma_x\sigma_E}$.

Now, consider an ordinary two dimensional spatial spreading resistance. A semicircular contact of radius r, has current flowing out toward a larger semicircle of radius R. The resistance of this configuration in ordinary space is $(\rho_\square/\pi)\log R/r$, with $\rho_\square$ the the resistance per square, of the two-dimensional medium. In our case we can take $R \sim kT$, and $r \sim \Delta E$, where ΔE is the width of the best resonance in Fig. 3.

Can we say something about σ_x and σ_E? The ordinary conductivity, σ, of the reservoir must be $\int \sigma_x dE$. Thus, $\sigma_x \sim \sigma/kT$, because the current flow is limited to an energy range ~kT about the Fermi-level. What about σ_E? It is related to the inelastic relaxation time τ_i. Let us assume that the typical inelastic process involves a phonon with energy $\hbar\omega$, transporting electrons over that energy range within the inelastic collision time τ_i. Let us also assume that we start with an electron population which has a serious deviation from thermal equilibrium, over such an energy range. Then ψ changes by kT, or so, over this range, and $\partial\psi/\partial E \sim kT/\hbar\omega$. Then

$$j_E = (\sigma_E/e)\partial\psi/\partial E \sim \sigma_E kT/e\hbar\omega. \tag{A.3}$$

The number of particles, per unit length, which have to be transported, to relax toward thermal equilibrium, is of order $\hbar\omega\partial n/\partial E$. Thus, the relaxation time becomes

$$\tau_i \sim (\hbar\omega\partial n/\partial E)/(j_E/e) = \frac{(\hbar\omega)^2 e^2}{kT\sigma_E}\partial n/\partial E, \tag{A.4}$$

or

$$\sigma_E \sim \frac{1}{\tau_i} e^2 \frac{\partial n}{\partial E} \frac{(\hbar\omega)^2}{kT}. \tag{A.5}$$

Thus

$$\rho_{\square}^{-1} = \sqrt{\sigma_x \sigma_E} = e \frac{\hbar\omega}{kT} \sqrt{\sigma \tau_i^{-1} \partial n / \partial E}. \tag{A.6}$$

The spreading resistance becomes

$$R = \frac{1}{e\pi} \left(\frac{kT}{\hbar\omega} \right) \sqrt{\tau_i (\partial E / \partial n) / \sigma} \log \frac{kT}{\Delta E}. \tag{A.7}$$

What about the validity of the *local* conductivities invoked above? This requires that the variations of carrier density occur on a scale which is large compared to that between the scattering or relaxation events. For motion along the energy coordinate that requires $\hbar\omega < kT$, where ω is the frequency of the typical phonon involved in the inelastic collisions. That may not be all that easily satisfied. In fact, the energy variation, *near* the resonant sample, must be much more striking and would really require $\hbar\omega < \Delta E$. This is most unlikely to apply. For motion along the spatial x direction, if we assume $\tau_{in} >> \tau_{el}$, which is likely, we are making a reasonable approximation.

REFERENCES

1. P. Erdös and R. C. Herndon: Adv. Phys. **31**, 65 (1982)
2. R. Landauer: Phys. Lett. **85A**, 91 (1981)
3. P. W. Anderson, D. J. Thouless, E. Abrahams, D. Fisher: Phys. Rev. B **22**, 3519 (1980)
4. P. D. Kirkman, J. B. Pendry: *The Statistics of 1 Dimensional Resistances,* preprint
5. R . Landauer: IBM J. Res. Dev. **1**, 223 (1957)
6. R. Landauer: Z. Physik B **21**, 247 (1975)
7. R. S. Sorbello: In *Macroscopic Properties of Disordered Media* ed. by R. Burridge, S. Childress, G. Papanicolaou (Springer, Heidelberg 1982) p. 251
   R. Landauer: Phys. Rev. B **16**, 4698 (1977)
   A. H. Verbruggen, H. Griessen, J. H. Rector: Phys. Rev. Lett. **52**, 1625 (1984)
   A. K. Das, R. Peierls: J. Phys. C **8**, 3348 (1975)
8. R. Landauer: Philos. Mag. **21**, 863 (1970)
9. A. Sommerfeld, H. Bethe: In *Handbuch der Physik*, ed. by A. Smekal (Springer, Berlin 1933), 2nd ed., Vol. 24, Part 2, p. 446
10. M. Büttiker: *Quantum Oscillations in Small Normal Metal Rings,* Supplement to Proceedings of Int. Conf. on Localization, Interaction and Transport Phenomena in Impure Metals, (Springer, Heidelberg, to be published)
11. M. Büttiker, Y. Imry, R. Landauer: Phys. Lett. **96A**, 365 (1983)
    M. Büttiker, Y. Imry, M. Azbel: *Quantum Oscillations in One-Dimensional Normal Metal Rings,* Phys. Rev. A, to be published
12. W. Apel: J. Phys. C: Solid State Phys. **16**, 2907 (1983)
    W. Apel, T. M. Rice: J. Phys. C: Solid State Phys. **16**, L271 (1983)
13. Y. Imry, M. Büttiker, R. Landauer, S. Pinhas: *Generalized Many Channel Conductance Formulae*, to be published

14. Y. Gefen, G. Schön: *Effect of inelastic processes on localization in one dimension,* preprint
15. R. Landauer: In *Electrical Transport and Optical Properties of Inhomogeneous Media*, ed. by J. C. Garland, D. B. Tanner (AIP, New York 1978) p.2, see Sec.9
16. H.-L. Engquist, P. W. Anderson: Phys. Rev. B **24**, 1151 (1981)
17. J. P. Carini, K. A. Muttalib, S. R. Nagel: Phys. Rev. Lett. **53**, 102 (1984)
18. G. Bergmann: Phys. Rev. B **28**, 2914 (1983)
19. T.-L. Ho: Phys. Rev. Lett. **51**, 2060 (1983)
E. Ben-Jacob, E. Mottola, G. Schön: Phys. Rev. Lett. **51**, 2064 (1983)
20. M. Büttiker, E. P. Harris, R. Landauer: Phys. Rev. B **28**, 1268 (1983)
21. H. A. Kramers: Physica **7**, 284 (1940)
22. M. Ya. Azbel: Solid State Commun. **45**, 527 (1983)
M. Ya. Azbel, A. Hartstein, D. P. DiVincenzo: Phys. Rev. Lett. **52**, 1641 (1984), and papers cited therein
23. S. Eränen, J. Sinkkonen, T. Stubb: Acta Polytechnica Scandinavica **1983**, 45 (1983). See also [22] and [16]
24. M. Büttiker: Unpublished notes
25. M. Büttiker: Private communication
26. A. D. Stone: Private communication
27. A. Fowler: Private communication
28. C. M. Soukoulis, J. V. José, E. N. Economou, Ping Sheng: Phys. Rev. Lett. **50**, 764 (1983)
F. Delyon, B. Simon, B. Souillard: Phys. Rev. Lett. **52**, 2187 (1984)
29. D. C. Langreth, E. Abrahams: Phys. Rev. B **24**, 2978 (1981). See also B. S. Andereck: J. Phys. C: Solid State Phys. **17**, 97 (1984)
30. M. Ya. Azbel: J. Phys. C: Solid State Phys. **14**, L225 (1981)

Interaction Effects in Weakly Localized Regime of Metallic Films

H. Fukuyama

Institute of Solid State Physics, The University of Tokyo, Roppongi, Minato-ku, Tokyo 106, Japan

Theories of interaction effects in weakly localized regime of metallic films are discussed including the higher order interaction effects, the inelastic scattering time (phase coherence time) and the effect of strong spin-orbit interactions. Such interaction effects have been characterized by two coupling constants, the BCS attractive interaction, $-\lambda_{ph}$, and the Coulomb repulsive interaction, μ, which have different cutoff energies.

1. Introduction

Mutual interactions between electrons are known to play essential roles on the localization phenomenon [1]. Such interaction effects are originally treated [2-5] in its linear order, g_i, and also in the linear order of the randomness parameter, $\lambda=\hbar/2\pi\varepsilon_F\tau$, ε_F and τ being the Fermi energy and the life time, respectively, which are independent parameters to characterize the system. Since the interaction can be strong in actual systems, as envisaged from the occurrence of the magnetic phase-transition or the superconductivity on one hand and the theoretical investigation yield exact predictions in Weakly Localized Regime (WLR) which have been checked experimentally with very high accuracy on the other hand [6,7], it is relevant to construct theories which take infinite order in interaction constant but in WLR. Such problem has recently been studied for the case of the repulsive Coulomb interaction based on ordinary diagram technique [8,9], which can be considered as an extention of theories of correlations in electron gas in the clean system [10] to those in random potential. The results for the contributions due to diffusion (i.e. g_1- and g_3- processes) are in accordance with those of FINKELSTEIN [11] and ALTSHULER and ARONOV [12], who employed different procedures. The contributions due to Cooperons (i.e. g_2- and g_4- processes) are shown to have characteristic features.

More generally, the interaction processes in metals will be characterized by two coupling constants, the BCS attractive interaction, $-\lambda_{ph}$ and the Coulomb interaction, μ, which have different cut-off energies; ω_D the Debye frequency for the former and ε_F for the latter.

In this paper we consider exclusively the metallic films, though most of the results has more generality, and the different roles played in WLR by λ_{ph} and μ will be clarified. The controversy over the inelastic scattering time, or the phase coherence time, τ_ε, due to electron-electron interaction in dirty metallic films is resolved. Effects of strong spin-orbit scattering are also discussed in the same context.

We take unit of $\hbar=k_B=1$.

2. Model

Our model is given by

$$H = H_0 + V \quad , \tag{1}$$

$$H_0 = \sum_i \frac{p_i^{\,2}}{2m} + u_0 \sum_{i,\ell} \delta(r_i - R_\ell) \quad , \tag{2}$$

$$V = \frac{1}{2} \sum_{i \neq j} [v(r_i - r_j) - \lambda_{ph}\, \delta(r_i - r_j)] \quad , \tag{3}$$

where v(r) is the Coulomb repulsive interaction

$$v(r) = e^2 / \varepsilon_0 r \equiv \sum v_B(q)\, e^{iqr} \quad , \tag{4}$$

and $-\lambda_{ph}$ is the BCS attractive interaction. The Coulomb interaction is screened, and the RPA screening in clean system, Fig.1, is charcaterized by the effective coupling constant, μ, given as an average of the scattering on the Fermi surface

$$\mu = N(0) \langle v_B(k-k') / [1 - v_B(k-k')\pi(k-k', 0)] \rangle_{|k|=|k'|=k_F} \quad , \tag{5}$$

where $N(0)$, $\pi(q,\omega_\ell)$ and k_F are the three-dimensional density of state per spin, the polarization function and the Fermi momentum, respectively. It is to be noted that μ as well as λ_{ph} involves a large momentum transfer in the scattering process. As a function of the energy-transfer, ω_ℓ, two parameters to characterize the system, μ and λ_{ph}, have different cut-off energies, ε_F and ω_D as schematically shown in Fig. 2.

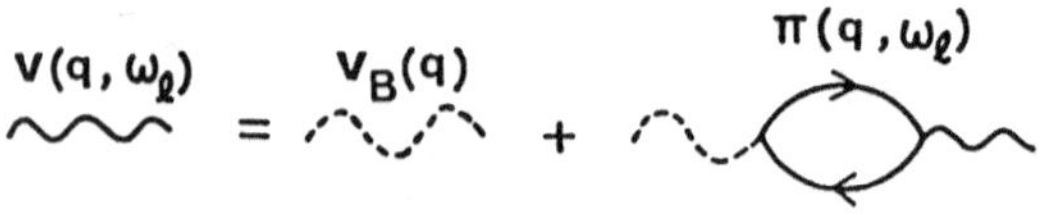

Fig. 1 RPA screening.

In metallic films in which we are interested in this paper, the typical thickness, b, is around several tens to hundred Å. For such b the quantization of the kinetic energy in the perpendicular direction is not important since $\varepsilon_F \gg (2\pi/b)^2/2m$ where m is the effective mass of electrons. Moreover the mean free path, ℓ, is usually $\ell < b$. In such circumstances the individual electronic states are essentially three-dimensional. On the other hand, the

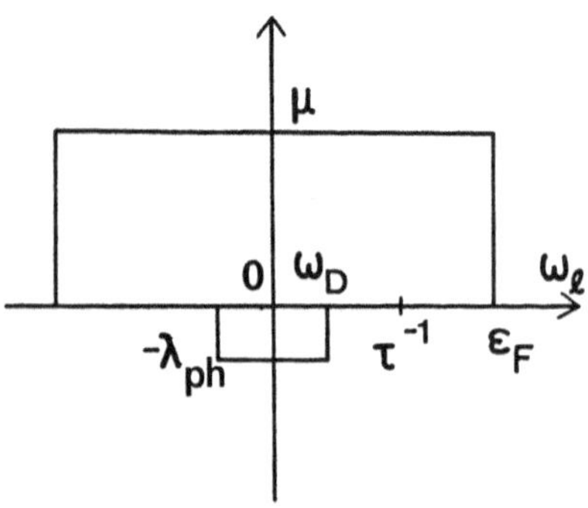

Fig. 2 Energy dependences of λ_{ph} and μ.

thermal diffusion length, L_T, given by $L_T=(D/T)^{1/2}$ where D and T are three-dimensional diffusion constant and the temperature, respectively, satisfies $L_T>b$. In this sense the diffusion is essentially two-dimensional. As regards the interaction processes, the dimensionality also depends on the amount of momentum transfer, q; i.e. for $q<(>)b$ the process is two(three)-dimensional.

Since $\varepsilon_F \gg \omega_D$ in ordinary metals, the life time, τ, due to elastic scattering can be either $\tau^{-1}>\omega_D$ or $\tau^{-1}<\omega_D$; the former is rather common and we confine ourselves to this case in the following as indicated in Fig.1.

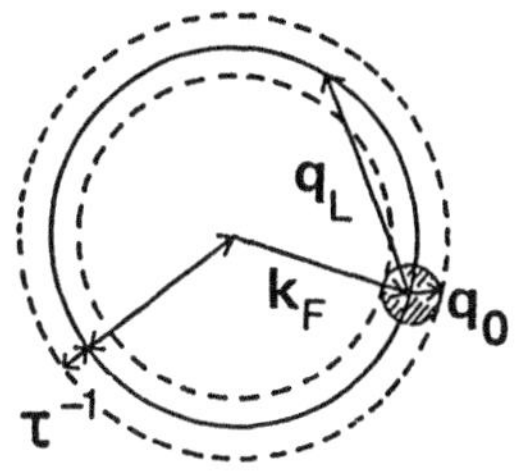

Fig. 3 The classification of processes with large momentum transfer, q_L, and small momentum transfer (shaded region).

In dirty metals the Fermi surface is smeared by the elastic scattering as schematically indicated in Fig.3. Due to the smearing effect the polarization function in (4) for $|q| \lesssim q_0 \equiv (D\tau)^{-1/2} = \sqrt{3}/\ell$, where $D=2\varepsilon_F\tau/3m$, is of diffusion type

$$\pi(q,\omega_\ell) = -2N(0)\,\frac{Dq^2}{Dq^2+|\omega_\ell|} \quad , \tag{6}$$

and the RPA-screened Coulomb interaction, $v(q,\omega_\ell)$, for such small transfer of momentum and energy is given by [13]

$$v(q,\omega_\ell) = \frac{1}{2N(0)}\,\frac{\kappa(q)^2}{q^2}\,\frac{Dq^2+|\omega_\ell|}{D\kappa(q)^2+|\omega_\ell|} \quad , \tag{7.a}$$

$$\kappa(q)^2 = \kappa_3^{\,2}\,(1-e^{-qb}) \quad , \tag{7.b}$$

where $\kappa_3^{\,2}=8\pi e^2N(0)/\varepsilon_0=4k_F/\pi a_B$, ε_0 and $a_B=\varepsilon_0/me^2$ being the dielectric constant and the effective Bohr radius. Here q is the momentum in the plane of films and in (7.a) only the uniform component along the perpendicular direction is retained, since it is relevant to the following discussions.

An important feature of the dynamically-screened Coulomb interaction given by (7.a) and (7.b) is that except for very small momentum transfer, $v(q,\omega_\ell)$ is essentially given by

$$v(q,\omega_\ell) = \frac{1}{2N(0)}\,\frac{Dq^2+|\omega_\ell|}{Dq^2} \quad , \tag{8}$$

since $\kappa_3^{\,2}$ is a large quantity. To be more explicit, the condition for the validity of (8) is

$$q_s \lesssim |q| \lesssim q_0 \quad , \quad \text{where} \tag{9.a}$$

$$q_s \sim (T/\varepsilon_F)(a_B/\ell b) \quad . \tag{9.b}$$

Even in dirty metals the matrix elements of interactions with large momentum transfer $|q|>q_0$ is not essentially affected by the elastic scattering, i.e. the smearing effect of the Fermi surface, and they are given by μ and λ_{ph}.

3. Conductivity

In the lowest order in interaction process the quantum corrections to the self-energy function have been characterized by the processes shown in Fig.4, where broken and double broken lines are diffuson and Cooperon, respectively. It is to be noted that only in the g_1-process is the momentum transfer small.

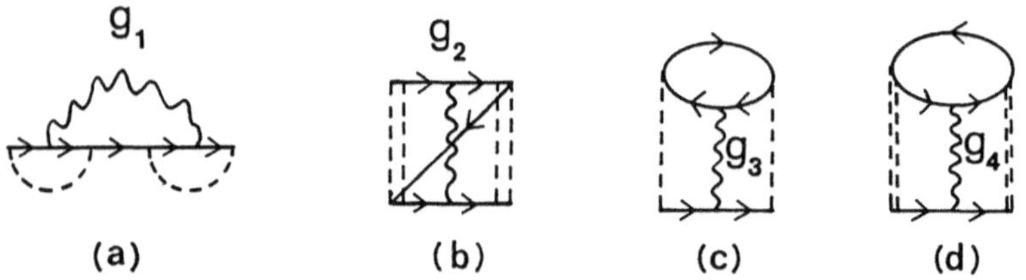

Fig. 4 Interaction processes in the lowest order of the interaction.

(i) g_1-process

The higher order interaction effect in the g_1-process is taken into account by considering the process shown in Fig.5, where the dotted line with a cross indicates the average of impurity potential and q_s and q_L are small and large momentum transfer, i.e. $q_s \lesssim q_0 \lesssim q_L \lesssim 2k_F$, respectively. Since the important contributions to vertex correction due to interaction in Fig.5(b) involves the large energy-transfer up to τ^{-1}, which is assumed to be much larger than ω_D, we can ignore λ_{ph} in this process. Hence the result of ref.9 applies by rewriting F as 2μ, and then the dynamically screened Coulomb interaction and the effective coupling constant are given by

$$v(q,\omega_\ell) = \frac{1}{2N(0)} \frac{D'q^2+|\omega_\ell|}{Dq^2} \quad , \tag{10.a}$$

$$g_1(q,\omega_\ell) \doteq \frac{(Dq^2+|\omega_\ell|)^2}{2Dq^2(D'q^2+|\omega_\ell|)} \quad , \tag{10.b}$$

$$D' = D(1-\mu) \quad . \tag{10.c}$$

These are valid for the region of (9.a)

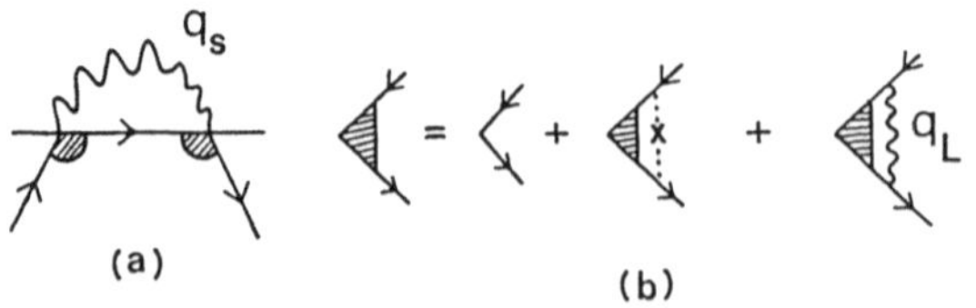

Fig. 5 The g_1-processes.

By use of (10.b) and repeating similar calculations as for the lowest order contribution, the quantum corrections to the conductance from the g_1-- process is given by

$$\sigma_1' / \sigma_\square = - \lambda^* g_1(\mu)\ \phi_2(T) \quad , \tag{11.a}$$

$$\phi_2(T) = 2\pi T \sum_{\omega_\ell > 0} \frac{1}{\omega_\ell} = \ell n \frac{1}{4\pi T\tau} \quad , \tag{11.b}$$

$$g_1(\mu) = \frac{1}{\mu} \ell n \frac{1}{1-\mu} \quad . \tag{11.c}$$

In (11.a) , $\sigma_\square$ and λ^* are defined by [14]

$$\sigma_\square = \frac{ne^2\tau}{m} d \equiv \frac{\varepsilon_F{}^*\tau}{\pi} e^2 \quad , \tag{12}$$

$$\lambda^* = [2\pi\ \varepsilon_F{}^*\ \tau]^{-1} \quad , \tag{13}$$

with

$$\varepsilon_F{}^* = \varepsilon_F\ (2k_F b\ /\ 3\pi) \quad . \tag{14}$$

(ii) g_3-process

The higher order contributions in the g_3-processes are given by the paramagnon types of processes [15-17] shown in Fig.6, where s and s' are spin indices. Since the interaction is spin-independent, s and s' are independent of each other. The first order contribution in these processes, Fig.6(a), is due to the static interaction, $\omega_\ell=0$, and then characterized by $\mu-\lambda_{ph}$, while the interactions in the higher order processes, i.e. Fig.6(b), (c) etc, involve large energy-transfer and then characterized by μ. Hence the contribution from the g_3-process is the sum of those of ref.9 and the contribution from the BCS interaction between antiparallel spins present in the lowest order process;

$$\sigma_3' / \sigma_\square = \lambda^* [2g_3(\mu) - \lambda_{ph}]\ \phi_2(T) \quad , \tag{15}$$

$$g_3(\mu) = 2\ (\frac{1}{\mu} \ell n \frac{1}{1-\mu} - 1) \quad . \tag{16}$$

(iii) g_2- and g_4- process

There exist cancellations of higher order contributions between the g_2- and g_4-process and the net result is due to the processes of the type shown in Fig.7, where the momentum transfer in the interaction is large. In this

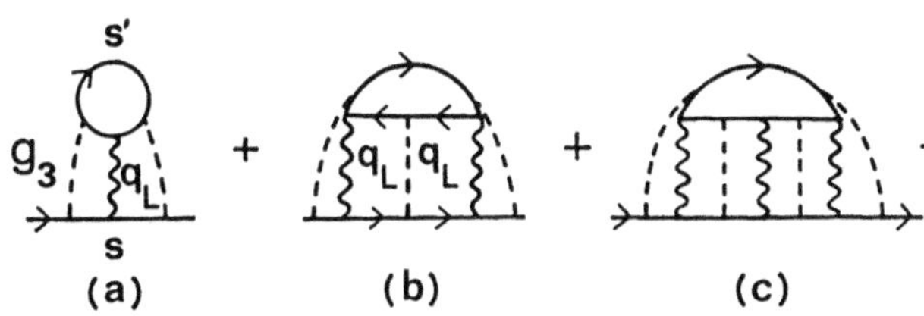

Fig. 6 The g_3-processes.

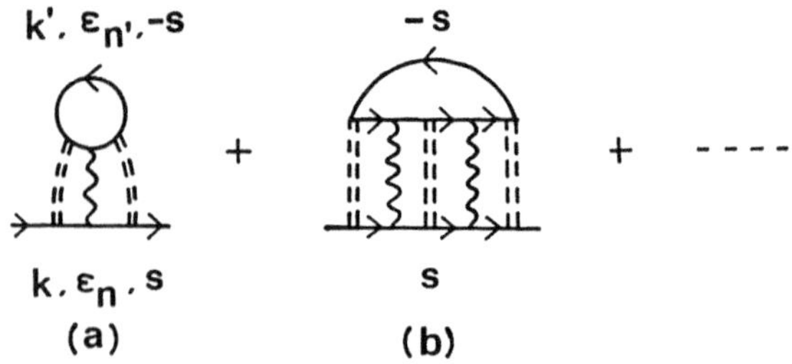

Fig. 7 Cooperon contributions to the self-energy function.

case, as was noted by ALTSHULER et al. [18], the effective interaction has strong dependences on q=k+k', and $\omega_\ell=\varepsilon_n+\varepsilon_{n'}$, defined as in Fig.7(a);

$$g_2 - 2g_4 \equiv - g_4(q,\omega_\ell) \quad , \tag{17.a}$$

$$g_4(q,\omega_\ell) = \begin{cases} -\lambda_s\,[1-\lambda_s\,\{\ln(1.14\omega_D/T)+\psi(\frac{1}{2}) \\ \qquad -\psi(\frac{1}{2}+\frac{Dq^2+|\omega_\ell|}{4\pi T})\}]^{-1} & (|\omega_\ell|<\omega_D) \\ -\mu\,[1+\mu\,\ln\varepsilon_F/\omega_D]^{-1} & (|\omega_\ell|>\omega_D) \end{cases} \tag{17.b}$$

where $\psi(z)$ is the di-gamma function and

$$\lambda_s = \lambda_{ph} - \mu^* \quad , \tag{18}$$

$$\mu^* = \mu\,[1+\mu\,\ln\varepsilon_F/\omega_D]^{-1} \quad . \tag{19}$$

Since the contributions from the region of $|\omega_\ell|<\omega_D$ are sensitive to T or the magnetic field, we consider those exclusively in the following. Equation (17.b) is valid for either $\lambda_s>0$ or $\lambda_s<0$, and may be rewritten as

$$g_4(q,\omega_\ell) = [\ln T_c/T + \{\psi(\tfrac{1}{2}) - \psi(\tfrac{1}{2}+\frac{Dq^2+|\omega_\ell|}{4\pi T})\}]^{-1} \quad (|\omega_\ell|<\omega_D) \tag{20}$$

where

$$T_c = 1.14\,\omega_D\,e^{-1/\lambda_s} \quad . \tag{21}$$

The energy variable of the Cooperon, $\mathcal{D}_\varepsilon$, coupled with $g_4(q,\omega_\ell)$ in the processes in Fig.7, is $(\varepsilon_n-\varepsilon_{n'})$ in contrast to $\omega_\ell=\varepsilon_n+\varepsilon_{n'}$ in $g_4(q,\omega_\ell)$; i.e.

$$\mathcal{D}_\varepsilon(q,\omega_{\ell'}) = \frac{1}{2\pi N(0)\tau^2}\;\frac{1}{[Dq^2+|\omega_{\ell'}|+\tau_\varepsilon(q,\omega_{\ell'})^{-1}]} \quad , \tag{22}$$

where $\omega_{\ell'}=\varepsilon_n-\varepsilon_{n'}$ and the life-time τ_ε of the Cooperon is introduced here phenomenologically, as was first introduced by ANDERSON et al. [19]. We indicated that τ_ε depends generally on q and $\omega_{\ell'}$ and as we will discuss later $\tau_\varepsilon(q,0)$ depends on q strongly. The difference on the energy dependences between $g_4(q,\omega_\ell)$ and $\mathcal{D}_\varepsilon(q,\omega_{\ell'})$ causes the complexity in the problem.

If $\ell n T_c/T >> 1$ and then q- and ω_ℓ- dependences in (20) can be ignored, the quantum corrections to the conductivity is given by

$$(\sigma_2' + \sigma_4') / \sigma_\square \simeq \lambda^* g_4(0,0) \bar{\Phi}_2(T) \quad . \tag{23}$$

where $\bar{\Phi}_2(T)$ is essentially same as $\phi_2(T)$, (11.b), but with the cut-off energy, ω_D, instead of τ^{-1}, i.e. $\phi_2(T) \sim \ell n\, \omega_D/4\pi T$. This result will be valid generally for non-superconducting system, i.e. $\lambda_s<0$, and for T not so low. On the other hand the results for $T\to 0$ but $\lambda_s<0$ are given by

$$(\sigma_2'+\sigma_4') / \sigma_\square \simeq \lambda^* \ell n \left[1 - \lambda_s \ell n\, \omega_D/T\right] \quad , \tag{24}$$

where we made use of the result of ref.9 with the replacement of ε_F by ω_D.

For superconducting system, $\lambda_s>0$, the contributions from g_2- and g_4-processes are qualitatively different, since the q- and ω_ℓ- dependences of $g_4(q,\omega_\ell)$ are to be carefully taken into account. If the temperature, T, is high enough, however, the contribution in the lowest order of the effective coupling constant is still given by (23), which takes contributions from the shaded region in Fig.8 uniformly. On the other hand, as, T gets close to T_c, the processes associated with $g_4(q,0)$, which signals the onset of superconductivity since $g_4(0.0)\to -\infty$ as $T\to T_c$ and used to be called the fluctuation propagator, yield such dominant contributions as those obtained by AZLAMAZOV-LARKIN [20] (σ_{AL}) and MAKI-THOMPSON [21] (σ_{MT}). In Fig.8 the energy region which yields such contributions is shown by a thick line. It is important to note that $(\sigma_2'+\sigma_4')$, (23), and σ_{AL} and σ_{MT} are all the quantum corrections to the conductivity lowest order in the randomness parameter, λ^*, derived from the self-energy correction of the type shown as Fig.7, but they have different regions of temperature for their applicability. So far to the knowledge of the author no complete investigation exists which interpolates between high temperature region and near T_c. It appears, however, that the magnetoresistance due to g_2- and g_4- process is appreciable even at relatively high temperature (e.g. $T\sim 10T_c$) if $\lambda_s>0$, and that the magnetoresistance at high temperatures will hence tell whether the system becomes superconducting or not at low temperatures.

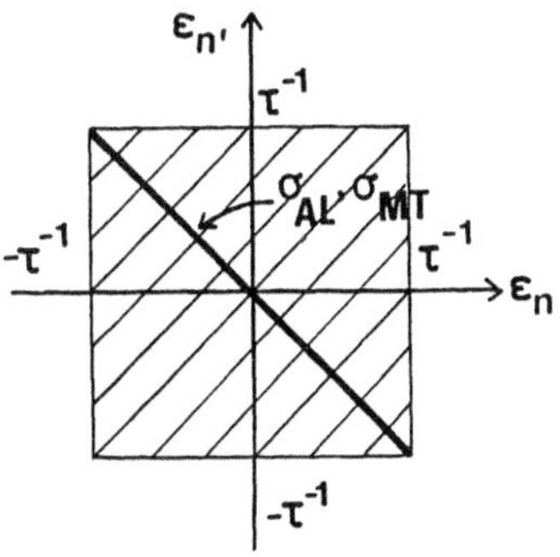

Fig. 8 Energy region for Cooperon contributions.

4. Inelastic Scattering Time

One of the most important physical quantities emerging from the study of the localization phenomenon is the inelastic scattering time, τ_ε, which was originally introduced by ANDERSON, ABRAHAMS and RAMAKRISHNAN [19], as a life time of Cooperon which determines the first order quantum correction to the conductivity. This τ_ε was assumed to be the same as the energy relaxation time, τ_e, of quasi-particle in disordered system. The latter, τ_e, due to Coulomb interaction,has been investigated some time ago by SCHMID [22], who found $\tau_e^{-1} \propto T^{d/2}$, d being the dimensionality. In the present context of the localization ABRAHAMS et al. [23] carefully examined τ_e on the Fermi surface and obtained

$$\tau_e^{-1} = \lambda^* \pi T \ln T_1/T \quad , \tag{25.a}$$

$$T_1 = 4(\varepsilon_F \tau)^2 D(\kappa_3^2 b)^2 \quad , \tag{25.b}$$

where κ_3^2 is introduced in (7.b). This result is obtained from the analysis of one-particle self-energy corrections of electrons in disordered metals. Since τ_ε is introduced via Cooperon, which is two-particle correlation function, it was not obvious that τ_ε is equal to τ_e. ABRAHAMS and the present author [24] evaluated τ_ε by diagrammatic analysis of the Cooperon and found that the life time of Cooperon, which is defined in Fig.9 where $\varepsilon_n(\varepsilon_n+\omega_\ell)<0$, has precisely the same value as (25.a) and (25.b) at Q=0 and $i\omega_\ell \equiv \omega \to 0$. LOPES DOS SANTOS [25] also confirmed this.

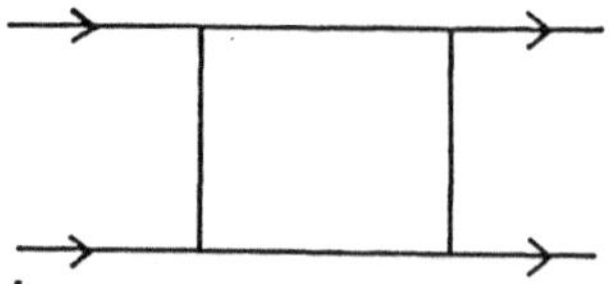

Fig. 9 The difinition of Q and ω_ℓ in the Cooperon.

On the other hand ALTSHULER et al. [26] claimed that τ_ε appearing in the quantum corrections to the conductivity is given by

$$\tau_\varepsilon^{-1} = \lambda^* \pi T \ln c \quad , \tag{26}$$

where c is a constant of the order of $\varepsilon_F{}^*\tau$. There exist a clear difference between(25.a) and (26) as regards the $\ln T$ factor and τ_e^{-1}, (25.a), is much larger numerically than τ_ε^{-1}, (26). Recently EILER [27] found the same result as (26) except for the numerical factor in c. The relationship between (25) and (26) has been a puzzle for some. This controversy can be solved as follows by noting [28] the strong Q-dependence of the life time of Cooperon as defined in Fig.9. By repeating the similar procedures as in ref.24 but by keeping Q finite, we obtain

$$\tau_\varepsilon^{-1}(Q) = -2N(0)\tau^2 \sum_q \int_{-\infty}^{\infty} dx \ \mathrm{Im}\ v_R(q,x) \ \{[f(x-\varepsilon)+f(x+\varepsilon)]\ \Gamma^R_{p-h}(qx) + 2n(x)\ \Gamma^R_{p-p}(q+Q,x)\} \tag{27.a}$$

$$\simeq -4N(0)\tau^2 \sum_q \int_{-\infty}^{\infty} dx \ \frac{\Gamma^R_{p-p}(q,x)}{\mathrm{sh}\beta x}\ \mathrm{Im}\ v_R(q-Q,x) \quad , \tag{27.b}$$

where $\beta=T^{-1}$, $f(x)=[e^{\beta x}+1]^{-1}$, $n(x)=[e^{\beta x}-1]^{-1}$ and Γ^R_{p-h}, Γ^R_{p-p} and $v_R(q,x)$ are retarded diffusion, Cooperon and the dynamically screened Coulomb interactions, respectively. (See ref.24 for more detail.) Since Γ^R_{p-p} contains τ_ε^{-1}, (27.b) is the self-consistent equation to determine τ_ε^{-1}. If Q=0, the q-integration is divergent from small q-region if $v(q,\omega_\ell)$, (8), is employed and this divergence due to $|q|\lesssim q_s$, q_s being defined by (9.b), is removed if (7.a) is employed. This leads to (25.a) with large T_1 reflecting very small q_s. However, the localization contribution to the conductivity, σ_L', is not solely given by $\tau_\varepsilon^{-1}(Q=0)$, since it is determined as a sum over Q as

$$\sigma_L' \propto \sum_Q \frac{1}{DQ^2+\tau_\varepsilon^{-1}(Q)} \quad . \tag{28}$$

This integration over Q may have the same effect as the cut-off of the contribution from the region of very small q in (27.b), $q < Q_0 \equiv (D\bar{\tau}_\varepsilon)^{-1/2}$, where $\bar{\tau}_\varepsilon$ is the effective value to be determined self-consistently. Hence q-integration in (27.b) is to be performed for $Q_0 \lesssim q$ and the result is given by [29]

$$\begin{aligned}\bar{\tau}_\varepsilon^{-1} &= \lambda^* \pi T \ln \bar{\tau}_\varepsilon T \\ &\simeq \lambda^* \pi T \ln 2\varepsilon_F^* \tau \quad .\end{aligned} \tag{29}$$

This is essentially the same result as (26).

The strong Q-dependence of τ_ε also implies its strong dependence on the magnetic field [28] but the analysis of magnetoresistance based on the constant τ_ε will not be modified seriously as discussed in ref.28.

Although $\tau_\varepsilon(Q=0)$ is not affected by the higher order interaction effects [28], $\bar{\tau}_\varepsilon$, (29), can depend on μ, but the straightforward manipulations by use of (10.a), instead of (18), indicates that there exist no contribution linear order in μ.

5. Effect of Strong Spin-Orbit Scattering

Spin-orbit scattering is known to cause interesting phenomena in the localization in films [6,30] and we will briefly examine here those problems discussed in the preceding sections in the presence of strong spin-orbit scattering by assuming $\tau_{s0} \sim \tau$, where τ_{s0} is the life time due to spin-orbit scattering.

As has been discsussed in detail elsewhere [31], the diffuson and the Cooperon depends on spin indices in this case. By repeating similar considerations as for the lowest order contributions we first find that the g_1-process yields the same result as in the absence of the spin-orbit interaction given by (11.a), while the contribution form the g_3-process is reduced by factor 4 compared with (15) as in the lowest order contribution [31,32]. On the other hand the Cooperon contribution, g_2- and g_4- processes, is reduced by factor 2 compared to that in the absence of the spin- orbit scattering.

In accordance with the invariance of the contribution from the g_1-process, which is due to the long wave length of the screened Coulomb interaction, the inelastic scattering time is seen also not be affected by strong spin--orbit scattering [29].

6. Summary

The higher order interaction effects have been discussed for the case of metallic films, which is neither two-nor three-dimensional in the sense that the one-particle state is three-dimensional whereas the diffusion is essentially two-dimensional. Characteristic roles played by the attractive BCS interaction, $-\lambda_{ph}$, and the repulsive Coulomb interaction, μ, have been clarified. The difference of the inelastic scattering times by ABRAHAMS et al. and by ALTSHULER et al. has been resolved by taking strong Q-dependence of the phase coherence time into account based on the diagrammatical analysis by ABRAHAMS and the present author.

References

1. P. W. Anderson: Physica **117&118B**, 30 (1983).
2. B.L. Altshuler and A.G. Aronov: Solid State Commun. **30**, 115 (1979).
3. B.L. Altshuler, A.G. Aronov and P.A. Lee: Phys. Rev. Lett. **44**, 1288 (1980).
4. H. Fukuyama: J. Phys. Soc. Jpn. **48**, 2169 (1980); Surface Sci. **113**, 489 (1982); Physica **117&118B**, 673 (1983).
5. P.A. Lee: Anderson Localization ed. by Y. Nagaoka and H. Fukuyama (Springer Verlag, 1982) p.62; H. Fukuyama: ibid p.89.
6. G. Bergmann: Z. Phys. B. **48**, 5 (1982), Phys. Rev. **B29**, 6114 (1984) ; to appear in Phys. Rev. C.
7. R.C. Dynes: Physica **109B&C**, 1857 (1982).
8. H. Fukuyama: Y. Isawa and H. Yasuhara: J. Phys. Soc. Jpn. **52**, 16 (1983).
9. Y. Isawa and H. Fukuyama: J. Phys. Soc. Jpn. **53**, 34 (1984).
10. For review, K.S. Singwi and M.P. Tosi in Solid State Physics ed. by F. Seitz, D. Turnbull and H. Ehrenreich (Academic Press, 1981) vol.36, p.177.
11. A.M. Finkelstein: Soviet Physics-JEPT **57**, 97 (1983).
12. B.L. Altshuler and A.G. Aronov: Solid State Commun. **46**, 429 (1983).
13. S. Maekawa, H. Ebisawa and H. Fukuyama: J. Phys. Soc. Jpn. **53**, #8 (1984).
14. H. Fukuyama: NATO Advanced Summer Institute on Percolation, Localization and Superconductivity ed. by A.G. Goldman and S. Wolf (Plenum Press, 1984).
15. N.F. Berk and J.R. Schrieffer: Phys. Rev. Lett. **17**, 433 (1966).
16. S. Doniach and S. Engelsberg: Phys. Rev. Lett. **17**, 750 (1966).
17. P. Fulde and A. Luther: Phys. Rev. **170**, 570 (1968).
18. B.L. Altshuler, A.G. Aronov, D.E. Khmelnitzkii and A. I. Larkin, Soviet Physics-JETP **54**, 4111 (1981).
19. P.W. Anderson, E. Abrahams and T.V. Ramakrishnan: Phys. Rev. Lett. **43**, 718 (1979).
20. L.G. Aslamazov and A.I. Larkin,: Phys. Lett. **26A**, 238 (1968).
21. K. Maki: Prog. Theor. Phys. **39**, 897 (1978); R.S. Thompson: Phys. Rev. **B1**, 327 (1970).
22. A. Schmid: Z. Physik **271**, 251 (1974).
23. E. Abrahams, P.W. Anderson, P.A. Lee and T.V. Ramakrishnan: Phys, Rev. **B24**, 6783 (1981).
24. H. Fukuyama and E. Abrahams: Phys. Rev. **B27**, 5976 (1983).
25. J.M.B. Lopes dos Santos: Phys. Rev. **B28**, 1189 (1983). In this paper it has been pointed out that the correct τ_e^{-1} is factor 2 smaller than eq.(25.a).
26. B.L. Altshuler, A.G. Aronov and D.E. Khmelnitzky: J. Phys. **C15**, 7367 (1982).
27. W. Eiler: to appear in J. Low Temp. Phys. **56**, #5/6 (1984).
28. H. Ebisawa and H. Fukuyama: J. Phys. Soc. Jpn. **52**, 3304 (1983).

29. H. Fukuyama: submitted to J. Phys. Soc. Jpn.
30. G. Bergmann: Z. Phys. **B48**, 5 (1982).
31. H. Fukuyama: J. Phys. Soc. Jpn. **51**, 1105 (1982).
32. G. Deutscher and H. Fukuyama: Phys. Rev. **B25**, 4298 (1982).

The Conductor Insulator Transition in Strongly Disordered Systems

W. Götze

Physik-Department der Technischen Universität München
D-8046 Garching, Fed Rep. of Germany, and

Max-Planck-Institut für Physik und Astrophysik
D-8000 München, Fed. Rep. of Germany

The basic concepts and approximations of the self consistent current relaxation theory for the particle motion in random potentials are reviewed. The scaling laws for the dynamical conductivity near the conductor insulator phase transition are discussed, the cross over from the transition dominated by quantum interference effects to the classical percolation transition is examined in particular. Results for the mobility and for the current excitations are compared with experimental data.

1. Introduction

Within the kinetic equation approach (KEA) a zero-temperature conductor is viewed as a Fermi gas, specified by the particle mass m, charge e and density n, moving in an array of scatterers, distributed randomly with density n_i and interacting with the particles via some potential u. Scattering processes relax the particle momentum with a characteristic decay rate $(1/\tau)$ and this leads to a finite dc conductivity $\sigma_o = e^2 n\tau/m$. Low frequency long wave length density excitations propagate as diffusion modes, specified by the diffusivity $D \propto \sigma$. Coulomb interactions can be incorporated by distinguishing properly between external and screened fields; diffusion modes are then replaced by damped plasmon excitations.

It is well known that the KEA is valid only in the limit of high dilution of the scattering centres $n_i \to 0$. The KEA becomes unvalid if the mean free path $\ell = v_F\tau$, where $v_F = k_F/m$ is the Fermi velocity, tends towards the particle wave length $\lambda_F = 2\pi/k_F$ or towards the spacing between the scattering centres $1/n_i^{1/3}$. The ultimate reason for the break down of the KEA is the occurence of a conductor-insulator phase-transition (CIT). For classical motion, the CIT is referred to as a percolation transition [1]; if the particle dynamics is dominated by quantum effects, the CIT is called an Anderson transition [2]. The disorder-induced CIT is of great academical interest, since it provides us with the simplest model of a fluid glass transition [3]. It is also of experimental interest, since it is the explanation of anomalies observed in doped semi conductors [4].

In this note I will review some results of the self-consistent current relaxation theory (SCCRT). This is an approximation approach developed in order to extend the KEA into the regime of strong disorder such, that predictions for experimental quantities can be made for the glass phase and for the CIT. I will concentrate on the mentioned model of noninteracting fermions; but some results for the classical Lorentz model and for Coulomb interaction effects will be touched upon also.

2. Density Fluctuations and Current Relaxation

Kubo's (impurity averaged) density propagator for wave vector $\vec{q}$ and frequency ω, $\Phi(q,\omega) = \Phi'(q,\omega) + i\Phi''(q,\omega)$, is an appropriate quantity in order to discuss particle motion in disordered systems. The density spectrum for $\omega \neq 0$, $\Phi''(q,\omega)$, is proportional to van Hove's dynamical structure factor $S(q,\omega)$. So, in principle $\Phi''(q,\omega)$ is measurable in a scattering experiment with a test particle transferring recoil momentum q and experiencing an energy loss ω to the electron system. Applying Mori's formalism [5] and observing the continuity equation for the density fluctuations $\rho(\vec{q})$ and currents $\vec{j}(\vec{q})$, one can write: $\Phi(q,\omega)=g(q)/[\omega+q^2 K_\ell(q,\omega)/g(q)]$. Here $g(q)$ is the wave vector-dependent compressibility and $K_\ell(q,\omega)$ is a certain correlation function formed with the longitudinal currents $j_\ell(q)=\vec{q}\vec{j}(\vec{q})/q$. If one approximates $K_\ell(q,\omega)\cong K_\ell(q=0,\omega)=K(\omega)$ one arrives at the Green-Kubo formula, connecting the density propagator for long wave length Φ_H with the compressibility g and the homogeneous current correlation function K:

$$\Phi_H(q,\omega) = - g/[\omega+q^2 K(\omega)/g]. \tag{1}$$

One can also show, that K is the $q\to 0$ limit of the transversal current correlation function $\Phi_T(q\to 0,\omega)=K(\omega)$ and so the (transversal) dielectric constant is given by $\varepsilon(\omega)-1=4\pi e^2 K(\omega)/\omega$. Consequently, the dynamical polarizability of the system reads

$$\chi(\omega) = e^2 K(\omega)/\omega, \tag{2}$$

while the dynamical conductivity, the current spectrum, is given by [6]

$$\sigma(\omega) = e^2 K''(\omega). \tag{3}$$

Applying Mori's reduction once more, one can write

$$K(\omega) = - (n/m)/[\omega+M(\omega)], \tag{4}$$

where $M(\omega)$ is the correlation function of the fluctuating forces experienced by the current due to the impurities. The current relaxation kernel $M(\omega)$ vanishes for $n_i=0$ and formula (4) [7] is introduced to separate the free particle motion from that nontrivial part due to randomness. Within the KEA one gets $M(\omega)=i/\tau$ and equ. (4) then is the known Drude formula. So $M''(\omega)$ generalizes the constant relaxation rate $1/\tau$ of the KEA to a general frequency dependent function. While in weakly disordered systems the current relaxation spectrum $M''(\omega)$ is a white noise spectrum, strong disorder will be reflected by a strong, even singular, frequency-dependence of $M''(\omega)$. Causality then requires the reactive part, $M'(\omega)$, to be non-zero.

Let us contemplate the relevance of $M(\omega)$ as a mean to characterize the conductor and the insulator phase. The dc conductivity $\sigma=\sigma(\omega=0)$ is non-zero if and only if $M''(\omega)<\infty$ for $\omega\to 0$ and then

$$\sigma = (ne^2/m)/M''(\omega) \quad ; \quad \omega\to 0 . \tag{5}$$

So a conductor is characterized by the low frequency relaxation spectrum to be finite. On the other hand, the static polarizability $\chi=\chi(\omega=0)$ is finite if and only if $0<\omega M'(\omega)<\infty$ for $\omega\to 0$, and then

$$\chi = (ne^2/m)/(\omega M'(\omega)) \quad ; \quad \omega\to 0 . \tag{6}$$

So an insulator is characterized by the low frequency relaxation kernel to exhibit a simple pole. Substitution of these two alternatives for the low frequency relaxation kernel, into equ. (1) yields two alternatives for the density propagator. In the conductor phase one gets

$$\Phi_H(q,\omega) = - g/[\omega+iq^2\sigma/(e^2 g)] . \tag{7}$$

Thus the density spectrum is regular. Density fluctuations, created at time zero, decay to zero if the time tends to infinity. The density fluctuations are ergodic variables in the conductor phase. On the other hand, in an insulator one finds

$$\Phi_H = -(1/\omega)f(q) \quad , \tag{8a}$$

$$f(q) = g/\left[1+q^2\chi/(e^2 g)\right] \quad . \tag{8b}$$

So for long times density fluctuations do not vary any more with time: $\langle\rho(q,t\to\infty)\rangle = f(q)\langle\rho(q,t=0)\rangle$. Inhomogenities, created at some time, do not disappear; the system does not approach a homogeneous equilibrium state. Connected with the singular behaviour of $M(\omega)$ in an insulator phase is a corresponding singular behaviour of $\Phi(q,\omega)$. The fluctuating forces exhibit a static contribution if and only if the density propagator exhibits a corresponding term. Density fluctuations and fluctuating forces are non-ergodic variables in the insulator phase.

Let us consider non-interacting electrons. Then $g(q\to 0)=\rho_F$, where ρ_F denotes the density of states at the Fermi level. Then one gets for the conductor a diffusion spectrum

$$\Phi_H''(q,\omega) = \rho_F Dq^2/\left[\omega^2+(Dq^2)^2\right] \quad , \tag{9a}$$

where diffusivity D and conductivity σ are connected by the Einstein relation

$$D = \sigma/(\rho_F e^2) \quad . \tag{9b}$$

On the other hand, in the hydrodynamic limit equ. (8a) yields purely static density fluctuations for the insulator phase

$$\Phi_H''(q,\omega) = \pi\delta(\omega)f(q) \quad . \tag{10a}$$

The formfactor $f(q)$ is characterized by a localization length r_o

$$f(q) = \rho_F/(1+q^2 r_o{}^2) \quad . \tag{10b}$$

There is an analogue of the Einstein relation, which now connects $r_o{}^2$ with the polarizability:

$$r_o{}^2 = \chi/(\rho_F e^2) \quad . \tag{10c}$$

Another formula, specifying the difference between conductors and insulators is obtained, if one calculates the quadratic radius of a density fluctuation, produced at time zero by a change of the local thermodynamic potential: $S(t)=\frac{1}{2}\langle(\Delta\gamma)^2(t)\rangle$. One finds for large times the asymptotic expressions

$$S(t) - S(t=0) = \begin{cases} Dt & \text{, conductor} \quad (11a) \\ r_o{}^2 & \text{, insulator} \quad . \quad (11b) \end{cases}$$

These results quantify the concepts of extended and localized motion respectively.

The results (9a), (9b) and (11a) for a conductor have been known for a long time. The preceeding formalism was proposed [8] in order to achieve corresponding general formulae for an insulator and to treat conductor and insulator within the same formal scheme. The formulae provide in particular a definition of a localization length with well specified statistical averaging procedures, and they also connect r_o with quantities like χ or $S(t)$ which can be measured in principle.

The discussion carried through above holds also for the classical Lorentz system after some modifications [9]. This is the model of a classical hard sphere or hard disk moving with fixed speed through an array of random point scatterers.

The discussion is to be modified, if Coulomb interaction between the particles is taken into account. The long range interaction implies $g(q)$ to vanish for $q \to 0$. For a bulk semi conductor with background DC ε one gets [10]

$$g(q\to 0) = q^2 \varepsilon/(4\pi e^2) \quad , \quad d=3. \tag{12a}$$

So equ. (7) describes a relaxation mode whose strength is strongly suppressed in the hydrodynamic limit $q \to 0$. If the electrons are bound to a plane surface so that a two-dimensional system is created, one gets

$$g(q\to 0) = q\varepsilon/(2\pi e^2) \quad , \quad d=2 \quad . \tag{12b}$$

Such model is applicable to MOS devices [11]. Then equ. (7) describes a peculiar mode. $\Phi_H''(q,\omega)/g$ is a symmetric central resonance curve, as discussed in connection with equ. (9a) for diffusion; but the half-width of the curve scales linearly with wave vector $\omega_H = \sigma 2\pi q/\varepsilon$, unlike the one for diffusion curves, which scales quadratically $\omega_H = Dq^2$. No simple physical interpretation of $f(q)$ is obvious in the case of a charged particle system.

From the preceeding discussion follows: the essence of an insulator is not the vanishing of the dc conductivity σ but rather the appearance of a pole in the relaxation kernel. Hence the signature of a CIT in a disordered system is the development of a low frequency singularity of the relaxation kernel, in particular the occurence of an important reactive part $M'(\omega)$. The pole of the relaxation kernel, its low frequency singularities and the dominance of its reactive part are in one to one correspondence with similar poles, singularities and reactive parts of the density propagator [8]. If one focuses on low frequency phenomena one can write [12]

$$K(\omega) = -(n/m)/M(\omega) \quad ; \quad |\omega/M(\omega)| \ll 1 \quad , \tag{13}$$

and eliminate the kernel $M(\omega)$ in favour of the current correlation function $K(\omega)$; the latter can be viewed as a frequency dependent diffusivity.

3. Selfconsistent Current Relaxation

The kernel $M(\omega)$ is the correlation function of the fluctuating forces F acting on the total particle momentum [5]. F turns out to be the sum of pair modes $\vec{q}U(\vec{q})\rho(\vec{q})^*$, consisting of fluctuations of the gradients of the random potential $\vec{q}U(\vec{q})$ for momentum $\vec{q}$ and density fluctuations of momentum $-\vec{q}$. A leading order factorization approximation thus yields

$$M(\omega) = \sum_{\vec{q}} w(q)\Phi(q,\omega) \quad , \quad \text{with} \tag{14a}$$

$$w(q) = q^2 \langle |U(\vec{q})|^2 \rangle/(dnm) \quad . \tag{14b}$$

For the model specified by impurity density n_i and impurity structure factor $s(q)$, which usually is replaced by unity, one gets $w(q) = |u(q)|^2 s(q) n_i q^2/(dnm)$. The contents of equ. (14a) is the following: the impurities transfer recoil momentum $\vec{q}$ to the electrons with a coupling amplitude $\vec{q}U(\vec{q})$. The density of states for the recoil process is proportional to van Hove's function $\Phi''(q,\omega)$. A golden rule treatment of the current decay rate yields $M''(\omega) \propto |\vec{q}U(\vec{q})|^2 \Phi''(q,\omega)$, where over all possible $\vec{q}$ has to be summed. Equivalently, one can argue that the current can decay into other modes. The simplest ones are pairs consisting of density fluctuations and random potential fluctuations, which act as static phonons.

Equations (14) formulate a self-consistency problem. The current relaxation kernel is given by the density propagator. But the kernel rules the current correlations, equ. (4), and via continuity equation the currents rule the time-dependence of the density fluctuations. Density fluctuations and current relaxations are to be calculated simultaneously. If the recoil spectrum in equ. (14a) is replaced by the one for a free fermion gas, $\Phi''(q,\omega) \cong \Phi_o''(q,\omega)$, the relaxation spectrum is practically a white one $M_o''(\omega) \cong 1/\tau$, and the rate $1/\tau$ is the KEA result. So the standard conductivity theory is reproduced as lowest order approximation. However, it was demonstrated /8,12/ that the feedback between current relaxation and density spectrum may be so effective, that equ. (14a) yields an insulator solution. In particular, the non-linear equations for $M(\omega)$ may lead to a CIT if parameters like n or n_i are shifted through some critical values. In the following section 4 some qualitative features of the CIT will be considered and the general discussion will be continued in sections 5 and 6.

4. The Critical Region for the Conductor Insulator Transition

Simple implications of the self-consistency equations can be obtained if one restricts oneself to small frequencies and to the neighbourhood of the CIT points. Motivated by the discussion of section 2 one defines implicitly a critical region by requiring three small parameters: $|\omega|$, $|1/M(\omega)|$ and $|\omega M(\omega)|$. Expansion in leading order in terms of these quantities is to be carried out for the self-consistency equations, which then simplify such that scaling laws can be read off. To proceed, we split off in equ. (14a) the generalized hydrodynamic form, equ. (1), for the propagator. The rest is to be approximated by a constant

$$M(\omega) = i\nu - \sum_{\vec{q}} W(q)/\left[\omega + q^2 K(\omega)/g(q)\right] \quad . \tag{15a}$$

Introducing the dimensionless frequency $\hat{\omega}=\omega/\nu$ and the dimensionless mobility $\hat{K}(\hat{\omega})=K(\omega)(\nu m/n)$, one finds

$$-i\hat{K}(\hat{\omega}) = 1 - F\left[\sqrt{\hat{\omega}/\hat{K}(\hat{\omega})}\right] \quad , \tag{15b}$$

$$F(z) = \sum_{\vec{q}} W(q)/\left[z^2 + (n\nu^2 q^2/mg(q))\right] \quad . \tag{15c}$$

Within the critical region one can expand in terms of

$$|\hat{\omega}/\hat{K}(\hat{\omega})| \ll 1 \quad . \tag{15d}$$

This expansion depends crucially on the $q \to 0$ asymptote of $W(q)$ and on the dimensionality d; I will consider several cases separately.

4a. The Percolation Transition for d=3

For the mentioned Lorentz system the evaluation of $W(q)$ is not trivial because of the singular interaction. But in qualitative agreement with equ. (14b) $W(q \to 0) \propto q^2$ and an upper q-cut-off is provided by the inverse particle diameter [9]. Hence $F(z)=F_o - z^2 F_1 + \ldots$, F_o, $F_1 > 0$; equ. (15b) reduces to a quadratic equation for $\hat{K}(\hat{\omega})$. One finds

$$\hat{K}(\hat{\omega}) = |\tau| S_{\pm}(\hat{\omega}/\tau^2) \quad . \tag{16a}$$

Here $\tau = F_o - 1$ denotes the separation parameter, the sign alternative refers to $\tau \gtrless 0$ respectively, and $S_{\pm}(z)$ is the uniquely determined causal solution of the elementary equation $S^2 \pm iS - iF_1 z = 0$. Hence

$$S_+(|z| \ll 1) = F_1 z \quad , \tag{16b}$$

$$S_-(|z| \ll 1) = i \quad , \tag{16c}$$

$$S_\pm(|z| \gg 1) = \sqrt{iF_1 z} \quad . \tag{16d}$$

The obvious implications are the following. For $|\omega|>\omega_c=\tau^2\nu$ a critical current spectrum increasing proportional to $\sqrt{\omega}$ is obtained. For $\tau<0$ the system is a conductor with a conductivity σ decreasing to zero linearly if $\tau\to0$. For $\tau>0$ the system is an insulator, where the localization length diverges like $1/\sqrt{\tau}$ if $\tau\to0$. The hydrodynamic regime is restricted to $|\omega|<\omega_c$ and ω_c slows down proportional to τ^2 if the critical point is approached. To discriminate between the conductor and the insulator phase, the system has to be examined on a time scale exceeding $1/\omega_c$.

Computer experiments done for the Lorentz system [13] are not carried out close enough to the critical point in order to test the scaling law (16a). For dimensionality d=2, F_1 diverges logarithmically; hence a scaling law holds only up to logarithmic corrections [9]. Computer data [14] support the corresponding results about the cross over of the long time tails connected with the critical spectrum [15]. Within the effective medium approach, applied to a classical random hopping model for the percolation problem [16], results equivalent to equs. (16) were obtained.

4b. The Anderson Transition for d=3

If one introduces the general phase space propagator $\Phi_{\vec{k},\vec{p}}(\vec{q},\omega)$, formed with Wigner's particle hole excitations $f_{\vec{k}}(\vec{q})=a^*_{\vec{k}+\vec{q}/2}a_{\vec{p}-\vec{q}/2}$, one can write the mode coupling equation (14a) as a triple sum

$$M(\omega) = \sum_{\vec{k}\vec{p}\vec{q}} w(q)\,\Phi_{\vec{k},\vec{p}}(\vec{q},\omega) \quad . \tag{17a}$$

The propagator exhibits low frequency long wave length singularities due to the diffusive motion of the density fluctuations $\rho(\vec{q})=\sum_{\vec{k}} f_{\vec{k}}(\vec{q})$, equ. (9a). These $q\to0$ singularities are strongly suppressed, since $w(q)$ vanishes in the small q limit, equ. (14b). But for non-interacting fermions, moving in a random potential, there holds a peculiar symmetry [17,18]:

$$\Phi_{\vec{k},\vec{p}}(\vec{q},\omega) = \Phi_{(\vec{k}-\vec{p}+\vec{q})/2,\,(\vec{p}-\vec{k}+\vec{q})/2}(\vec{p}+\vec{k},\omega) \quad . \tag{18}$$

Hence equ. (17a) can be rewritten into the equivalent form [18]:

$$M(\omega) = \sum_{\vec{k}\vec{p}\vec{q}} w(\vec{k}+\vec{p})\,\Phi_{\vec{k},\vec{p}}(\vec{q},\omega) \quad . \tag{17b}$$

This shows, that there is another singularity hidden in $M(\omega)$ due to the regime $\vec{p}+\vec{k}\to0$ in equ. (17a) or, equivalently, due to $q\to0$ in equ. (17b). This singularity is not suppressed by a vanishing w. To extract this singularity one pulls out the hydrodynamic contribution in equ. (17b) and lumps the rest into a constant. One obtains again equ. (15a), with some complicated vertex $W'(q)$ instead of $W(q)$. The main point is that W' approaches some non-zero constant in the long wave length limit $W'(q\to0)=W_0\neq0$, as opposed to $W(q)$, which vanishes in that limit [18]. Consequently, the asymptotic expansion of equ. (15c) changes to $F(z)=F_0-F_1'z+\ldots$, $F_1'\propto W_0>0$. Thus equ. (15b) yields the scaling law

$$\hat{K}(\hat{\omega}) = |\tau|\,S_\pm'(\hat{\omega}/|\tau|^3) \quad . \tag{19a}$$

The notation is chosen in analogy to the one used in equs. (16). $S'(z)$ is the causal solution of the equation $S'^2 \pm iS' - iF_1'\sqrt{z}S' = 0$. Hence

$$S_+(|z| \ll 1) = F_1'z \quad , \tag{19b}$$

$$S_-(|z| \ll 1) = i \quad , \tag{19c}$$

$$S_\pm(|z| \gg 1) = \sqrt[3]{-iF_1'^2} \quad . \tag{19d}$$

So the scaling law for quantum systems differs from that obtained for classical particles; the result (19a) is compatible with a scaling law originally derived by Wegner [19] for the Anderson model. The SCCRT yields approximate expressions for Wegner's unspecified scaling exponent and for the scaling function. Like for the classical system one gets a linear variation of σ with τ. But the critical frequency scales differently, $\omega_c \propto |\tau|^3$, and, as a result, the critical spectrum varies like $\sqrt[3]{\omega}$. The localization length diverges like $1/|\tau|$.

The reason for the difference between scaling laws for classical and quantum systems, reflected by equs. (16a, 19a) are wave mechanical interference effects, which were found [20] to dominate the multiple scattering events. Equivalently, these effects alter the low frequency asymptote of $\sigma(\omega)$ for the conductor [17], which in the present frame is produced by the leading asymptote of $F(z)-F_o$. The preceeding analysis can be extended to arbitrary dimensionality d. Wegner scaling is found for $2<d<4$. For $d=4$ logarithmic corrections enter the scaling equations. $d=4$ is an upper critical dimensionality in the sense, that for $d>4$ the scaling exponents become independent of d and the scaling law agrees with the result (16) found for classical systems. The results of this subsection are due to BELITZ, GOLD and GÖTZE [21]. The connection of Wegner scaling with interference relaxation was also discussed by ABRAHAMS and SHAPIRO [22] within a different frame. The preceeding results have subsequently been published also by VOLLHARDT and WÖLFLE [23].

4c. The Cross Over from the Anderson Transition to the Percolation Transition [21]

To study the classical limit of a zero temperature Fermi system the following points are to be observed ($\hbar$ is written explicitly only in this subsection). The Fermi velocity $v_F=(\hbar k_F/m)$ is to be kept fixed and it adopts the role of the particle speed in the classical system. The density $n=(mv_F)^3/(6\pi^2\hbar^3)$ diverges in the $\hbar\to 0$ limit and so one better studies the dimensionless mobility $\hat{K}$ or the relaxation kernel M, which refer to single particles, rather than the conductivity, which increases with n. The random potential fluctuations are to be specified by some characteristic cut off length L_o, since point scatterers cannot be used in classical systems. The crucial parameter then is the ratio of De Broglie wave length $\lambda_F=2\pi h/(mv_F)$ and fluctuation extension

$$\xi = 2L_o/\lambda_F \quad . \tag{20}$$

The classical limit is equivalent to the case where potential fluctuations are long ranged compared to the particle wave length: $\xi \gg 1$ [24]. The problem consists of a careful determination of W' in equ. (15a) as function of q and ξ, so that the ξ dependence of the coefficients in $F(z)=F_o-F_1'z-F_1z^2+\dots$ can be determined.

For the asymptote of the mobility one finds

$$\sigma = \sigma_M|\tau|f(\xi) \quad . \tag{21}$$

Here the separation parameter is the relative difference of the Fermi energy ε_F from the edge value: $\tau=(\varepsilon_F-\varepsilon_c)/\varepsilon_c$. $\sigma_M=e^2k_F/(6\pi^2\hbar)=e^2v_F/(6\pi^2\hbar^2)$ is the conductivity scale, which is of the order of the one called minimal conductivity by MOTT [4,25]. $f(\xi)=1.4$ for $\xi \ll 1$, so that in the limit of short-ranged scatterers the length parameter L_o does not enter the theory. In agreement with a reasoning of PEPPER [26] the theory yields an increase of the conductivity scale for long-ranged fluctuations, since $f(\xi)=2.5\xi$ for $\xi \gg 1$. This conductivity enhancement then leads the classical limit for the mobility: $\sigma/n=\sigma_M|\tau|$ $2.5\cdot\xi/n=5|\tau|e^2 L_o/(m^2\pi v_F)$.

For the dimensionless dynamical conductivity one gets the two parameter scaling law

$$\hat{K}(\hat{\omega}) = |\tau| S_\pm(\hat{\omega}\tau_o(\xi)/|\tau|^3; \hat{\omega}/\tau^2) \quad , \tag{22}$$

where $\tau_o(\xi)=F_1'^2/F_1$ tends to a constant for $\xi \ll 1$ and vanishes in the classical limit, $\xi \gg 1$, like $\tau_o=1.9/\xi^4$. Hence the dynamics is dominated by interference effects according to equ. (19a) if $|\tau|<\tau_o$, while the scaling law for percolation, equ. (16a), holds for $|\tau|>\tau_o$. The range of validity of Wegner scaling shrinks to zero in the limit $L_o/\lambda_F\to\infty$.

4d. Weak Localization

Dimensionality d=2 is a lower critical one for the discussion of the preceeding two subsections, since $F(z)=-F_o'\log z+\ldots$, $F_o'\propto W_o>0$, diverges for small frequencies logarithmically. A critical region in the sense of equ. (15d) is still defined and the self consistency equ. (15b) reduces to

$$-i\hat{K}(\hat{\omega}) = 1 + F_o'\log|\hat{\omega}/\hat{K}(\hat{\omega})| \quad . \tag{23}$$

There is no solution with a non-zero dc conductivity and thus there cannot exist a true metallic conductor in d=2 [20].

A dynamical scaling law cannot be derived from equ. (23). But one can introduce a frequency ω^*, separating the low frequency regime of hydrodynamic dynamics from the regime $|\omega|>\omega^*$ of critical dynamics. Obviously,

$$\hat{K}(\hat{\omega}) = i[1 + F_o'\log(\hat{\omega}/i)] \quad , \qquad |\omega|>\omega^* \quad , \tag{24a}$$

$$\omega^* = \nu \exp[-1/F_o'] \quad . \tag{24b}$$

For $|\omega|<\omega^*$, $\hat{K}(\hat{\omega})=\hat{\chi}\hat{\omega}$, and the susceptibility can be converted into a localization length with equ. (10c) to get

$$\log r_o^2 = 1/F_o' + \ldots \quad . \tag{24c}$$

Hence the critical regime is identical with the regime of weak disorder: $F_o' \ll 1$. Then ω^* is an exceedingly small frequency and the formula for the critical current correlations [17] covers the whole range of experimental relevance. The localization length, equ. (24c) [27], is exceedingly large compared to the natural microscopic length scales. For long-ranged random potentials F_o' decreases: $F_o'\to 0$ for $\xi\to\infty$, and all the mentioned deviations from the true metallic behaviour disappear in the classical limit [21].

4e. The Quasi Transition in d=2

For large z one can expand in equ. (15c) $F(z)=(A/z^2)-(B/z^4)+\ldots$, where A and B increase linearly with W. Thus the self-consistency equation (15b) leads to the expansion of the polarizability for strong disorder: $1/\chi\propto A-(B/A)$. This is a Curie-Weiss law for the glass polarizability. Extrapolating this result into the regime of weak coupling one expects therefore a polarization catastrophe, i.e. a transition to a conductor. Fluctuations bend the $1/\chi_o$ versus W-curve and shift the transition downward [28]. In d=2 the fluctuations prevent a true instability, but there occurs a change from a $\chi\propto r_o^2$ of a size given by the microscopic scales to an exponentially big one, equ. (24c). A $1/r_o$ versus coupling plot shows a rather abrupt transitive from a well-defined insulator to one, where $1/r_o$ cannot be made visible on a linear scale, Fig. 1 [21].

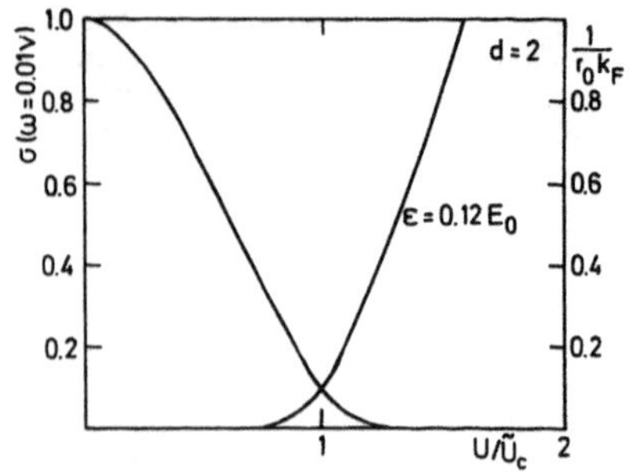

Fig. 1

The dc conductivity vanishes. But if one considers the conductivity for a small frequency, say $\bar{\omega}=0.01\nu$, one observes the following. In the regime of strong disorder $\bar{\sigma}=\sigma(\bar{\omega})$ is very small and cannot be seen on a linear plot on scale $\sigma_o=ne^2/m\nu$. If the regime of weak localization is entered, ω^* will abruptly fall below $\bar{\omega}$ and $\bar{\sigma}$ will increase. In this regime $\bar{\sigma}$ depends on $\bar{\omega}$ very weakly only, Fig. 1.

Hence the SCCRT predicts a quasi transition from a true insulator to a quasi metallic state. Under many experimental conditions this quasi transition cannot be distinguished from a true CIT. The quasi transition becomes the sharper the larger the potential fluctuation range L_o relative to the particle wave length λ_F. In the classical limit the quasi transition converges towards the true percolation transition. It should be emphasized that the quasi transition from strong to weak disorder does not depend on having a cross over to classical percolation. The quasi transition from strong to weak localization occurs also e.g. in WEGNER'S [29] n-orbital model, provided the parameters are chosen properly [30].

5. The Mobility

The lowest order approximation $M_o(\omega)$ of the SCCRT is obtained by substituting the density propagator $\Phi_o(q,\omega)$ of the particle system without impurities into equs. (14). For non-interacting particles, one finds the relaxation spectrum to be practically a white one $M_o''(\omega)\cong M_o''(\omega=0)=1/\tau$. Then

$$1/\tau = \sum_{\vec{q}} w(q)\,\Phi_o''(q,\omega=0) \qquad (25)$$

turns out to be the KEA result for the current relaxation rate. Since the recoil spectrum is obtained as superposition of particle hole excitations, the recoil process can be reinterpreted in terms of scattering processes. Formula (25) then yields the variational solution of the Boltzmann equation for the mobility. If Coulomb interaction is incorporated, equ. (25) yields the Mott-Jones formula, i.e. the KEA result for scattering from screened impurities [7].

The essential point of the self-consistency concept then is the following. Due to interaction, the particles diffuse rather than fly around with the Fermi velocity v_F. Hence, they stick much longer near the scattering centre; this is reflected by the long wave length low frequency recoil spectrum for diffusion to be much larger than the corresponding spectrum for free flight

$$\Phi_H''(q,\omega=0) = \rho_F/[Dq^2] \gg \rho_F\pi/[2v_F q] = \Phi_o''(q,\omega=0) \quad . \qquad (26)$$

As a result, the transfer of recoil is more efficient $M''(\omega=0)>M_o''(\omega=0)$, and hence the true conductivity is lower than the KEA result $\sigma<\sigma_o$, equ. (5). The same conclusion holds if Coulomb interactions are incorporated. Obviously, there is a positive feed back: the decrease of the diffusivity, $D<D_o$, leads to a further increase of the recoil spectrum, which implies a further suppression of the mobility. For some critical coupling the mobility is zero, the particle gets trapped, the CIT is obtained [8]. This continuous decrease to zero of the diffusivity is seen elementarily if one uses the simplified equ. (15a) for $\omega=0$, observing equs. (5), (9b) and (4): $D/D_o=1-A$, where D_o is the KEA diffusivity and

$$A = \sum_q W(q)g(q)^2 m/(nq^2) \quad . \tag{27}$$

The criterion for the CIT then reads A=1; the number A is the relevant dimensionless coupling constant of the theory. For A<1 the system is a conductor, for A>1 it is an insulator.

For calculations, intended to extend from the regime of the KEA to the limit of strong disorder, an approximation for the density propagator in terms of $M(\omega)$ is needed in equ. (14a), which is valid for all q. The simplified equ. (15a) cannot be used, since asymptotic expansions for $z\to 0$, $q\to 0$ have been performed there. To proceed, the collision-rate approximation, known from the theory of gases was generalized to [8]

$$\Phi(q,\omega) = \Phi_o(q,\omega+M(\omega))/\left[1+M(\omega)\Phi_o(q,\omega+M(\omega))/g(q)\right] \quad . \tag{28}$$

This formula can be obtained by approximating the infinite q dependent matrix of relaxation kernels, obtained within MORI'S [5] scheme, by one q independent matrix element $M(\omega)$ [28]. Originally Φ_o and g were used as free gas correlation functions; then Coulomb interactions have been incorporated by evaluating these functions within the random phase approximation including local field corrections [31,32]. For $qv_F/|\omega+M(\omega)| \ll 1$ the approximation for $\Phi(q,\omega)$ reduces to equ. (1). The numerical results to be quoted are based on the coupled self-consistency equations (14,28).

It has to be noticed, that equ. (28) exhibits two defects. First, it is not compatible with the requirements imposed by the symmetry (18). As a result, quantum interference effects are lost, if equ. (28) is substituted into equ. (14). As a consequence, the quasi transition in d=2 is treated as a true CIT, if the theory is applied e.g. to MOS-systems. The scaling laws following from equ. (18,28) [12,28] are not the correct ones discussed in section 4b. Second, in deriving equ. (28) it was anticipated that all relaxation kernels are of the same order. In particular it is implied, that all kernels diverge at the CIT. For the Lorentz system it was shown, however, that all kernels, except the one for the current, remain finite at the percolation point [33]. Approximating finite kernels by infinite ones leads to wrong asymptotic behaviour for some quantities at the CIT. It would be desirable to eliminate this artifact, even though it might not strongly influence the figures to be shown. A recent claim by BECKER and KELLER [34] about improvements in the derivation of the SCCRT seems unjustified to me, since their work does not reproduce correctly the known result [17] for the quantum interference correction to $\sigma(\omega)$ (equ. (24a) for $\xi=0$); since momentum cut-offs and ξ-dependences are not specified; and since their work implies the existence of two singular points for certain sets of parameters instead of one CIT.

In Fig. 2 [35], D/D_o for the d=3 Lorentz system is shown as function of the dimensionless scatterer density $n^*\propto n_i$ in comparison with the molecular dynamics data of BRUIN [13]. The dashed line is the simplified result discussed above in connection with equ. (27). For the percolation point A=1 one gets [9], $n_c^*=9/4\pi\cong 0.72$ while Monte Carlo work yields $n_c^*=0.81\pm 0.05$ [36]. For d=2 the theory misses the critical point by about a factor of two, but D/D_o plotted as function of n/n_c agrees perfectly with the data [15]. Successful comparison of computer data for the diffusivity done for the Anderson model [37] with the results of the SCCRT has been reported also [38].

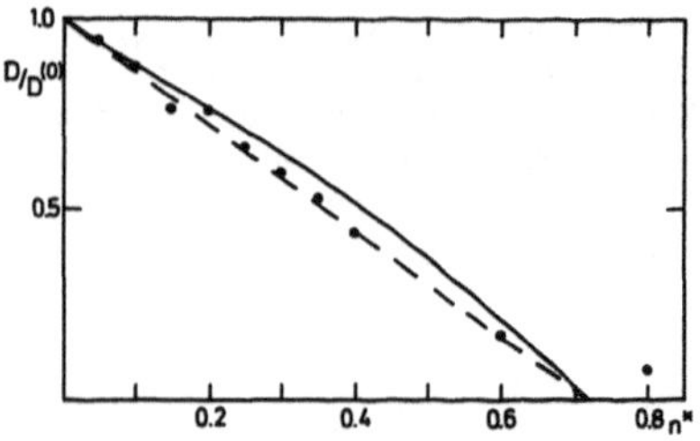

Fig. 2

In Fig. 3 [39] experimental values [40] of the mobility of a Si-100-inversion layer are shown to fall below the KEA result (dotted curve), if the electron density n falls below 2 $10^{12}cm^{-2}$. The scatterers are provided by Na^+ ions and all parameters of this device are so well known, that the theoretical curves (full and dashed) contain only the uncertainty about the distance between impurity and electron layer (1.34 A and 0.00 A). The sodium doped Si-100-device is such an important system to check dynamical theories for strong disorder, since n and n_i can be varied independently and reproducably over more than one decade. In Fig. 4 data of KOCH and MAZURE [41], done for accumulation layers are shown for the CIT points (the quasitransition points from a quasi conductor to an insulator in the sense of section 4e). A mathematical model of an electron system in d=2 misses the transition by about a factor 2 (dotted line). But if the known form factors for the layer thickness are incorporated (dashed line for ions in the layer, full line for ions 21 Å above the electron layer) the theory accounts for the data.

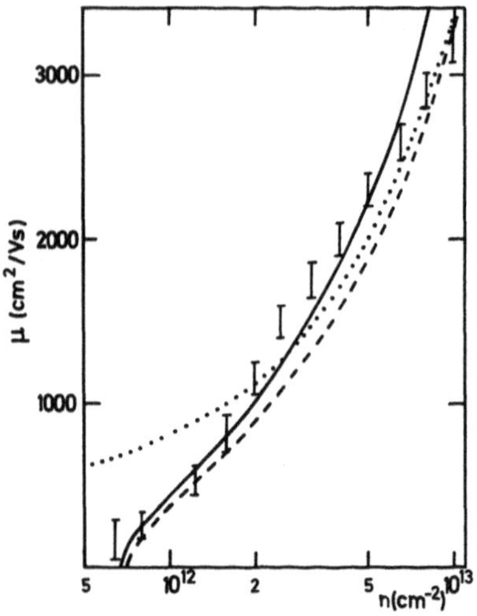

Fig. 3

For d=3 systems with ion impurities of valence Z the transition criterion A=1 can be cast into the form used by MOTT [42] in his discussion of electron localization in an isolated self-consistently screened Coulomb potential: $n_c^{1/3}a^*=f$; here a* is the effective Bohr radius and $f\sim 0.2$. For ions desolved in rare gas crystals the SCCRT yields [31] n_c=2.8, 4.4 and 4.2 (in $10^{22}cm^{-3}$) for Hg:Xe (Z=2), La:Ar (Z=3), Sn:Xe (Z=4), while the corresponding experimental values are 5.9 [43], 2.5 [44], 2.8 [45].

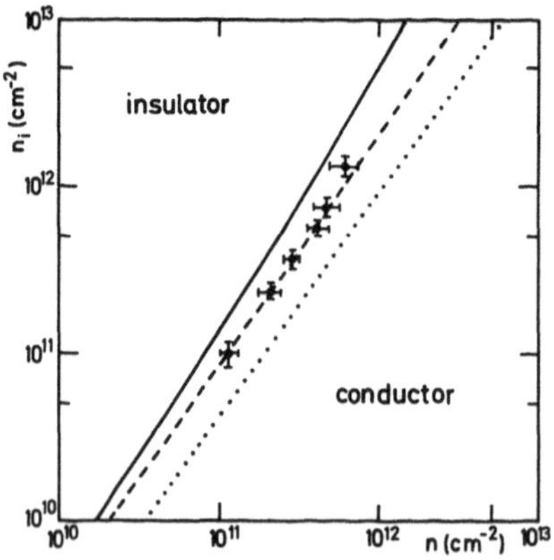

Fig. 4

6. The Current Correlations

The relaxation enhancement is important only for small frequencies. For large ω the density spectrum is given by the one of the electron system without impurities. Hence $M''(\omega)$ is well approximated by the lowest order result $M_o''(\omega)$, equ. (25). This statement needs some reservation, since for large frequencies anomalies due to a dynamical break down of screening require a self-consistent treatment of plasmon dynamics and electron impurity scattering. This is automatically achieved by the SCCRT [31,32]; but since the experimental implications are not directly connected with localization I will not report the details [39]. So one understands the existence of a localization precursor in the sense, that $K''(\omega)$ rises with increasing frequency from its suppressed dc value up to some maximum at $\omega \neq 0$, before it decreases like the Drude Lorentzian. Consequently, the velocity correlation function $K(t)$, the Fourier back transform of $K''(\omega)$, will exhibit some indication of oscillatory motion. In particular, it will change its sign, while the KEA predicts the positive monotonous exponential $K_o(t)=\exp-(t/\tau)$. In Fig. 5 [35] the results of the theory for the d=3 Lorentz system are compared with the molecular dynamics data [13]. The area under $K(t)$ yields the diffusivity, Fig. 2. Notice, that no fit parameter enters the figures 2 and 5.

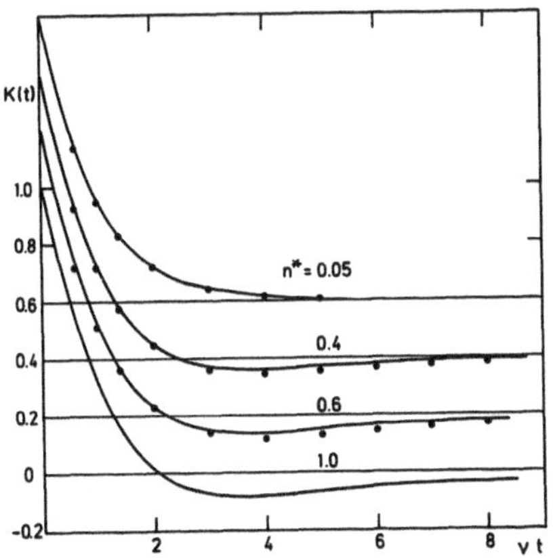

Fig. 5

Measurements of $\sigma(\omega)$, equ. (3), near a CIT have first been carried out by ALLAN, TSUI and DE ROSA [46] for Si-MOS systems. Extension of this work is shown in Figs. 6, 7 [47]. To analyse the data (full curves in Fig. 6, dots in Fig. 7) the electron density was used as an adjustable parameter; but it was always close to the one determined by measurement. Since the random potential was unknown, a model, specified by a strength parameter U and a range parameter $q_o=2\pi/L_o$, was used. So all the theoretical curves shown (dashed curves in Fig. 6 and full curves in Fig. 7) are fixed by two fit parameters U, q_o.

The Arrhenius plot Fig. 7 exhibits the characteristic change from metallic, essentially temperature independent, conductivities to thermally activated ones due to a shift of the Fermi energy ε_F through the mobility edge ε_c [48]. Careful analysis of such plots by MOTT and collaborators [49,50,4] originally was the only means to detect a CIT in Anderson's sense. A whole series of such plots was analyzed with the conductivity versus Fermi-energy curves obtained within the SCCRT [51]. Since the theory reproduces Mott's conductivity scale and PEPPER'S [26] results concerning the conductivity enhancement due to long ranged potential fluctuations, the data could be described reasonably. It was concluded also, that Arrhenius plots are not sensitive enough to check details of the theory. Whether one uses the continuous σ versus ε_F curves of the SCCRT [51] or some ad hoc function with e.g. a discontinuity involved, cannot be discriminated experimentally in most cases. Arrhenius plots are a challenge of the theory only if they are accompanied by other experimental information, as demonstrated e.g. in connection with Figs. (6, 7).

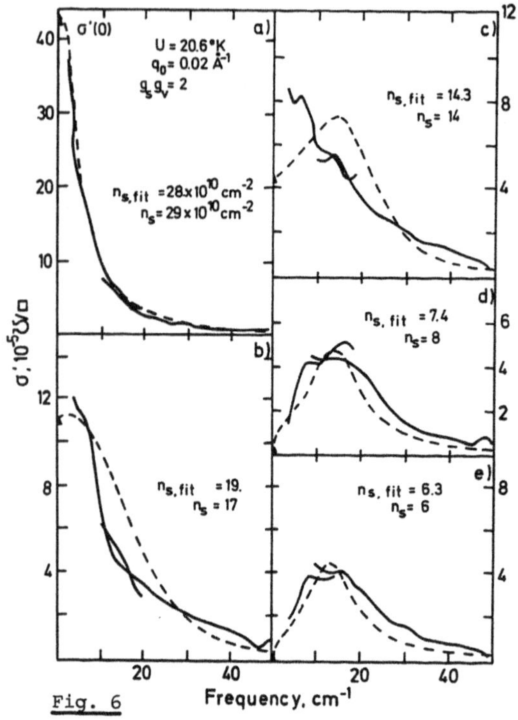

Fig. 6

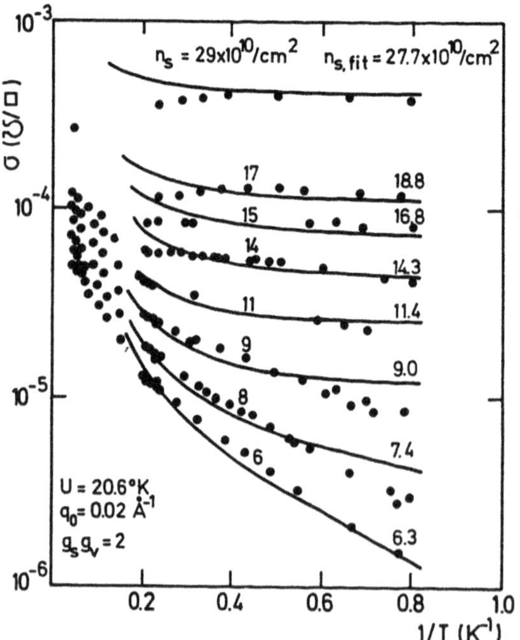

Fig. 7

The Si-100 accumulation layer experiments by Koch and Mazuré are so challenging, since all parameters of the system are known. Fig. 8 [52,39] clearly shows the change of σ(ω) from a Drude curve with high frequency screening anomalies to one with a low frequency quasi gap if the Fermi energy is shifted through the mobility edge. So the SCCRT cannot only produce the position of the CIT point for this system, Fig. 4, but also the dynamics for intermediate and large frequencies.

7. Acknowledgements

I thank Drs. D. Belitz and A. Gold for cooperation and discussions.

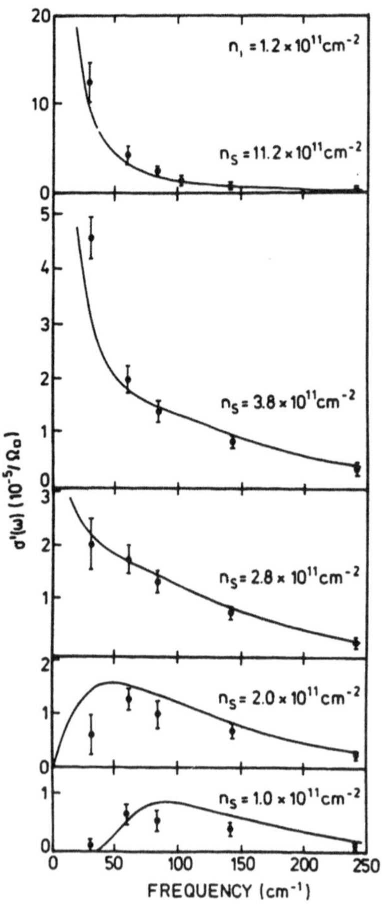

Fig. 8

8. References

1 S. Broadbent and J. Hammersley: Proc. Camb. Phil. Soc. 53 , 629 (1957)
2 P.W. Anderson: Phys. Rev. 109 , 1492 (1958)
3 P.W. Anderson: Comm. Sol. State Phys. 2 , 193 (1970)
4 N.F. Mott and E.A. Davis: "Electronic processes in non-crystalline materials" (Clarendon Press, Oxford 1979)
5 H. Mori: Progr. theor. Phys., Kyoto 33 , 423 (1965)
6 R. Kubo: J. Phys. Soc. Japan 12 , 570 (1957)
7 W. Götze and P. Wölfle: Phys. Rev. B 6 , 1226 (1972)
8 W. Götze: Solid State Comm. 27 , 1393 (1978)
9 W. Götze, E. Leutheusser and S. Yip: Phys. Rev. A 23 , 2634 (1981)
10 P.C. Martin: Phys. Rev. 161 , 141 (1967)
11 T. Ando, A.B. Fowler and F. Stern: Rev. Mod. Phys. 54 , 437 (1982)
12 W. Götze: J. Phys. C 12 , 1279 (1979)
13 C. Bruin: Phys. Rev. Lett. 29 , 1670 (1972) - Physica 72 , 261 (1974)
14 B.J. Alder and W.E. Alley: J. St. Phys. 19 , 341 (1978)
15 W. Götze, E. Leutheusser and S. Yip: Phys. Rev. A 25 , 533 (1982)
16 Odagaki T. and M. Lax: Phys. Rev. B 24 , 5284 (1981)
17 L.P. Gorkov, A.I. Larkin and D.E. Khmelnitskii: JETP Lett. 30 , 228 (1979)
18 P. Prelovsek: in "Recent Developments in Condensed Matter Physics", J.T. Devreese edit., Plenum Press, New York and London, Vol. II, 191 (1981)
19 F. Wegner: Z. Phys. B 25 , 327 (1976)
20 E. Abrahams, P.W. Anderson, D.C. Licciardello and T.V. Ramakrishnan: Phys. Rev. Lett. 42 , 673 (1979)
21 D. Belitz, A. Gold and W. Götze: Z. Phys. B 44 , 273 (1981)
22 B. Shapiro and E. Abrahams: Phys. Rev. B 42 , 4889 (1981)
23 D. Vollhardt and P. Wölfle: Phys. Rev. Lett. 48 , 699 (1982) - P. Wölfle and D. Vollhardt: in "Anderson Localization", Y. Nagaoka and H. Fukuyama edit., Springer, Berlin (1982)
24 J. Ziman: J. Phys. C 1 , 1532 (1968)
25 N.F. Mott: Phil. Mag. 22 , 7 (1970)
26 M. Pepper: Proc. Roy. Soc. London A 353 , 225 (1977)
27 D. Vollhardt and P. Wölfle: Phys. Rev. Lett. 45 , 842 (1980) - Phvs. Rev. B 22 , 4666 (1980)
28 W. Götze: Phil. Mag. 43 , 219 (1981)
29 F. Wegner: Phys. Rev. B 19 , 783 (1979)
30 D. Belitz and W. Götze: Phys. Rev. B 28 , 5445 (1983)
31 A. Gold and W. Götze: Helv. Phys. Acta 56 , 47 (1983)
32 A. Gold and W. Götze: Sol. State Comm. 47 , 627 (1983)
33 W. Götze and E. Leutheusser: Z. Phys. B 45 , 85 (1981)
34 K.W. Becker and J. Keller: Z. Phys. B 54 , 283 (1984)
35 W. Götze, E. Leutheusser and S. Yip: Phys. Rev. A 24 , 1008 (1981)
36 J. Kertész: J. de Physique 42 , L 393 (1981)
37 P. Prelovsek: Phys. Rev. Lett. 40 , 1596 (1978) - Phys. Rev. B 18 , 3657 (1978) - Solid St. Comm. 31 , 179 (1979); J. Stein and U. Krey: Z. Phys. B 37 , 13 (1980) - Solid St. Comm. 36 , 951 (1980)
38 P. Prelovsek: Phys. Rev. B 23 , 1304 (1981)
39 A. Gold and W. Götze: to be published; A. Gold: PhD thesis, Technische Universität München, May 1984
40 A. Hartstein, T.H. Ning and A.B. Fowler: Surf. Sci. 58 , 178 (1976)
41 F. Koch and C. Mazuré: Private communication (1984)
42 N.F. Mott: Proc. Roy. Soc. 34 , 416 (1949)
43 O. Cheshnovsky, O.U. Even and J. Jortner: Phys. Rev. B 25 , 3350 (1982)
44 R. Römer, F. Siebers and H. Micklitz: Sol. State Comm. 27 , 881 (1980)
45 R. Ludwig, T. Paul and H. Micklitz: Z. Phys. B 47 , 31 (1982)
46 S.J. Allen, D.C. Tsui and F. DeRosay: Phys. Rev. Lett. 35 , 1359 (1975)

47 A. Gold, S.J. Allen, B.A. Wilson and D.C. Tsui: Phys. Rev. B 25 , 3519 (1982)
48 L. Banyai: Physique des Semiconducteurs, Hulin edit., Dunot, Paris, p. 417 (1964)
49 N.F. Mott: Phil. Mag. 13 , 989 (1966)
50 N.F. Mott, M. Pepper, S. Pollit, R.H. Wallis and C.J. Adkins: Proc. Roy. Soc. London A 345 , 169 (1975)
51 D. Belitz and W. Götze: Phil. Mag. B 43 , 517 (1981) - A. Gold and W. Götze: J. Phys. C 14 , 4049 (1981) - A. Gold: J. Phys. C 18 , L 815 (1982)
52 A. Gold, W. Götze, C. Mazuré and F. Koch: Sol. State Comm. 49 , 1085 (1984)

Recent Developments in the Metal-Insulator Transition

G.A. Thomas and M.A. Paalanen

AT&T Bell Laboratories, 600 Mountain Avenue, Murray Hill, NJ 07974, USA

A comparison of new experimental and theoretical results confirms earlier indications that Coulomb interactions play an important role in the behavior of disordered systems. These results point toward a new scenario for theoretical discussions of the problem--particularly in systems with apparently anomalous critical behavior. To solve the problem in the end, a full theory will be needed, including both localization and interactions. However, the solution may look like the existing, non-interacting, scaling theory of localization, with strong spin flip scattering (produced by the interactions). This approximation correctly describes the critical exponent of the conductivity in the regime within 1% of the metal-insulator transition.

1. Introduction

Historically, one of the clear expositions of the metal-insulator transition was made by MOTT[1]. In his terms, recent results indicate that the transition is both of the Mott and Anderson types, in that both Coulomb interactions and disorder play important roles. In his videotaped lecture for the Santa Cruz conference this summer, Mott suggested[2] that in doped semiconductors disorder plays a crucial role. He asked the conference participants if the transition in Si:P, for example, occurred in a conduction band, where disorder might dominate over interactions, or not. Anderson answered in his talk[3] (given before Mott's tape) that he felt it was a Mott transition in the sense that it occurred in an impurity band, where interactions are important. He noted, however, that the approach to the transition could be understood within his scaling theory provided that interactions need only be included to produce strong spin flip scattering. (PAALANEN, et al. and RUCKENSTEIN, et al.[4] independently presented a similar suggestion at both Santa Cruz and at the semiconductor conference in San Francisco based on new NMR results.)

We will summarize selected results here, and argue that they show strong Coulomb effects generally, and thus that the Fermi energy lies in an impurity band at what is a Mott-Anderson transition. The recent results also confirm Mott's characteristic conductivity[2] as an accurate indicator of the region near the transition where anomalous electronic diffusion occurs.

2. Motivation: the Critical Exponent Puzzle

One way of describing our motivation for the continued intensive study of disordered, three-dimensional systems is the apparently fundamental disagreement between theory and experiment illustrated in Figure 1a. The figure shows[4] the conductivity as a function of carrier density for Si:P, and indicates that, as the measured conductivity (open circles) goes to zero, it does so smoothly and with a critical exponent of order 1/2. The values of conductivity at zero-temperature must be obtained from extrapolations to T=0 K of measurements at low temperatures (here, of order 10mK, or 10^{-4} of the Fermi temperature of 100K). The values of density were obtained from the concentration of P atoms in the Si crystals. As illustrated in Figure 1b., each P donates its one extra valence electron to the electronic system, while the P ions form the random potential in which the electrons move and interact. We define a metal here as a substance having a non-zero conductivity at T=0K, and find that the metallic state occurs at densities above $n=3.75\ 10^{18}$ cm^{-3} in Si:P. (At this point, the donors would be 65Å apart if they were ordered in a cubic lattice.)

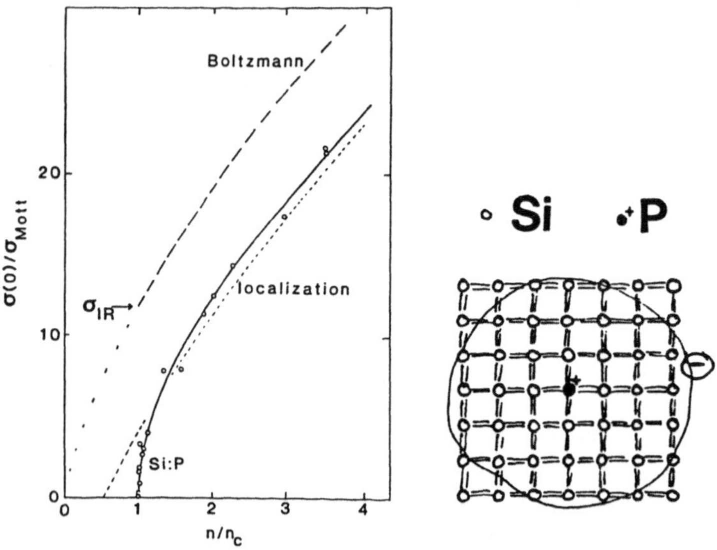

Fig. 1a: Discrepancy between experiment and theory in Si:P [17];
Fig. 1b: Schematic of a P donor in Si

The usual Boltzmann conductivity[1] is shown as the long dashes, for which the electrons in their Fermi distribution are imagined to scatter weakly with a mean free path l:

$$\sigma_B = ne^2 l/(\hbar k_F) \ . \tag{1}$$

(Including screening, in the Born approximation, l is roughly constant in Si:P and the decrease in the conductivity arises from the decreasing density.) This picture becomes unreasonable below the Ioffe-Regel conductivity, at which the Fermi wave length becomes equal to l. However, even with weak scattering and without electron-electron interactions, coherent back-scattering tends to localize the electrons and reduces the conductivity[2,3,5,6] below the Boltzmann curve, as indicated by the short dashes, calculated from

$$\sigma = \sigma_B[1-3/(k_F l)^2] \ . \qquad (2)$$

If this perturbation theory of localization is extended beyond the usual region of applicability, it extrapolates to a metal-insulator transition with critical behavior of the following form:

$$\sigma = (n/n_c-1)^1 \ . \qquad (3)$$

So we have the result:

exponent=1
(non-interacting, scaling theory
of localization[2,3,5,6])

Significant recent theoretical advances in our understanding of the Coulomb interactions were reported at Santa Cruz by LEE and by CASTELLANI[12]. This new calculation of Coulomb interactions with disorder, but neglecting a full treatment of localization, gives the same exponent as the non-interacting theory:

exponent=1
(disordered interacting theory
without localization[11,12])

This result is proposed to apply to the universality classes of disordered interactions in the presence of a large magnetic field or of strong spin scattering, where localization effects may not be as important as the interactions.

This critical exponent 1 provides a simple way of contrasting the theoretical and experimental behaviors. The solid line fit[4] to the experimental points in Figure 1 gives:

exponent=1/2
(one electron/scattering site, experiment,
within 0.1% of n_c)

Si:P	AT&T Bell	1980-83[4]

The Mott characteristic conductivity[1,2] used for normalization in the figure is less than the Ioffe-Regel value by a factor arising from the reduction, due to localization processes, in the effective density of states that enters the conductivity. (In Si:P, this reduction is 1/13.)

Several important recent results confirm the experimental findings shown in Figure 1:

exponent= 1/2
(one electron/ scattering site)

Ge:Sb	Tokyo	1981[7]
Si:P	Tokyo	1982[7]
Si:P	Cornell	1984[8]
Si:As	Cornell	1984[8]
Si:P+As	Cornell	1984[8]
Si:Sb	Cambridge	1984[9]

The confirmation of this critical behavior to as close to the transition as 0.1% in relative density has come from measurements[4] very near the transition under uniaxial stress, and from studies of the dielectric susceptibility, as illustrated in Figure 2. The right axis and solid points show the square of the conductivity as a function of relative density, in agreement with those in Figure 1. The left hand axis and open circles show the inverse of the dielectric susceptibility, and indicate a linear critical behavior. Recent unpublished results[10] confirm these results in the insulator. Simple dimensional considerations suggest (and calculations presented by KAWABATA[11] indicate) that this dependence should give half the characteristic exponent, and thus we conclude from these results also the value:

exponent= 1/2
(one elec./scat. site, diel. sus.)

Si:P	AT&T Bell	1980,83[4]
Si:P	Rochester	1984[10]
Si:As	Rochester	1984[10]

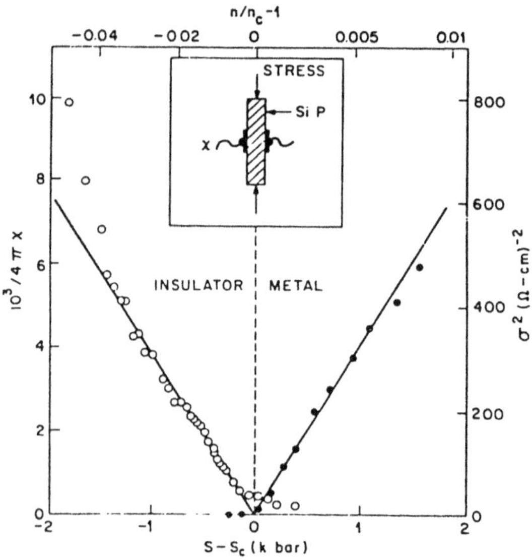

Fig. 2: Dielectric susceptibility and conductivity very near the transition in Si:P, both indicating exponent 1/2 [4,7]

There are, however, a range of systems in which the observed exponent is 1, an example of which[4,7] is shown in Figure 3a. The data are plotted with the same axes as Figure 1, with the open circles again the results for Si:P, which are in agreement with the results for Ge:Sb. The solid circles are measurements for special Ge:Sb samples in which additional scattering centers were added beyond one per electron, by putting in equal amounts of Sb and B. These two impurities exchange an electron, and contribute charged scattering centers, but no carriers at low T, as illustrated schematically in Figure 3b.

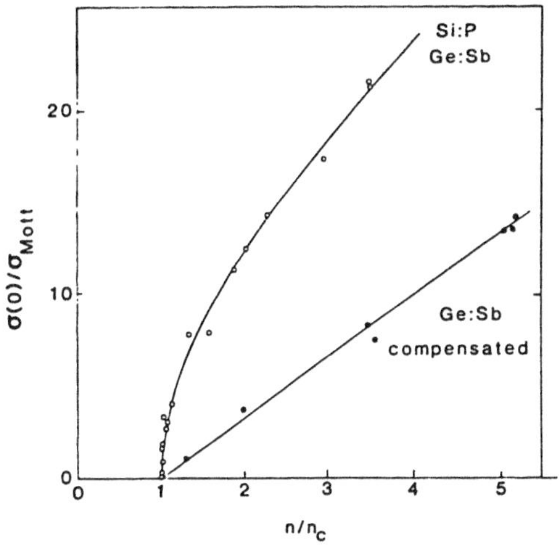

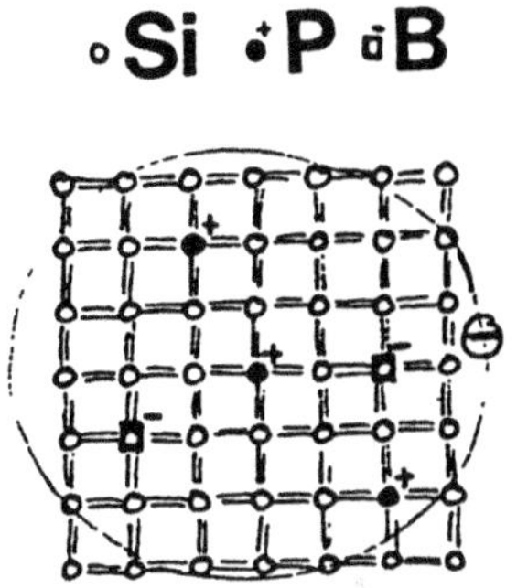

Fig. 3a: Change in critical behavior [7] from one scattering site/conduction electron to several (compensated)

Fig. 3b: Schematic of compensated Si:P

This addition of compensating impurities increases the scattering and thus reduces the conductivity at fixed carrier density n (by reducing l), as shown in the figure. The exponent changes:

exponent=1
(several scattering sites/elec.)

Ge:Sb(comp.)	Tokyo	1982[7]

More recent results give the same exponent in a variety of systems where there are several scattering centers per electron. However, none of the results giving exponent 1 resolve the region within 1% of the critical point. Particularly detailed measurements were summarized by DYNES[13] at Santa Cruz and also by NISHIDA[14] (as illustrated in Figure 4a) and by others[15,16] on amorphous alloys of a semiconductor and a metal:

exponent=1
(several scattering sites/elec.)

a-Si:Au	Tokyo	1982,84[14]
a-Si:Nb	AT&T Bell	1982,84[13]
a-Si:Al	"	1984[13]
a-Ge:Au	Illinois	1984[15]
a-Si:Mo	Stanford	1984[16]

These amorphous alloys are illustrated schematically in Figure 4b. Here, the critical concentration of the metal is about 10%, with an average spacing between the metal atoms of around 2 lattice spacings. The open circles in Figure 4a indicate that a large field does not seem to affect the exponent in these alloys.

The same exponent is also found in compound semiconductors which are doped at a series of concentrations and unavoidably

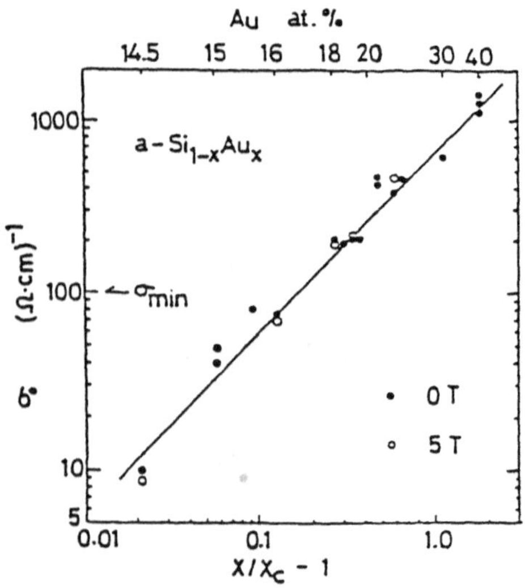

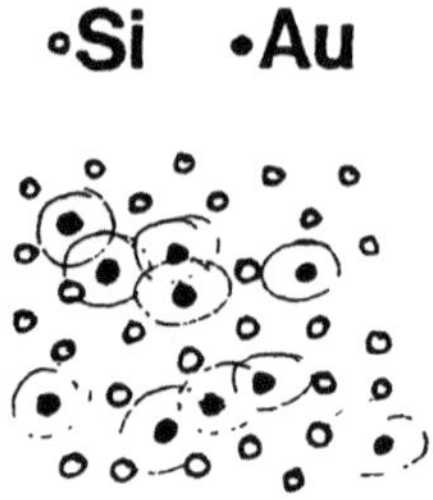

Fig. 4a: New evidence for exponent 1, both at zero and large magnetic field, in samples of aSi:Au [14];
Fig. 4b: Schematic of an amorphous alloy

compensated, as studied by MORITA and co-workers [17]:

exponent=1
(several scattering sites/elec.)

GaAs:n	Tohoku	1984[17]
InSb:n	"	1984[17]

Here, n indicates some impurities that donate electrons to the system.

Another case in which particularly nice data are available is that in which a sample near the transition is driven through it by a large applied magnetic field as discussed by VON MOLNAR [18] and MANSFIELD [19]:

exponent=1
(several scatt. sites/elec., large field)

Gd_3S_4:v	IBM	1983,84[18]
InSb:n	London	1984[19]

Here, v indicates vacancies which free carriers in the system. Figure 5a shows these results of von Molnar, in which the applied field reduces the degree of disorder, turning an insulator into a metal. In Figure 5b, the results for InSb are shown as a function of T for a series of fields. In this case, the field primarily shrinks the electron's wave function, turning a metal into an insulator. Other similar measurements[9] on InSb:n and InP:n as a function of field appear to us to give consistent results, and these as well as most of the above (particularly the Si:P and the studies in large field) indicate anomalous

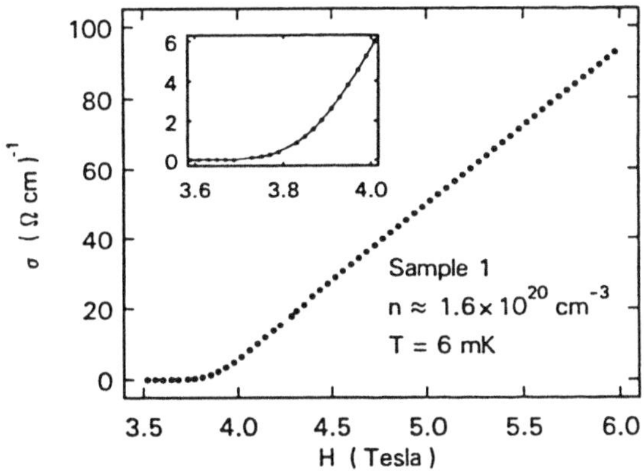

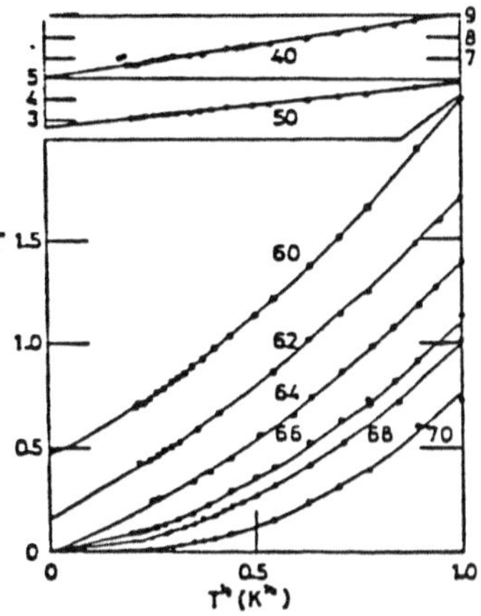

Fig. 5a: New evidence of a continuous transition with exponent 1 in Gd_3S_4:v as a function of magnetic field [18];

Fig. 5b: Example of anomalous diffusion below Mott's characteristic conductivity in a metallic sample of InSb:n [19]

diffusion for conductivities below Mott's characteristic value. This anomalous diffusion has been analyzed[7,18,19] as a correction to the T=0K conductivity of the form $AT^{1/2}$, with A positive and increasing as the transition is approached, as illustrated in Figure 5b. This increasing slope corresponds to a decreasing diffusion temperature, i.e. to a greater sensitivity of the electrons' motion to T as they become more nearly localized.

3. Scenario for Discussing the Exponent Puzzle

Two years ago, during the Semiconductor conference in Montpellier, Anderson discussed the exponent puzzle informally over dinner at a fine restaurant in St. Martin de Londre. He wondered if there could somehow be strong spin flip scattering in the cases where the exponent was 1/2. In an essentially one-line calculation on a piece of scrap paper (the napkins were cloth), he showed how the second order correction to the beta function in perturbation theory could give 1/2. For this to happen, however, the full theory, when it was eventually solved, had to be equivalent to a situation in which the Coulomb interactions entered only to make the collective spin motion quasi-static compared to the electronic motion. These spins would then kill the first order correction to beta and resurrect the second order term from its value of zero without this time reversal symmetry-breaking.

This beta function, shown in Figure 6, is now relatively well known[3,5,11], as the logarithmic derivative of the dimensionless conductance with respect to the log of the length scale. In the non-interacting scaling theory of localization, it is constructed as a function of the conductance, g, which is assumed to be the only relevant scaling parameter. (It is humorous to note that in the original sketch of the beta function by ABRAHAMS et al.[3,20] the slope of the curve near the metal-insulator transition can be measured from the figure to

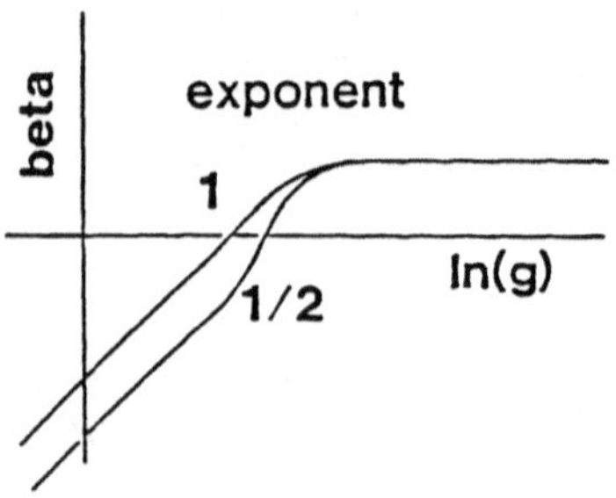

Fig. 6: Sketch of beta function for exponents 1 and 1/2

give an exponent 1/2. A sketch for both exponents is shown in Figure 6.)

A qualitative consideration of the physical meaning of the beta function requires an idea of the conductivity on a short, as opposed to a macroscopic, length scale. This variation of the conductivity with length scale can be observed directly in the frequency-dependent conductivity. Figure 7a shows the conductivity in the insulator and Figure 7b shows it in the metal, both near the transition. The length within the scaling theory can be written as L or equivalently as $(D/f)^{1/2}$, where D is the electronic diffusion constant and f is the frequency. This equivalence of scaling in length or in frequency was shown by GORKOV et al.[21] in an early version of the scaling theory. The behavior on a short length scale can thus be investigated at finite frequency, from which point scaling to infinite length is moving to zero frequency.

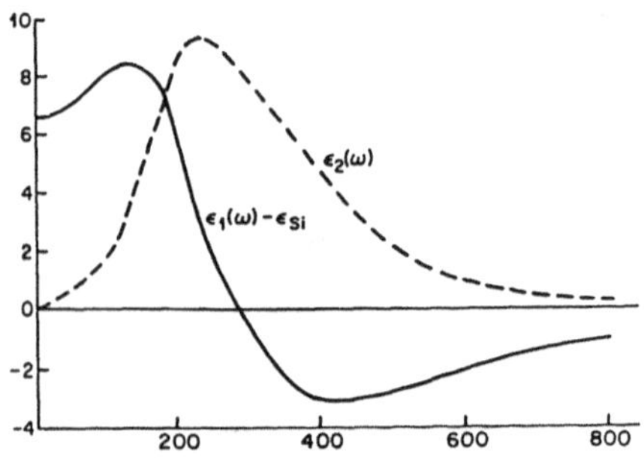

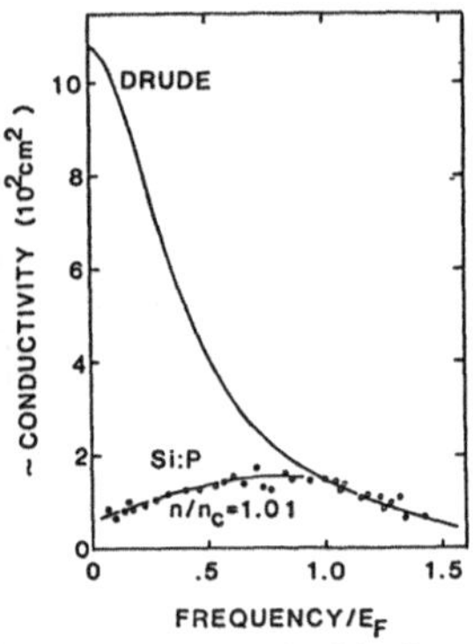

Fig. 7a: Frequency-dependent conductivity and dielectric susceptibility in insulating Si:P

Fig. 7b: Frequency-dependent conductivity in metallic Si:P compared to Drude behavior [23]

In Figure 7a, an experimental measurement of the behavior in insulating Si:P shows that the conductivity at finite f scales to 0 as f goes to 0. Here, beta is just barely negative on a short length scale, and thus scales toward decreasing conductance as the length scale becomes long. (The figure also shows that the dielectric constant scales to a finite value, aside from glassy contributions at very low frequencies which are not shown here, but were discussed by BHATT [21] at Santa Cruz.)

In Figure 7b, recent unpublished results for a metallic sample[23] show that here, where the finite f conductivity is higher and its slope is smaller than in the insulator, the conductivity goes to a finite value as f goes to 0. Figure 7b also shows qualitatively that there is a very large deviation from ordinary metallic (Drude) behavior in this sample, since it is near the transition. These and similar data could be used, in principle, to test the scaling hypothesis that the conductivity times $(D/f)^{1/2}$ is the only necessary parameter to describe the beta function, and thus determine whether a given sample will be either a metal or an insulator. However, the critical exponent discussed above already tells us that some refinement of the original theory is necessary. At Santa Cruz, GOETZE [24] also emphasized the importance of studying the far-infrared conductivity in order to understand the metal - insulator transition, and he presented a self-consistent, approximate description of the problem.

Within a non-interacting model, it is not at all clear where the strong spin scattering could come from in a system such as Si:P in the metallic state, since the only spins are on the conduction electrons. With strong short-range interactions, however, spins that act almost as if they were localized can occur. One way in which this can happen was discussed at the Santa Cruz conference by RICE[25], as well as by ANDERSON[3] and by PAALANEN[4], and at the Semiconductor Conference in San Francisco by RUCKENSTEIN [4], based on the Brinkman-Rice analysis of Gutzwiller's approximate solution to the Hubbard Hamiltonian[1]. Within this picture, an electron's effective mass can be greatly enhanced if its neighboring sites are occupied by other electrons to which it feels a strong repulsion. In effect, it can quiver in place, but it can be unable to hop if the repulsion is strong enough. Within this model the spin diffusion is slowed compared to the charge diffusion by a small Stoner factor, but the principal effect on all the properties comes from an enhanced effective mass. Rice emphasized, however, that clusters of spins (which are not present in periodic systems) may play a more important role in systems with nearly ideal randomness such as Si:P.

4. Selected Results Showing Strong Correlations (recent and recently reconsidered)

Although the Brinkman-Rice model has only been treated for an ordered lattice, it shows qualitatively how the effective density of states entering the magnetic susceptibility and the specific heat can be strongly enhanced. This enhancement occurs at the same time that the frequency-dependent conductivity is suppressed near low f[1], as seen for example in Figure 7b. RICE also discussed[25] an extension of this theory to finite T, showing that the specific heat enhancement grows rapidly as the temperature is lowered below a reduced Fermi temperature (reduced, that is, by the enhanced effective mass). Ruckenstein suggested that we should work toward a generalization of the Gutzwiller approximation that includes the randomness.

The qualitative features that we can then look for experimentally in a system that might exhibit the Brinkman-Rice type of effects are the following:

1. sensitive to one electron/site
2. low f conductivity suppressed at the same time that
3. low T specific heat enhanced and
4. low T magnetic susceptibility enhanced

We have already anticipated the first of these factors in the classification of samples above, noting that the exponent 1/2 occurs in systems with one electron per scattering site, and that a distinct change occurs with the controlled addition of extra sites, as seen in Figure 3.

We have also noted above the low f depression of the far-infrared conductivity shown in Figure 7. At much lower densities than those shown in Figure 7, the spectrum has been explained quantitatively[7] in terms of clusters of electrons, with the energies shown to be strongly dependent on the Coulomb forces. Qualitatively similar behavior must occur near the transition, but with very large clusters occurring with increasing probability.

At the Santa Cruz conference, SASAKI and KAMIMURA [26] as well as RICE [25] reconsidered the third factor above as evidenced by measurements of the specific heat in Si:P, illustrated in Figure 8. The important feature here is the sharp upturn in the effective density of states C/T at low T. The roughly linear slope at high T in this plot as a function of T^2 is partly due to the phonons. However, part of the variation in C/T above 1K is due to the electrons whose contribution is probably slowly temperature-dependent,and possibly similar to the behavior presented by RICE[25] for periodic systems. The values of C/T in Figure 8 are normalized to the free electron density of states, gamma, which is about 0.03mJ/K for this sample. In previous analysis of this data, the free electron intercept of the linear region above about 2K was emphasized--a fact suggesting a lesser

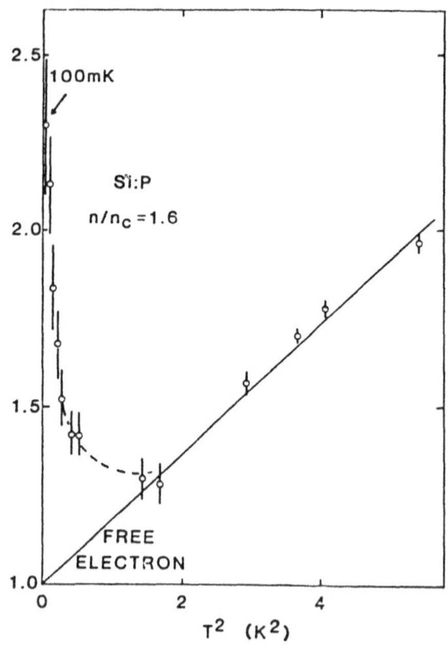

Fig. 8: Anomalous specific heat at low temperature in metallic Si:P [7]

role for interactions and a transition in a conduction band. The lowest T behavior, however, is certainly most important in estimating the T=0K properties. In Mott's terms, both this enhanced density of states for all excitations and the reduced one for current carriers (Fig. 7), are indications of strong correlations in an impurity band[1].

The fourth expectation noted above for a Brinkman-Rice-like system involves the magnetic susceptibility. Following the Santa Cruz conference, results have been written up by SARACHIK and co-workers[27] showing a strong enhancement of the magnetic susceptibility at low temperatures in Si:B as illustrated in Figure 9. These results confirm earlier studies showing qualitatively similar behavior in metallic samples of Si:P by SASAKI and co-workers[1,7]. Measurements have been extended to lower T in insulating samples of Si:P by ANDRES et al.[7,25,28], and are shown for a barely insulating sample in Figure 10. At still lower densities, analysis[7,25] of the magnetic susceptibility shows that clusters of spins with a wide distribution of antiferromagnetic interaction strengths accounts quantitatively for the suppression of the susceptibility below the Curie curve. Presumably, qualitatively similar behavior occurs near the transition, but with large clusters being relatively important. In going to the metallic side, the finite T susceptibility does not change drastically, but we speculate that its zero T value may become finite.

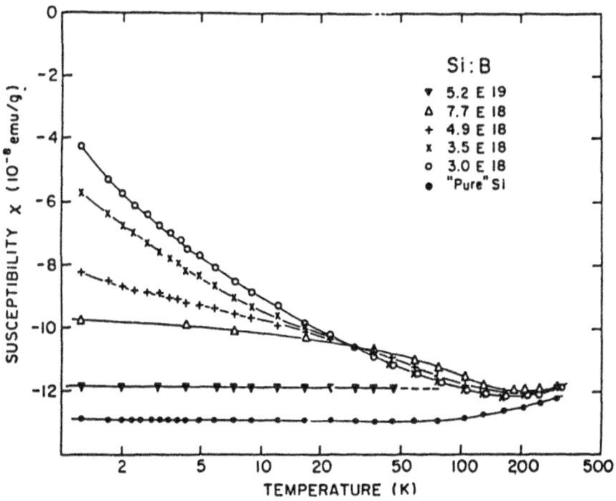

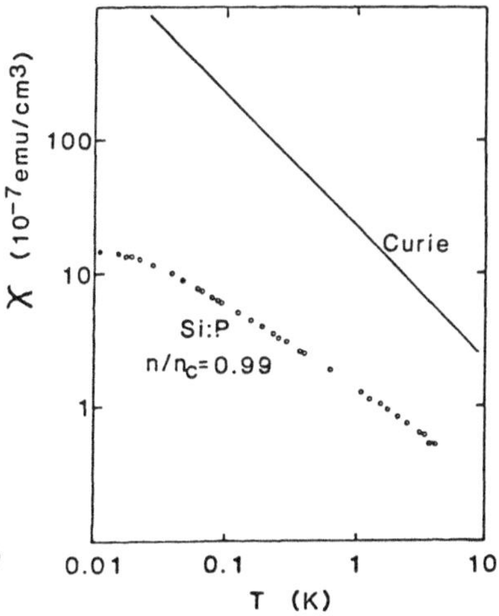

Fig. 9: Anomalous magnetic susceptibility at low temperature in metallic Si:B [27]

Fig. 10: Magnetic susceptibility in barely insulating Si:P at low temperature

A further indication of the presence of spins with quasi-localized character was presented at the Santa Cruz conference by PAALANEN et al.[4]. They showed that in barely metallic samples of Si:P the ^{29}Si nuclear spins relaxed over an order of magnitude faster, as a result of their interaction with their environment, than they would in the presence of non-interacting electrons. The temperature-dependence of this enhanced rate is approximately $T^{1/2}$, which indicates a distinct difference from non-interacting spins (either delocalized, as in the Korringa model, or localized). This spin-lattice relaxation time is found to get even shorter as the frequency (and applied magnetic field) are decreased. This dependence is roughly linear, but with a finite intercept at zero frequency in a metallic sample. This behavior appears to be related to the apparent slowing down of the spin diffusion. Rice emphasized that the variation of the Zeeman splitting should enter, since it is larger than T in much of the experimental region.

5. Conclusion

In the systems with strong correlations discussed above, we imagine that the electrons in a dirty metal encounter three sorts of scattering as they try to move around. They scatter strongly, of course, from the random potential. They scatter from the spins, which can appear quasi-static to them, because of the slow collective spin motion. Finally, they interact with the other electrons' Coulomb fields. A satisfactory explanation of the critical exponent that describes the metal-insulator transition will probably require a complete theory. However, the approach to the transition can be explained with spin scattering in a non-interacting model.

Helpful discussions and important collaborative interactions leading to this paper have come particularly from A.E. Ruckenstein, and also from E. Abrahams, P.W. Anderson, R.N. Bhatt, P.A. Lee, T.M. Rice, T.F. Rosenbaum, W. Sasaki, and others as indicated in the references.

References

1.N.F. Mott, The Metal-Insulator Transition,(Taylor and Francis, London).
2.N.F. Mott, Solid State Electronics, in press(proceedings of the Santa Cruz Conference); N.F.Mott and M.Kaveh, J. Phys. C15, 1707 (1982); N.F.Mott, Proc. Roy. Soc. London APS382PS, 1 (1982); see also Phil. Mag., in press(proc. Bar Ilan Conf.,1983).
3.P.W. Anderson, Sol St. Elec., in press.
4.M.A. Paalanen, A.E. Ruckenstein, and G.A. Thomas, ibid.; A.E.Ruckenstein, M.A.Paalanen and G.A.Thomas, Proc.ICPS17, San Francisco,1984.
5.H. Fukuyama, ibid.; See also S.Hikami, in Anderson Localization ,Springer, 1982, p.15 and other articles therein.
6.R.N. Bhatt, ibid.; R.N.Bhatt and T.V.Ramakrishnan, Phys. Rev.B28,6091(1983).
7.A review with extensive references is given by R.F. Milligan et al., Phys. Rev. B27,7509 (1983); and G.A.Thomas, Phil.Mag, in press.
8.P.Newman and D.F.Holcomb, Phys. Rev., in press.
9.A. Long and M. Pepper, Solid State Elec., in press; G.Biskupski, et al., J.Phys.C17,L411 (1984).
10.T.G. Castner, private comm. (1984).
11.A. Kawabata, Sol.St. Elec., in press.
12.C. Castellani and P.A.Lee, ibid.
13.R.C. Dynes, ibid.; D.G.Bishop, E.G.Spencer and R.C.Dynes, preprint (1984).

14.N. Nishida, et al., Sol. St. Elec., in press. 15.J.M. Mochel, private comm. (1984).
16.T. Yoshizumi et al., Solid State Elec., in press.
17.S. Morita et al., ibid.
18.S. vonMolnar et al.,ibid.;S.Washburn et al.,Phy.Rev.B,in press
19.R. Mansfield et al. ibid.
20.E. Abrahams et al., Phys. Rev. Lett.42, 673 (1979).
21.L.P. Gorkov et al., Soviet Phys. JETP Letters, 30, 228 (1980).
22.R.N. Bhatt, Solid State Elec., in press.
23.J.B. Mock and G.A.Thomas, unpublished.
24.W. Goetze, ibid.
25.T.M. Rice, ibid.; see also R.N.Bhatt, and P.A.Lee, Phys Rev. Lett.48,344 (1982); W.F.Brinkman and T.M.Rice, Phys. Rev.B2, 4302 (1970).
26.W. Sasaki and H. Kamimura, oral remarks (1984); T.Takimori and H.Kamimura, J.Phys.C16,5167 (1983).
27.M.A. Sarachik et al., private comm. (1984).
28.K.Andres, et al., Phys. Rev.B24,244 (1981); Similar effects are observed in CdS:n, R.E.Walstedt et al., J. Appl. Phys. 50, 1700 (1979).

The Scaling Theory of Localisation

A. MacKinnon

Imperial College, Department of Physics, Blackett Laboratory
Prince Consort Road, London SW7 2BZ, United Kingdom

While the application of scaling ideas to the behaviour of non-interacting electrons in disordered systems has provided some notable successes, especially in 2-dimensions, it has also thrown up some new questions and failed to provide consistent answers to others. Using mainly numerical data the, wide range of applicability of the scaling concept will be demonstrated, as well as its limitations.

We still do not understand the nature of the states close to a mobility edge. The short range fluctuations in the wave functions may play a critical role in transport properties, but are largely ignored by the scaling theory.

1. Introduction

Since the first formulations of the scaling theory of localisation by WEGNER [1] and ABRAHAMS et al. [2] considerable progress has been made in our understanding of the physics of transport phenomena in disordered systems. Several conferences and workshops have taken place [3,4,5] and review articles have appeared [6,7] . The numerical simulation of such systems has also advanced [8-11] producing a convincing demonstration of the validity of the scaling ansatz.

In the meantime, several new questions have emerged. Among these are the effects of electron-electron interaction [7] , of time-reversal symmetry and spin-orbit coupling [12,13] , of finite temperatures [14] and of polaron formation [15] and of magnetic fields.

In the absence of electron-electron interaction and magnetic fields three universality classes have been predicted [12,13] :

a) Orthogonal, due to normal potential scattering.

b) Unitary, due to lack of time reversal-symmetry, magnetic impurities, etc.

c. Symplectic, due to spin-orbit coupling.

Only the latter should display an Anderson/Metal-Insulator Transition in 2-dimensions.

Recently, there has been some interest in the short range behaviour of the wave functions close to the critical region [16,17] . On the one hand we can learn the limitations of the scaling theory for higher temperatures [9]; while, on the other hand, the possibility of large scale fluctuations which could fundamentally alter the critical behaviour cannot be ignored [15]. Here the fashionable concept of "fractals" [18] has been introduced [16].

2. A Model Quintet

In this paper we shall consider 5 different tight binding models belonging to 3 different universality classes.

The general form of the Hamiltonians on a two-dimensional lattice may be written

$$H = \sum_{ii' \; jj' \sigma\sigma'} H^{\sigma\sigma'}_{ii' \; jj'} \tag{1}$$

where i,i' and j,j' denote the lattice sites and σ,σ' ($=\pm 1$) are spin indices. There are two constraints on this form:

$$H^{\sigma\sigma'}_{ii' \; jj'} = (H^{\sigma'\sigma}_{i'i \; j'j})^{*} \tag{2}$$

due to hermiticity, and

$$H^{-\sigma,-\sigma'}_{ii' \; jj'} = \sigma\sigma' \, (H^{\sigma\sigma'}_{ii' \; jj'})^{*} \tag{3}$$

from the symmetry properties of the spin-orbit coupling.

In particular we consider the "Rectangular" and "Lloyd" [19] models with

$$H = W \sum_{ij} \varepsilon_{ij} \, |ij\rangle\langle ij| + V \sum_{iji'j'} |ij\rangle\langle i'j'| \tag{4}$$

where $i' = i \pm 1 \quad j' = j$

or $\quad i' = i \quad j' = j \pm 1$

The parameters ε are independent random variables distributed according to the distributions

$$p(\varepsilon) = \begin{cases} 1 & -\tfrac{1}{2} < \varepsilon < \tfrac{1}{2} \\ 0 & \text{otherwise} \end{cases} \qquad \text{"Rectangular"} \tag{5a}$$

$$p(\varepsilon) = \pi^{-1}(\varepsilon^2 + 1)^{-1} \qquad \text{"Lloyd"} \tag{5b}$$

We shall only treat those states at the band centre ($E=0$) leaving only 1 independent parameter, W/V. The above models contain purely diagonal disorder, whereas the three remaining have purely off-diagonal disorder. In these cases we write

$$H = \sum_{ijj'\sigma\sigma'} f(\phi^{\sigma\sigma'}_{ijj'}) \, |ij\sigma\rangle\langle ij'\sigma'| + V \sum_{ii'j\sigma} |ij\sigma\rangle\langle i'j\sigma| \tag{6}$$

where $i' = i \pm 1$ and $j' = j \pm 1$. The function $f(\phi)$ takes one of the forms:

$$f(\phi^{\sigma\sigma'}_{ijj'}) = V\cos\phi \, \delta_{\sigma\sigma'} \qquad \text{"Orthogonal"} \tag{7a}$$

$$= V\cos\phi \qquad \text{"Unitary"} \tag{7b}$$

$$= V\exp(i\phi) \qquad \text{"Symplectic"} \tag{7c}$$

where ϕ is a random variable between 0 and 2π.

In these 3 cases there is again only a single free parameter, the energy, E/V. In all the above models the variable V defines the units of energy and is irrelevant to the nature of the states.

Similar models have been the subject of earlier numerical work [21,22]. As defined the 3 models with off-diagonal disorder are very anisotropic: in the "i" direction the matrix elements are constants and only couple like spins, whereas in the "j" direction they have random phases and couple like and unlike spins equally.

This property makes it particularly easy to study the localisation properties of the models, while, hopefully, retaining those aspects of the Hamiltonian which determine its universality class.

It is worth noting at this point that the latter 3, off-diagonal, models are local gauge invariant [20] , in the j direction whereas the former 2, diagonal, models are not. We should expect the local gauge invariant models to obey scaling laws for smaller system sizes than the others, since their behaviour cannot be sensitive to boundary conditions [9] .

3. The Long Strip Method

The properties of the above Hamiltonians have been studied using the Long Strip Method [9-11] . The system is considered to be finite in direction 'j' with periodic boundary conditions and width, M. In the "i" direction the system is essentially infinite, so that we deal with a very long cylinder of circumference M.

We can rewrite any of the Hamiltonians to be considered in the form

$$V_{ii+1}\,\phi_{i+1} = (E-H_i)\phi_i - V_{ii-1}\,\phi_{i-1} \tag{8}$$

where the index 'i' refers to the "slices" of the cylinder and all quantities are defined in the M-dimensional subspace of the circumference.

Equ. (8) may be written as a transfer matrix

$$\begin{bmatrix}\phi_{i+1}\\ V_{i+1i}\phi_i\end{bmatrix} = \begin{bmatrix}V_{ij+1}^{-1} & 0\\ 0 & V_{i+1i}\end{bmatrix} \begin{bmatrix}(E-H_i) & -I\\ I & 0\end{bmatrix} \begin{bmatrix}\phi_i\\ V_{ii-1}\phi_{i-1}\end{bmatrix} \tag{9}$$

The time required to iterate such a form will be dominated by the inversion, V_{ii+1}^{-1}. By confining ourselves to models such that V_{ii+1} etc.may be replaced by a unit matrix, I, we may use the simpler formula

$$\begin{bmatrix}\phi_{i+1}\\ \phi_i\end{bmatrix} = \begin{bmatrix}(E - H_i) & -I\\ I & 0\end{bmatrix} \begin{bmatrix}\phi_i\\ \phi_{i-1}\end{bmatrix} \tag{10}$$

To study localisation we require the behaviour of the wave function on long length scales. This is given by the product

$$T_L = \prod_{i=1}^{L} \begin{bmatrix}(E - H_i) & -I\\ I & 0\end{bmatrix} \tag{11}$$

whose eigenvalues rise, or fall, exponentially with L on average. We require, in particular, the eigenvalue, t^L, which rises or falls most slowly: that is the contribution with the longest growth/decay length. Several methods for calculating this are discussed by MACKINNON and KRAMER [9] .

The localisation length on the cylinder of width M, λ_M, is defined as

$$\lambda_M^{-1} = \lim_{L\to\infty} \frac{-1}{2L} \ln |t^L|^2 \tag{12}$$

In the analysis that follows, the renormalised localisation length

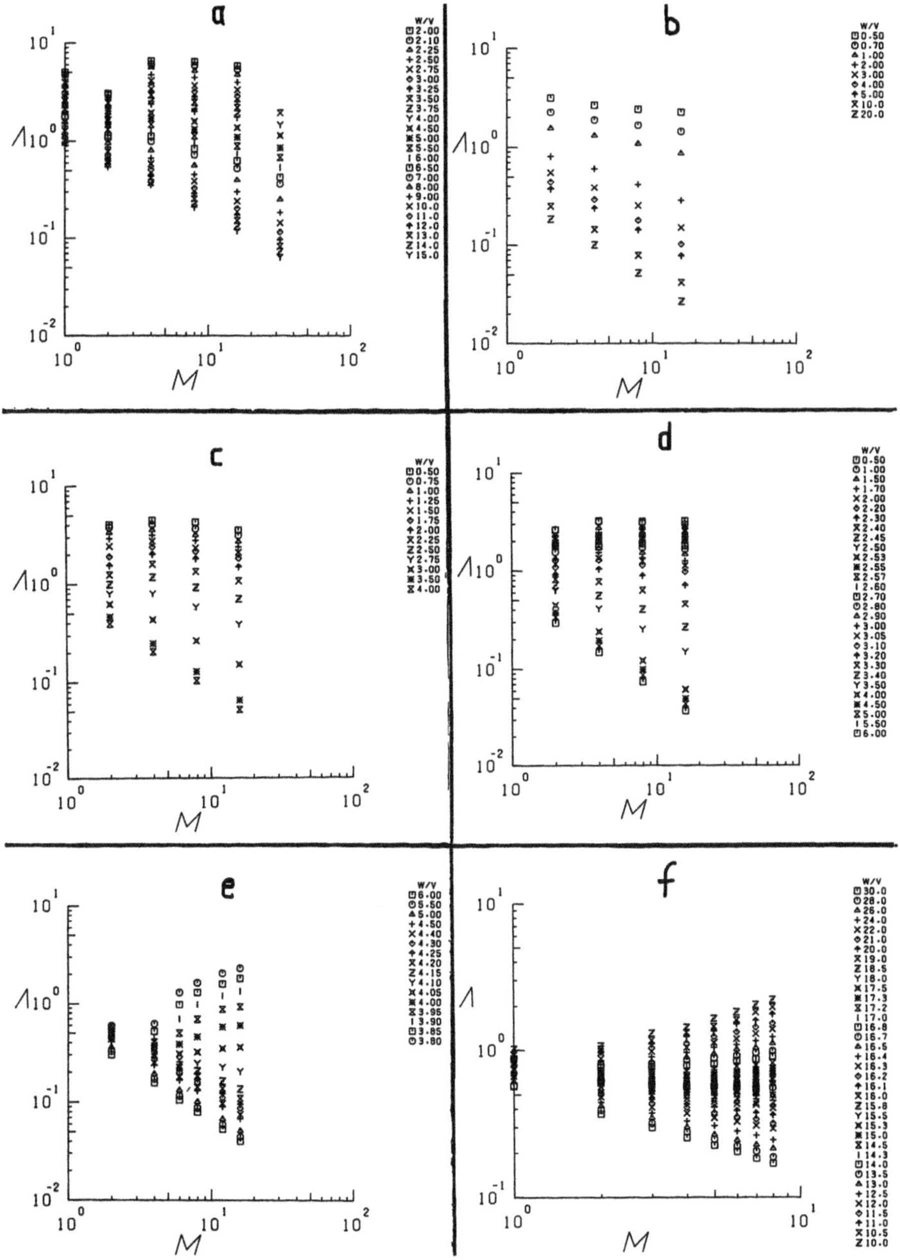

Figure 1
The renormalised localisation length, Λ, vs. circumference of the cylinder, M, for the Rectangular (a), Lloyd (b), Orthogonal (c), Unitary (d) and Symplectic (e) models in 2-dimensions and for the Rectangular model in 3-dimensions (f). The symbols refer to different values of disorder (W/V) or energy (E/V) as listed in the upper right of each figure.

$$\Lambda = \lambda_M / M \tag{13}$$

plays a decisive role. If we consider the cylinder as a one-dimensional chain of MxM "block spins", then Λ is the localisation length on the length scale of such a chain.

Fig. 1 shows the raw data for our 5 Models, together with that of the "Rectangular" Model in 3-dimensions.

Except for some deviations for small M, the data for the "Rectangular" (1a) "Lloyd" (1a) and "Orthogonal" (1c) models show that Λ always falls with increasing M. By contrast in the "Symplectic" (1e) and "3-d Rectangular" (1f) models the behaviour changes from Λ falling with increasing M for small Λ to Λ rising with increasing M for large Λ. In the "Unitary" (1d) case the behaviour for large Λ is unclear.

These tendencies may be interpreted as follows: falling Λ implies λ_M tending to become smaller than M and we might expect it to converge to $\Lambda = \lambda_\infty / M$ where λ_∞ is the localisation length of the two-dimensional system. Falling Λ is therefore a symptom of localisation.

When Λ rises by contrast λ_M is tending to become larger than M. This may be interpreted as a signature of extended states.

Therefore the "Rectangular", "Lloyd" and "Orthogonal" Models have only localised states, the "Symplectic" and "3-d Rectangular" Models have a transition and the "Unitary" model is undecided.

4. Scaling

a) General

In order to extract more information from our data, we make use of the ideas of the renormalisation group [1,2] . Let us start from the ansatz

$$\frac{d\ln\Lambda}{d\ln M} = \chi\,(\ln\Lambda). \tag{14}$$

This implies that the change in Λ when the scale of M is altered depends on Λ alone, and not on disorder, W or energy, E, or M separately. The general solution of (14) can be written

$$\Lambda(M,W) = f\{\xi(W)/M\} \tag{15}$$

where the constant of integration $\xi(W)$ is a characteristic length which depends on W(or E) but not on M. Before applying these ideas to our data let us consider some limiting cases.
i) For small Λ, as discussed above λ_M converges to its value in the 2-d or 3-d system. Thus

$$\Lambda(M,W) = \xi(W)/M \tag{16a}$$

$$\chi(\ln\Lambda) = -1 \tag{16b}$$

and we can identify $\xi(W)$ with λ_∞.
ii) For large Λ, perturbation theory gives

$$\begin{aligned}\Lambda(M,W) &= (M/\xi(W))^{d-2}\\ \chi(\ln\Lambda) &= d-2\end{aligned} \tag{17a}$$

where d is the dimensionality. In this case $\xi(W)$ is more difficult to identify (see sect. 6 below)

iii) From the behaviour of $\chi(\ln\Lambda)$ close to a fixed point $\chi(\ln\Lambda_c)=0$ we can calculate the critical indices of $\xi(W)$. Expanding χ around $\chi=0$ gives

$$\frac{d\ln\Lambda}{d\ln M} = \chi'(\ln\Lambda - \ln\Lambda_c) \qquad (18)$$

whose solution can be written

$$\ln\Lambda = \ln\Lambda_c + \{M/\xi(W)\}^{\chi'} \qquad (19)$$

This function must be analytic in W for all finite M. Thus

$$\xi(W) \propto (W-W_c)^{-\nu} \qquad (20)$$

where $\nu = 1/\chi$, is the only compatible form for ξ on both sides of the transition.

b) Analysis of the Data

How can we test whether our data are compatible with the scaling ansatz? One possible question asks: is it possible to choose one set of $\xi(W)$'s such that all the data fall on a single curve in accordance with (15)? Using a least squares procedure [9] to choose the $\xi(W)$'s we obtain the diagrams in Fig. 2. Not only has the procedure worked well (except 2b) but we have a bonus: not only is (15) fulfilled but we obtain the same functions $f(\chi)$ for the 2-dimensional "Rectangular", "Lloyd" and "Orthogonal" Models (Fig. 2a) and also for the 3-dimensional "Rectangular" and "Lloyd" Models (Fig. 2d). This result confirms that these models belong to the same "Orthogonal" [12,13] universality class.

The results for the "Unitary" model (Fig. 2b) are too noisy to be conclusive, whereas the behaviour of the "Symplectic" (Fig.2c) model is more reminiscent of the 3-d model. This is in agreement with previous predictions for such models [12,13]. In particular, there is clearly an Anderson transition in the "Symplectic" model, in spite of its 2-dimensional character.

5. Deviations From Scaling

The curves in Fig. 2 were calculated using only data for $M > 4$ (except 2c, $M > 6$). For smaller M either the lattice constant becomes significant or the phase correlation length/mean free path is comparable with M. In such cases the data are sensitive to boundary conditions and deviate from scaling [9]. Thus the phase correlation length marks a limit of scaling behaviour. The results of the theory are valid for temperatures such that the inelastic scattering length [14] is larger than this. At higher temperatures we return to classical, Drude, behaviour with the electrons moving ballistically between inelastic scattering processes.

It is important to note that the theory remains valid on length/temperature scales which are shorter than the characteristic length, $\xi(w)$ (15), in the theory. Only in this way can we describe the phenomena of weak localisation and the critical indices where an Anderson transition exists. In these cases $\xi(w)$ diverges and M is much smaller. For example, in the 2-d "Rectangular" model with W=2, $M \simeq 10$ and $\xi(W) \simeq 10^6$.

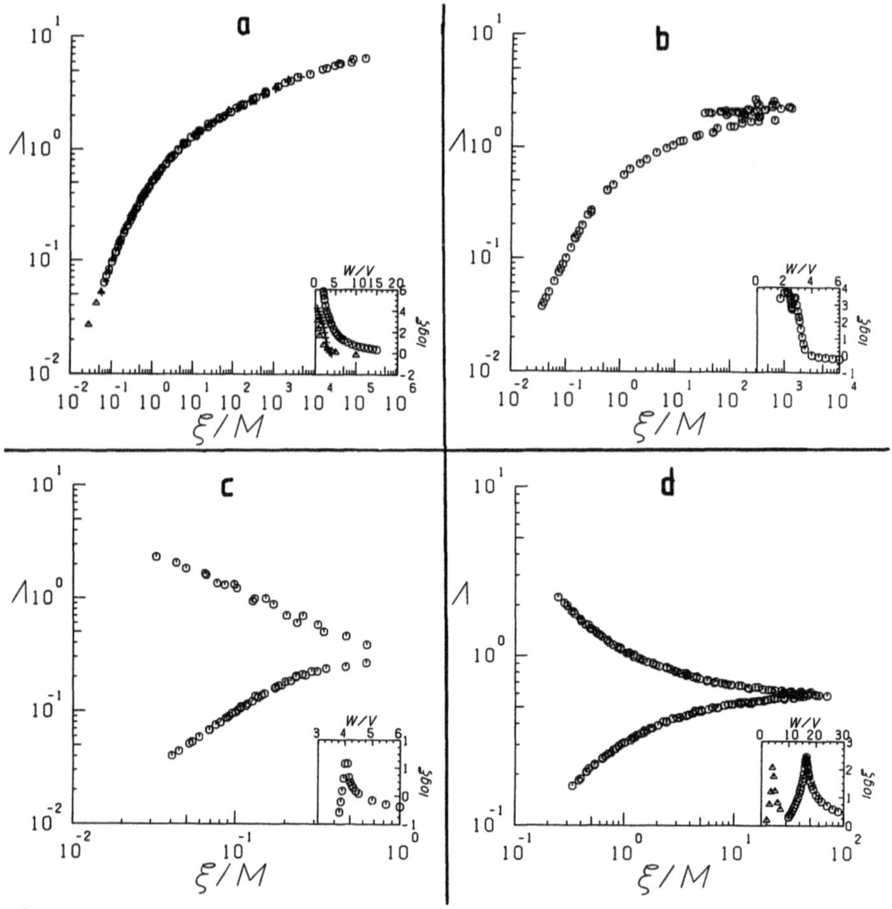

Figure 2
The scaling function, Λ vs ξ(W)/M where ξ(W) is the characteristic length determined by least squares. The figures refer to:

(a) Rectangular (○), Lloyd (△) and Orthogonal (+) models in 2-d.
(b) Unitary model in 2-d.
(c) Symplectic model in 2-d.
(d) Rectangular (○) and Lloyd (△) models in 3-d.

An exception to this is the "Symplectic" model, (Figs. 1e and 2c) where the tendency towards extended states, often called weak anti-localisation, only manifests itself when M is larger than a characteristic length related to the spin-orbit scattering. Below this, the behaviour is reminiscent of localisation (see Fig. 1e). This is in agreement with analytical theory on this case [13] . It implies, however, that although the behaviour for large Λ can be described by a one parameter scaling law, [12] , the critical behaviour requires a more complex analysis which would require much more data than is presented here.

6. Short Range Behaviour

It has been suggested [15,16] that at lengths much shorter than the characteristic length, ξ, but longer than the phase correlation length the states behave in a fractal manner. In analogy with their analysis we can interpret

$$d^* = \chi(\ln \Lambda)+2 \tag{21}$$

as an effective (fractal) dimension. This concept is particularily applicable to the region around $\chi=0$, which corresponds to short length scales. It is a curious result of the scaling concept that the critical index, ν, of the length, ξ, which characterises the behaviour at long length scales can be calculated using (18-20) from the opposite extreme, the limit of short length scales.

This implies that there is no relevant length scale below, ξ, until we reach the phase correlation length, which in the above models is comparable with the lattice constant. Exactly at the transition, $\xi=\infty$, the behaviour must be fractal, in the sense that it is independent of the length scale [17]. Presumably the short range behaviour is similar on both sides of the transition. We thus have fractal behaviour up to the localisation length or up to a length scale above which the states are 3-dimensional and uniform. We can identify this length with $\xi(W)$ on the extended side. A simple picture [23] for the fractal states is obtained by considering a cube of side, L. A number αL^2 of conduction channels enters the cube but only αL leave the opposite face. The probability that the electrons emerge from a given point is negligible; they emerge from a fraction αL^{-1} of the face, giving a "speckle" pattern. This picture illustrates the important point that the states must be highly non-uniform to behave fractally, so that they could be classified by, for example, a vanishing participation ratio, $\{ L^3 \int |\psi|^4 \}^{-1}$. This non-uniformity is essential to understanding the properties of real mobility edges, since it can lead, for example, to polaron formation [15], which may fundamentally alter the critical behaviour.

The above discussion is, however, not valid for the 2-d "symplectic" case since its short range behaviour is not simply related to the long range, as already noted.

Conclusions

We have considered data from several different model Hamiltonians which were chosen both to represent various universality classes and also for ease of calculation using the long strip algorithm. The universality of the scaling curves for the "orthogonal" class, in both 2-d and 3-d has been demonstrated by deriving a common curve for the scaling function (15) for the "Rectangular", "Lloyd" and "Orthogonal" models. At the same time we have seen that the behaviour of the "Unitary" and "Symplectic" models is different, and that the latter displays a clear metal insulator transition. Its critical behaviour is more complicated, however, and requires further study.

The short range behaviour of the eigenstates close to the transition has recently excited interest because of the possibility of fractal behaviour and polaron formation. More work is needed before the details are understood, however.

Acknowledgments

I should like to thank B. Kramer, J.B. Pendry and P.D. Kirkman for many useful discussions,and the British Petroleum Venture Research Unit for financial support.

References

1. F.J. Wegner: Z. Phys. 25, 327 (1976).
2. E. Abrahams, P.W. Anderson, D.C. Licciardello, T.V. Ramakrishnan: Phys. Rev. Lett. 42, 673 (1979).
3. Y. Nagaoka, H. Fukuyama (eds) "Anderson Localisation" (Springer) (1982).
4. This Volume.
5. International Seminar on Localisation in Disordered Systems at Johnsbach, GDR, to be published by Teubner-Texte zur Physik, Leipzig (1984).
6. G. Bergmann: Physics Reports 107, 1 (1984).
7. H. Fukuyama: to be published.
8. A. MacKinnon, B. Kramer: Phys. Rev. Lett. 47, 1546(1981).
9. A. MacKinnon, B. Kramer: Z. Phys. B. 53, 1 (1983).
10. J.L. Pichard, G. Sarma: J. Phys. C14, L127 and L617 (1981).
11. J.L. Pichard: Thesis, Universite de Paris-Sud (1984).
12. R. Oppermann, K. Jungling: Phys. Lett. 76A, 449 (1980).
13. S. Hikami, A.I. Larkin, Y. Nagaoka: Prog. Theor. Phys. 63, 707 (1980).
14. P.W. Anderson, E. Abrahams, T.V. Ramakrishnan: Phys. Rev. Lett. 43, 418 (1979).
15. M.M. Cohen, E.N. Economou, C.M. Soukoulis: Phys. Rev. Lett. 51, 1202 (1983).
16. C.M. Soukoulis, E.N. Economou: Phys. Rev. Lett. 52, 565 (1984).
17. H. Aoki: J. Phys. C16, L205 (1983).
18. B.B. Mandelbrot: "Fractals: Form, Chance, and Dimension" (Freeman, San Francisco) (1977).
19. P. Lloyd: J. Phys. C2, 1717, (1969).
20. R. Oppermann, F. Wegner: Z. Phys. B 34, 327 (1979).
21. C.M. Soukoulis, I. Webman, G.S. Crest, E.N. Economou: Phys. Rev. B26, 1838(1982).
22. U. Krey, W. Maass, J. Stein: Z. Phys. B49, 199(1983).
23. J.B. Pendry: J. Phys. C, in print (1984).

Anderson Transition and Nonlinear σ- Model

Franz Wegner

Institut für Theoretische Physik, Ruprecht-Karl-Universität
D-6900 Heidelberg, Fed. Rep. of Germany

1. Anderson Transition

A particle (e.g. an electron) moving in a random one-particle potential may have localized and extended eigenstates depending on the energy of the particle. The energy E_c which separates the localized states from the extended states is called the mobility edge. Extended states can carry a direct current whereas localized states are bound to a certain region and can move only with the assistance of other mechanisms (e.g. phonon-assisted hopping). Thus the residual conductivity is expected to vanish for Fermi energies E in the region of localized states, and to be nonzero for E in the region of extended states. This transition from an insulating behaviour to a metallic one is called Anderson transition.

This problem can be mapped onto a field theory of interacting matrices. The critical behaviour near the mobility edge will be discussed. The theory has a G(m,m) symmetry which, for finite frequency, breaks to a G(m) x G(m) symmetry. Depending on the potential, G stands for the unitary, orthogonal and symplectic group. Due to the replica trick m equals 0. The replica trick can be circumvented by using fields composed of commuting and anticommuting components. Then one deals with unitary graded and unitary orthosymplectic symmetries.

I refer to lectures given in Les Houches [1], Sanda-Shi [2], Trieste [3], and Sitges [4]. Most of the material presented here can be found in ref. [4] and in the original papers [5,6,7]. A few remarks concerning developments for interacting systems on similar lines are added.

2. Mapping on a Static Problem and Continuous Symmetry

Consider a one-particle tight-binding model

$$H = \sum_{rr'} f_{rr'} \, |r\rangle\langle r'| \tag{1}$$

where the ket $|r\rangle$ stands for an atomic orbital at site r of a d dimensional Bravais lattice. Then the Green's function

$$\mathcal{G}(r,r', z_p) = \langle r| \, (z_p - H)^{-1} \, |r'\rangle \tag{2}$$

can be expressed as expectation value over the field ϕ

$$\mathcal{G}(r,r', z_p) = s_p \, \langle \, \phi^*_{pa}(r') \, \phi_{pa}(r) \, \rangle \tag{3}$$

with respect to the "density"

$$\prod_p \{\det(z_p - f)\}^m e^{-\mathcal{H}} \quad \text{where} \tag{4}$$

$$\mathcal{H} = c \Sigma s_p (z_p \delta_{rr'} - f_{rr'}) \phi_{pa}(r) \phi^*_{pa}(r')$$

$$= -c \operatorname{tr}(\phi\ s\ \phi^+ f) + c \operatorname{tr}(\phi\ s\ z\ \phi^+). \tag{5}$$

Here $c = 1$, $p = 1,2$ is the energy index, $s_p = \pm i$, $z_p = E - is_p\omega/2$, E real, $\operatorname{Im} \omega > 0$, and $a = 1,2,\ldots m$ is the replica index. In the matrix formulation ϕ is a $2m \times N$ matrix, the columns are labelled by the energy and replica indices, the rows by the N lattice vectors r. The matrices s and z are diagonal with elements s_p and z_p, resp. The factors s guarantee the convergency of the integrals.

In order to get rid of the determinant in (4) we may formally choose $m = 0$. This is called the replica trick. Although this means literally that no degrees of freedom are left and (3) becomes meaningless, one can in practice do the calculation for general m and finally set $m = 0$. Diagrammatic expansions are well-defined for $m = 0$.

A mathematically and conceptually clean way is to add anticommuting components to ϕ [7,8]. Then the integral over the anticommuting components yields the determinant. Thus we may choose ϕ to be a $4 \times N$ matrix with N rows

$$\phi(r) = (S_1(r),\ S_2(r),\ \zeta_1(r),\ \zeta_2(r)) \tag{6}$$

where S_p are complex, ζ_p anticommuting components. We denote the set of graded matrices

$$X = \begin{pmatrix} a & \zeta \\ \eta & b \end{pmatrix} \tag{7}$$

by $M(n_1, m_1, n_2, m_2)$ where the blocks a, ζ, η, b are $n_1 \times n_2$, $n_1 \times m_2$, $m_1 \times n_2$, $m_1 \times m_2$ matrices and a, b are even, ζ, η odd elements of the graded algebra. Thus $\phi \in M(N, 0, 2, 2)$, $s, z \in M(2, 2, 2, 2)$, $f \in M(N, 0, N, 0)$. For an elementary introduction to graded matrices and groups see [9]. Working with these matrices all traces have to be read as graded traces.

$\mathcal{H}$ is invariant under linear transformations $\phi \to \phi U$ with $U \in U(m) \times U(m)$ and $U \in UPL(1,1) \times UPL(1,1)$, resp. In the limit $\omega \to 0$ the symmetry group is $U(m, m)$ or the pseudo-unitary graded subgroup of $UPL(2, 2)$ obeying $UsU^+ = s$. Thus the contribution proportional to ω, $\operatorname{tr}(\phi\phi^+)$, breaks this larger symmetry. The expectation value of the symmetry breaking term

$$\sum_p < \phi_{pa}(r)\phi_{pa}(r) > = i(\mathcal{G}(r,r,E + \omega/2) - \mathcal{G}(r,r,E - \omega/2)) \tag{8}$$

is proportional to the density of states ρ in the limit $\omega \to 0$, thus playing the role of the order parameter.

3. Composite Variables and Nonlinear σ-Model

Now let us consider the ensemble average over the random potentials H. Suppose the matrix elements f are Gaussian distributed with

$$\overline{f_{rr'}} = 0,\ \overline{f_{rr'} f_{r''r'''}} = \delta_{rr'''}\delta_{r'r''} M_{r-r'} . \tag{9}$$

This model is called local-gauge invariant since the distribution of the Hamiltonians H is invariant under gauge transformations

$$|r\rangle \rightarrow \exp(i\phi_r)\ |r\rangle\ . \tag{10}$$

Since these transformations are unitary we call it a unitary ensemble. Accordingly the only nonvanishing one-particle Green's function is

$$\overline{\mathcal{G}(r,r',\ z_p)} = \delta_{rr'}G(z_p) = s_p\ \langle\phi^*_{pa}(r')\phi_{pa}(r)\rangle \tag{11}$$

and the only nonvanishing two-particle Green's functions $K(z_1,\ z_2)$ are

$$\begin{aligned} K(r,r',\ z_1\ z_2) &= \overline{\langle r|(z_1 - H)^{-1}|r'\rangle\langle r'|(z_2 - H)^{-1}|r\rangle} \\ &= \langle\phi^*_{1a}(r')\phi_{1a}(r)\phi^*_{2b}(r)\phi_{2b}(r')\rangle \end{aligned} \tag{12}$$

$$\begin{aligned} K''(r,r',\ z_1,\ z_2) &= \overline{\langle r|(z_1 - H)^{-1}|r\rangle\ \langle r'|(z_2 - H)^{-1}|r'\rangle} \\ &= \langle\phi^*_{1a}(r)\phi_{1a}(r)\phi^*_{2b}(r')\phi_{2b}(r')\rangle \end{aligned} \tag{13}$$

where the average is taken with respect to

$$\begin{aligned} \overline{\exp(-\mathcal{H})} = \exp\{&-c\ \mathrm{tr}(\phi\ s\ z\ \phi^+) \\ &+ c/2\ \sum_{rr'} \mathrm{tr}(\phi(r)\ s\ \phi^+(r')\phi(r')\ s\ \phi^+(r))\}\ . \end{aligned} \tag{14}$$

By means of a Gaussian transformation we can transform [6] to composite variables Q (effectively $Q(r) \sim \sqrt{s}\ \phi^+(r)\phi(r)\ \sqrt{s}$) where $Q(r)$ is a 2m x 2m matrix and $Q(r) \in M(2,\ 2,\ 2,\ 2)$, resp., so that

$$G(z_p) \sim \langle\ Q^{pp}_{aa}(r)\ \rangle \tag{15}$$

$$K(r,r',\ z_1,\ z_2) \sim \langle\ Q^{12}_{ab}(r)\ Q^{21}_{ba}(r')\ \rangle \tag{16}$$

$$K''(r,r',\ z_1, z_2) \sim \langle\ Q^{11}_{aa}(r)\ Q^{22}_{bb}(r')\ \rangle \tag{17}$$

which is averaged with respect to exp(-L) with

$$\begin{aligned} &L = 1/2\ c\ \Sigma\ w_{rr'}\ \mathrm{tr}(Q(r)\ Q(r')) + c\ \sum_r \mathrm{tr}\ \ln(z - Q(r))\ , \\ &\sum_{r'} w_{r-r'}\ M_{r'-r''} = \delta_{rr''}\ . \end{aligned} \tag{18}$$

Even if local-gauge invariance is absent, then a similar transformation to matrices Q can be performed [10].

The saddlepoints of L are matrices Q with eigenvalues

$$\lambda_p = 2E_0^{-2}\{z_p - s_p(E_0^2 - z_p^2)^{1/2}\}\ , \quad E_0^2 = 4\sum_r M_r\ . \tag{19}$$

If instead of the Hamiltonian (1) we choose a model Hamiltonian [11]

$$H = n^{-1/2} \sum_{r\alpha,r'\beta} f_{r\alpha,r'\beta}\ |r\alpha\rangle\ \langle r'\beta| \tag{20}$$

with n orbitals per lattice site $\alpha = 1,\ 2,\ \ldots n$, and

$$\overline{f_{r\alpha,r'\beta}} = 0 , \quad \overline{f_{r\alpha r'\beta}\, f_{r''\gamma r'''\delta}} = \delta_{r'''}\delta_{r'r''}\delta_{\alpha\delta}\delta_{\beta\gamma} \; M_{r-r'} , \tag{21}$$

when we obtain again the Lagrangian (18) but with $c = n$. In the limit $n \to \infty$ the saddlepoints become very sharp and

$$G(r\alpha, r'\beta, z_p) = \delta_{rr'}\delta_{\alpha\beta}\lambda_p . \tag{22}$$

The eigenstates are in the interval $(-E_0, +E_0)$ obeying a semicircle law.

For large but finite n the density of states obeys a scaling law near E_0, $\rho(E)/n = n^{-\xi}\, \bar{\rho}((|E| - E_0)n^{2\xi})$ with $\xi = 2/(6 - d)$ for $d < 4$, $\xi = 1$ for $d > 4$ [12,13].

The zerodimensional model has been originally introduced by WIGNER [14], see the books [15,16] and the review [17], as a statistical model of nuclei. The present method has now returned to nuclear physics and has been used in the investigation of fluctuation properties of nuclei and nuclear reactions [18].

(If Q contains only commuting components, then the saddlepoint argument is correct. If Q contains also anticommuting components, then the UPL-symmetry is used additionally.) Starting from this limit one can perform a systematic expansion in powers of n^{-1} which was useful to obtain information on the mobility edge behaviour [19].

The contour of the integrals over Q can be made to run through the non-compact saddlepoint manifold obtained from the diagonal matrix with eigenvalues λ_p obtained by pseudo-unitary transformations. If we assume that the fluctuations of the eigenvalues of Q are irrelevant similar to the fluctuations of the length of the vector in the n-vector model, then Q obeys

$$(Q - \lambda_1)(Q - \lambda_2) = 0 . \tag{23}$$

From $Q^{pp} \approx \lambda_p$ for small Q^{12}, Q^{21} one obtains from (23)

$$Q^{11} = 1/2(\lambda_1 + \lambda_2) + \{(1/2(\lambda_1 - \lambda_2))^2 - Q^{12}\, Q^{21}\}^{1/2} , \tag{24}$$

and similarly for Q^{22}. Thus the "longitudinal" components Q^{11}_{ab}, Q^{22}_{ab} can be expressed in terms of the "transversal" components Q^{12}_{ab}, Q^{21}_{ab}. The matrix elements Q^{12}_{ab} are independent. The Taylor expansions of (24)

$$Q^{11}_{ab} = \lambda_1\delta_{ab} - \frac{1}{\lambda_1-\lambda_2} \sum_c Q^{12}_{ac}\, Q^{21}_{cb} + \ldots \tag{25}$$

yields

$$< Q^{11}_{ab} > = \lambda_1\delta_{ab} \tag{26}$$

since the summation runs over $m = 0$ equal contributions in the case of the replica trick, and $< Q^{11}_{a2}\, Q^{11}_{2a} > = - < Q^{11}_{a1}\, Q^{11}_{1a} >$ due to the UPL(1,1) symmetry. Therefore

$$G(r, r, z_p) = \lambda_p \tag{27}$$

holds. The transverse fluctuations do not affect the averaged one-particle Green's function which thus shows no critical behaviour at the mobility edge. This is a particular feature of the $m = 0$ component problem.

We may construct an effective interaction L_{eff} for the matrices Q. Any local potential of Q obeying the full pseudo U(m,m) symmetry and UPL(2,2) symmetry is a function of λ_1 and λ_2 only and thus a constant. Any interaction containing one gradient can be expressed as a surface integral. The most simple interaction containing two gradients is

$$L^0_{eff} = 1/4\ \hat{K} \int d^d r\ \mathrm{tr}(\nabla Q(r) \nabla Q(r))\ . \tag{28}$$

It is the only interaction containing two derivatives ∇ obeying rotational and inflection symmetry. To (28) the symmetry breaking term has to be added

$$L_{eff} = L^0_{eff} + \frac{i\omega}{4v} \int d^d r\ \mathrm{tr}(s\ Q(r)) \tag{29}$$

where v is the volume per lattice site. The harmonic approximation which is exact in the limit $n \to \infty$ yields for (29) the two particle Green's function

$$K(q,\ z_1\ z_2):\ = \sum_{r,\beta} e^{iqr}\ K(0\alpha,\ r\beta,\ z_1,\ z_2) \sim \frac{\rho(E)}{-i\omega + Dq^2} \tag{30}$$

in the hydrodynamic limit, and thus a diffusion type behaviour. Therefore the Goldstone mode due to the continuous symmetry of $U(m,m)/U(m) \times U(m)$ corresponds to the diffusion mode in the random potential.

4. Mobility Edge Behaviour

In order to obtain the power laws near the mobility edge we now use the well-known results from conventional critical phenomena. The analogies with an isotropic ferromagnet at temperature T, magnetic field h, correlation length ξ, and with magnetization m, and transverse and longitudinal susceptibilities $\chi_\perp$ and $\chi_\parallel$, are listed in the first two rows below

$E - E_c$	$\hat{=}$	$T - T_c$	1
$-i\omega$	$\hat{=}$	h	$\beta+\gamma = d\nu$
q	$\hat{=}$	q	ν
ξ	$\hat{=}$	ξ	$-\nu$
ρ	$\hat{=}$	m	$\beta = 0$
K	$\hat{=}$	$\chi_\perp$	$-\gamma = -d\nu$
K''	$\hat{=}$	$\chi_\parallel$	$-\gamma = -d\nu$

The localization length ξ indicates the range over which the wavefunction decays. The third column shows how these quantities scale, that is, if $E - E_c$, $-i\omega$, and q are multiplied by factors b^1, $b^{\beta+\gamma}$, b^ν, then ξ, ρ, K, K'' are rescaled by factors $b^{-\nu}$, $b^{-\gamma}$, resp. Since $\beta = 0$, and the scaling law $2\beta + \gamma = d\nu$ holds, all exponents can be expressed by ν. Thus, e.g.,

$$K(b^\nu q,\ b^{d\nu}\omega,\ b(E - E_c)) = b^{-d\nu}\ K(q,\omega,\ E - E_c)\ . \tag{31}$$

The conductivity σ can be expressed by the diffusion constant D and by K as

$$\sigma \sim \rho D \sim \rho^2 \partial K^{-1}/\partial q^2\ \ ; \tag{32}$$

thus it scales like

$$\sigma \sim b^s \sim (E - E_c)^s \tag{33}$$

with

$$s = 2\beta + \gamma - d\nu = (d - 2)\nu \ . \tag{34}$$

Similarly the polarizability diverges as one approaches the mobility edge like

$$\chi \sim \sigma/\omega \sim b^{s-d\nu} \sim (E_c - E)^{-2\nu} \ , \tag{35}$$

compare [20].

All the scaling laws agree with the prediction from real-space renormalization for the homogeneous fixed point ensemble [21].

The model (28 - 29) is very similar to the nonlinear σ-model on matrices with unitary symmetry U(2m)/U(m) x U(m) for m = 0. The "pseudo"-symmetries yield the same diagrams; only some overall signs are flipped. The W-function and thus the critical exponent s has been calculated in $d = 2 + \varepsilon$ dimensions [22-24] yielding

$$s = 1/2 + O(\varepsilon) \ . \tag{36}$$

This holds for systems in which time-reversal invariance is broken but isotropy and inversion symmetry are maintained on average.

For time-reversal invariant systems two universality classes known may be represented by the model (21) with [11]

$$\overline{f_{r\alpha r'\beta} f_{r''\gamma r'''\delta}} = (\delta_{rr'''}\delta_{r'r''}\delta_{\alpha\delta}\delta_{\beta\gamma} + \delta_{rr''}\delta_{r'r'''}\delta_{\alpha\gamma}\delta_{\beta\gamma})M_{r-r'} \tag{37}$$

and by a model with a spin-dependent potential [22]

$$H = n^{-1/2} \Sigma f_{r\alpha\sigma r'\beta\sigma'} |r\alpha\sigma\rangle \langle r'\beta\sigma'| \tag{38}$$

with $\alpha = 1, 2, \ldots n/2$ and

$$\overline{f_{r\alpha\sigma r'\beta\sigma'} f_{r''\gamma\sigma''r'''\delta\sigma'''}} = (\delta_{rr'''}\delta_{r'r''}\delta_{\alpha\delta}\delta_{\beta\gamma}\delta_{\sigma\sigma'''}\delta_{\sigma'\sigma''} + \delta_{rr''}\delta_{r'r'''}\delta_{\alpha\gamma}\delta_{\beta\delta}\sigma\sigma'\delta_{\sigma,-\sigma''}\delta_{\sigma',-\sigma'''})M_{r-r'} \ . \tag{39}$$

Using the replica trick the models are governed by orthogonal O(m,m)/O(m) x O(m) and symplectic Sp(m,m)/Sp(m) x Sp(m) symmetries, resp., with m = 0. The orthogonal case yields [19,23,24,25]

$$s = 1 + O(\varepsilon^4) \ . \tag{40}$$

In the symplectic case the W-function does not show a zero up to four-loop order in the physical region.

For two-dimensional systems the d.c. conductivity vanishes at all energies in the orthogonal and unitary case [19,22,26,27]. In the symplectic case there may be a region where the conductivity behaves better than ohmic [28]. In all cases one finds for σ of a square of length L in the quasi-metallic region

$$\sigma = \sigma_0 - \frac{e^2\alpha}{\pi^2 h} \ln\left(\frac{L}{a}\right) + \ldots \tag{41}$$

with $\alpha = 1, 0, -1/2$ in the orthogonal, unitary and symplectic case, respectively.

The time-reversal invariant system can also be treated by introducing fields ϕ and Q composed of commuting and anticommuting components. Then the matrices are of the form (7) where the block a contains real elements, b 2 x 2 submatrices of quaternion form, ζ and η pairs of adjoint Grassmann variables. Then the underlying symmetry is the unitary orthosymplectic group. The system is still described by $\mathcal{H}$, (5), and L, (18), with c = 1/2, n/2 for model (21), (37) and c = -1/2, -n/2 for model (38-39). (For details see [7].) Although the paths of integration over the Q-matrices are different the saddlepoints are the same and the 1/n expansion of one system can be obtained from the other by changing the sign of n. This symmetry relation has been first obtained by OPPERMANN and JÜNGLING [22] on a diagrammatic basis, and is related to the fact that manifolds $O(2m_1 + 2m_2)/O(2m_1) \times O(2m_2)$ and $Sp(-m_1 - m_2)/Sp(-m_1) \times Sp(-m_2)$ yield the same low temperature expansion [24].

Moreover, there is a second way to handle the unitary ensemble by introducing $\phi \in M(0, N, 2, 2)$. It also yields (5) and (18), but with c = -1, -n. Thus the saddlepoint expansion is invariant under the change of 1/n into -1/n. This has also been observed in[22] and is related to the formal equivalence of $U(m_1 + m_2)/U(m_1) \times U(m_2)$ and $U(-m_1 - m_2)/U(-m_1) \times U(-m_2)$ [29].

Although the formal expansions in powers of the inverse coupling constant $\hat{K}^{-1}$ (28) differs only by sign factors from the compact symmetry, the phase transition for the system of noncompact symmetry is quite different [6,30,31]. For $\omega \to 0$ the local expectation values $\overline{Q^n(r)}$, $n > 1$, remain finite only in the region of extended states. They diverge as one approaches the mobility edge reflecting strong amplitude fluctuations. The critical dimensions of these operators exceed the dimensionality of the lattice. They determine the critical exponents of the inverse participation ratio [32,33].

Recently Efetov [34] has argued that contributions beyond perturbation theory may become decisive. He predicts a jump of the metallic conductivity at the mobility edge for the Anderson model on a Cayley tree.

5. Strong Magnetic Field

In a magnetic field the system is described by the unitary symmetry which does not show an indication for extended states in two dimensions. In a strong magnetic field the quantized Hall effect, however, predicts a metallic state. Recently it has been pointed out [35] that a violation of inflection symmetry created by the magnetic field yields an additional term

$$\int d^d r \; tr(Q(r)[\partial_x Q(r), \partial_y Q(r)]) \tag{42}$$

which should allow for the Hall current. The authors use anticommuting fields ζ and the replica trick. Then there is no need for convergency guaranteeing factors s, and the symmetry group becomes compact. Then (42) describes topological charges, and the Hall conducting phase is supposed to be a phase of condensed topological excitations.

Restricting to the Hilbert space of the lowest Landau level for free electrons the density of states can be calculated exactly for electrons moving in a white-noise potential [36] or more generally in a system of uncorrelated point-like scatterers [37]. The components S(r) and $\zeta(r)$ of the field ϕ can be composed to a superfield $\psi(r,\theta) = S(r) + \theta\zeta(r)$, θ being Grassmannian. The effective Lagrangian exhibits supersymmetry in the (r,θ) superspace which reduces the problem from a two-dimensional to a zero-dimensional one. A similar reduction for higher Landau levels or two-particle Green's functions is not known.

However, the value of one-, two- and three-particle diagrams is uniquely determined by the number of Euler trails through the diagram which allows a speedy evaluation. An estimate of the conductivity in the band centre [38] yields $\sigma_{xx} = 1.4\ e^2/\pi h$. 1/n expansion starting out from a translational invariant saddlepoint have not given an indication for a transition to a conducting regime [39].

6. Interacting Electrons

The 1/n expansion [40] and the field theoretic formulation [41] in terms of fields ϕ and matrices Q can be carried over to the problem of interacting electrons. The Fermi statistics requires the fields ϕ to be anticommuting. The energy index p now labels the Matsubara frequencies. I quote only some results and refer the reader to the review by CASTELLANI and DI CASTRO [4] and some of the original papers [40-43].

If spin is not conserved and the long-range Coulomb interaction is taken into account, then the conductivity exponent s obeys s = 1 to lowest order in $\varepsilon = d - 2$. If the interaction is of short range, then one obtains in the spin flip case (magnetic impurities) s = 1/2, in a homogeneous magnetic field s = 1, and in a spin-orbit potential antilocalization shows up as in the non-interacting case.

If spin is conserved, then the renormalization group equations show initially a strong enhancement of the magnetic susceptibility [43]. Then the region of validity of the renormalization group equations is left. Thus the nature of the metal-insulator transition is not completely clear in this case. It may be that it is governed by slowly growing and decaying magnetic clusters [44].

References

1 D.J. Thouless, p.5, E. Abrahams, p.9, F. Wegner, p.15, Phys.Reports 67 (1980)
2 D.J. Thouless, p.2, F.J. Wegner, p.8, S. Hikami, p.15, P. Wölfle and D. Vollhardt, p.26, in Y. Nagaoka, H. Fukuyama (eds.) "Anderson Localization", Springer Series in Solid-State Sciences 39 (1982)
3 F. Wegner, Lecture Notes in Physics 201, 454 (1984)
4 F. Wegner (i), C. Castellani, C. di Castro (ii), Lecture Notes in Physics (VIII Sitges Conf. on Applications of Field Theory to Statistical Mechanics, 1984)
5 F. Wegner, Z. Phys. B35, 207 (1979)
6 L. Schäfer, F. Wegner, Z. Phys. B38, 113 (1980)
7 F. Wegner, Z. Phys. B49, 297 (1983)
8 K.B. Efetov, Zh. Eksp. Teor. Fiz. 82, 872 (1982); JETP 55, 514 (1982)
9 V. Rittenberg, M. Scheinert, J. Math. Phys. 19, 709 (1978)
10 A.M. Pruisken, L. Schäfer, Phys. Rev. Lett. 46, 490 (1981); Nucl. Phys. B200 [FS4], 20 (1982)
11 F.J. Wegner, Phys. Rev. B19, 783 (1979)
12 B.V. Bronk, J. Math. Phys. 5, 215 (1964)
13 K. Ziegler, Z. Phys. B48 (1982); Phys. Lett. 99A, 19 (1983)
14 E. Wigner, Ann. Math. 62, 548 (1955); 67, 325 (1958)
15 M.L. Mehta, Random Matrices and the Statistical Theory of Energy Levels (Academic Press, New York, London 1967)
16 C.E. Porter, Statistical Theory of Spectra: Fluctuations (Academic Press, New York, London 1965)
17 T.A. Brody, J. Flores, J.B. French, P.A. Mello, A. Pandey, S.S.M. Wong, Rev. Mod. Phys. 53, 385 (1981)

18 J.J.M. Verbaarschot, H.A. Weidenmüller, M.R. Zirnbauer, Phys. Rev. Lett. 52, 1597 (1984)
H.A. Weidenmüller, Ann. of Physics 158, 120 (1984)
J.J.M. Verbaarschot, M.R. Zirnbauer, Ann. of Physics 158, 78 (1984)
19 R. Oppermann, F. Wegner, Z. Phys. B34, 327 (1979)
20 T.F. Rosenbaum, K. Andres, G.A. Thomas, R.N. Bhatt, Phys. Rev. Lett. 45, 1723 (1980) and references 6 and 7 therein
21 F.J. Wegner, Z. Phys. B25, 327 (1976)
22 R. Oppermann, K. Jüngling, Phys. Lett. 76, 449 (1980); Z. Phys. B38, 93 (1980)
23 S. Hikami, Prog. Theor. Phys. 64, 1466 (1980); Phys. Rev. B24, 2671 (1981)
24 E. Brézin, S. Hikami, J. Zinn-Justin, Nucl. Phys. B165, 528 (1980)
25 S. Hikami, Nucl. Phys. B215 [FS7], 555 (1983)
26 E. Abrahams, P.W. Anderson, D.C. Licciardello, T.V. Ramakrishnan, Phys. Rev. Lett. 42, 673 (1979)
27 L.P. Gorkov, A.I. Larkin, D.E. Khmel'nitskii, Pis. Zh. Eksp. Teor. Fiz. 30, 248 (1979); JETP Lett. 30, 228 (1979)
28 S. Hikami, A.I. Larkin, Y. Nagaoka, Prog. Theor. Phys. 63, 707 (1980)
29 F.J. Wegner, Nucl. Phys. B180 [FS2], 77 (1981)
30 G. Parisi, J. Physics A14, 735 (1981)
31 A. Houghton, A. Jevicki, R.D. Kenway, A.M.M. Pruisken, Phys. Rev. Lett. 45, 394 (1980)
32 F. Wegner, Z. Phys. B36, 209 (1980)
33 A.M.M. Pruisken, preprint
34 K.B. Efetov, Pis'ma Zh. Eksp. Teor. Fiz. (1984), in print and preprint
35 H. Levine, S.B. Libby, A.M.M. Pruisken, Phys. Rev. Lett. 51, 1915 (1983); Nucl. Phys. B FS , 30, 49, 71 (1984)
A.M.M. Pruisken, Nucl. Phys. B235 [FS11], 277, and present volume
36 F. Wegner, Z. Phys. B51, 279 (1983)
37 E. Brézin, D.J. Gross, C. Itzykson, Nucl. Phys. B235 FS11 , 24 (1984)
38 S. Hikami, Phys. Rev. B29, 3726 (1984), and present volume
39 T. Streit, J. de Physique Lett. 40, L713 (1984), and present volume
S. Hikami, Prog. Theor. Phys. 72, 722 (1984)
40 R. Oppermann, Z. Phys. B49, 273 (1983)
R. Oppermann, J. Phys. Soc. Jap. 52, 3554 (1983)
M. Ma, E. Fradkin, preprint
41 A.M. Finkel'stein, Zh. Eksp. Teor. Fiz. 84, 168 (1983); JETP 57, 97 (1983)
42 A.M. Finkel'stein, Pis. Zh. Eksp. Teor. Fiz. 37, 436 (1983); JETP Lett. 37, 436 (1983)
B.L. Altshuler, A.G. Aronov, Sol. St. Comm. 46, 429 (1983)
C. Castellani, C. di Castro, P.A. Lee, and M. Ma, Phys. Rev. B30, 527 (1984)
C. Castellani, C. di Castro, G. Forgacs, and S. Sorella, preprint
43 R. Oppermann, Sol. Stat. Comm. 44, 1297 (1982)
C. Castellani, C. di Castro, P.A. Lee, M. Ma, S. Sorella, and E. Tabot, Phys. Rev. B30, 1596 (1984)
A.M. Finkelstein, Z. Phys. B56, 189 (1984)
44 P.W. Anderson, this volume

Localization and Superconductivity

G. Deutscher, A. Palevski, and R. Rosenbaum

Department of Physics and Astronomy, Tel Aviv University, Ramat Aviv
Tel Aviv 69978, Israel

The interplay between electron localization and superconductivity is discussed for two-dimensional and three-dimensional systems. In the two-dimensional case (thin films) theory and experiments on T_c depression, critical field and the effect of superconductivity fluctuations are reviewed and compared. In the three-dimensional case (percolative, granular, and amorphous systems near the metal-insulator transition) experiments on the quenching of superconductivity and on the anomalous magnetoresistance are reviewed with reference to current theoretical ideas.

1. Introduction

From the study of granular metals [1][2] it has been known for many years that superconductivity disappears near the metal-insulator (M-I) transition. For the case of thin films, it was observed that the relevant parameter for the decrease of the critical temperature T_c was the resistance per square $R_\square$ rather than the normal state resistivity ρ_n [3], while for thick granular films it was shown that the grain's size was an important parameter in its own right: for a given value of the resistivity, the depression of T_c is more pronounced for smaller grains [4]. Both of these observations could be understood in terms of thermodynamic fluctuations of the superconducting order parameter, known to depress the (mean field) critical temperature. For thin films, $R_\square$ was known from the theory of ASLAMAZOV and LARKIN [5] to be the relevant scaling parameter for these fluctuations in two dimensions, while for thick granular specimens these fluctuations can become *zero-dimensional* in character when the effective superconducting coherence length becomes of the order of the grain's size [6], which explains the importance of this parameter.

Recently, disorder has been shown to lead to electron localization in a manner that depends critically on dimensionality [7]. In particular, theory predicts, and experiments verify [8], that in 2D even a weak disorder (i.e. a value of $R_\square$ small compared to $\hbar/e^2$) is sufficient to induce electron localization and to produce an observable logarithmic increase of the resistance at low temperatures, and at the same time a depression of T_c linear in $R_\square$. More generally, it is now possible to view the depression of T_c as being one of the phenomena associated to electron localization.

In order to make a critical comparison between theory and experiment, it is necessary to distinguish between the 2D and the 3D case. In 2D, we are mostly interested in the detailed verification of the predictions of weak localization theory. The effects are small, but the theories, based on a perturbation approach, are supposed to be very accurate in the considered limit of weak disorder. The samples are thin films with very well defined and homogeneous physical properties (smooth and continuous films, with a mean free path usually limited by diffuse reflections at the surfaces, and typically $R_\square$ < 10 $\Omega/\square$, very much below the characteristic value for strong localization $\sim$ 30K $\Omega/\square$). Superconducting properties in the weakly localized 2D regime are now well understood, we briefly review them in the next section.

In the 3D case, localization effects are in general immeasurably small, unless samples are in the vicinity of the M-I transition, with resistivities larger than 1.10^{-4} Ωcm. This poses a theoretical problem, because the perturbation approach is not valid near the transition - and also an experimental one, because usual metals and alloys simply *do not have* resistivities in the range that we require. Doped semiconductors may of course be used for the study of normal state properties near the M-I transition, but we do not have that resource if we wish to study the disappearance of superconductivity. The only way high normal state resistivities may be achieved in metals is by mixing them with an insulating element. This may be done in a number of ways, but immediately leads to problems of homogeneity. It will therefore be necessary to first review some of the properties of inhomogeneous metals before discussing some of the latest developments on the interplay between strong localization and superconductivity.

2. The 2D Case: Weak Localization and Superconductivity

We review, successively, the T_c depression in zero field and the behaviour of H_{c2}, and the anomalous magnetoresistance above T_c.

2a. T_c-Depression and Modified Upper Critical Field H_{c2}

MAEKAWA et al. [9] predict that

$$\ln \frac{T_c}{T_{co}} = - \frac{R_\square}{\pi^2\hbar/e^2} \left[\left(\ln \frac{\hbar D_o K_3^2}{k_B T_c \tau_o}\right) \left(\ln \frac{\hbar}{k_B T_c \tau_o}\right)^2 - 2/3 \left(\ln \frac{\hbar}{k_B T_c \tau_o}\right)^3 \right] \tag{1}$$

where T_c and T_{co} are respectively the depressed and "bulk" value of the critical temperature, D_o the coefficient of diffusion, τ_o the electron scattering time and K_3 the inverse screening length in 3D. For usual metallic films, the expression inside the brackets is of the order of 10 to 100, and we have roughly

$$\frac{T_{co} - T_c}{T_{co}} \simeq R_\square / 10^3 \ \Omega. \tag{2}$$

For a film with $R_\square \simeq 100\Omega$, (2) predicts $(T_{co}-T_c)/T_c \simeq 1.10^{-1}$, which is the right order of magnitude (Fig.1) although a more detailed analysis seems to reveal some discrepancies [10]. However, there are many difficulties which complicate a detailed and meaningful comparison between theory and experiment: i) at small $R_\square$ values at which films are smooth and homogeneous ($R_\square$ < 10Ω) the predicted depression in T_c ($\sim 1.10^{-2}$K) is of the order of the width of the transition in thin films and of the reproducibility of the transition itself, which may be affected by stresses, etc.; ii) at high $R_\square$ values ($R_\square$ > 100Ω) one deals generally with semicontinuous films [11] for which the meaning of quantities such as D_o and τ_o that enter (1) is not at all straightforward [12]. High $R_\square$ films, presumably continuous, can be prepared by quench condensation onto liquid He cooled substrates, but in

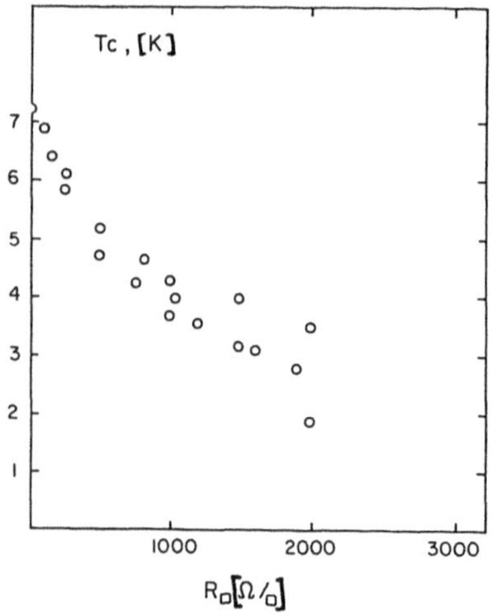

Fig.1: T_c depression in thin films as a function of the resistance per square (After [3])

that case the conditions of deposition may by themselves substantially enhance or depress T_c, again making comparison with experiment difficult; iii) the vortex unbinding Kosterlitz-Thouless transition [13] adds a further complication.

The same comments could apply to the behaviour of H_{c2}. Theory [9] predicts a temperature-dependence somewhat steeper than the regular Type II behaviour, an effect indeed sometimes observed experimentally [10]. Again, for small values of $R_\square$ the predicted effects are small, and in high $R_\square$ films inhomogeneity (percolation) effects play a predominant role [14].

2b. Anomalous Magnetoresistance above T_c

An anomalous (negative) magnetoresistance (MR), occurring at low magnetic fields, applied perpendicular to the plane of a thin metallic film [8], has been of the most convincing proof of the existence of electron localization in 2D, in excellent agreement with theory [15]. A characteristic feature of this MR is that it crosses over from an H^2 to a $\ell n\ H$ dependence at the characteristic field H_0 necessary to destroy the phase-coherence of the localized states. The value of this field is determined by the spatial extension of these states, which in the simplest case of infinite spin orbit and spin flip scattering times is given by the inelastic diffusion length $L_{in}=(D\ \tau_{in})^{\frac{1}{2}}$, where D is the coefficient of diffusion and τ_{in} the inelastic scattering time. In that case, $H_0 = \phi_0/4\pi L_{in}{}^2$. A detailed review of this anomalous MR, including the effects of spin orbit interaction that may inverse its sign, is presented by G. Bergmann in the Proceedings of LT17 [16]. In this section, we wish to review briefly the modifications to this anomalous MR when the metallic film becomes a superconductor at some lower temperature T_c, i.e. when superconducting fluctuations are present.

The theory of LARKIN [17] predicts essentially that the functional field dependence of the anomalous MR remains the same as in the absence of superconducting fluctuations, but with a positive sign and a temperature-dependent amplitude factor $\beta(T/T_c)$ which diverges at T_c. The function β does not contain any adjustable parameter, hence the verification of Larkin's theory is a very fine and decisive test of our current understanding of the interplay between superconductivity and weak localization. This experimental

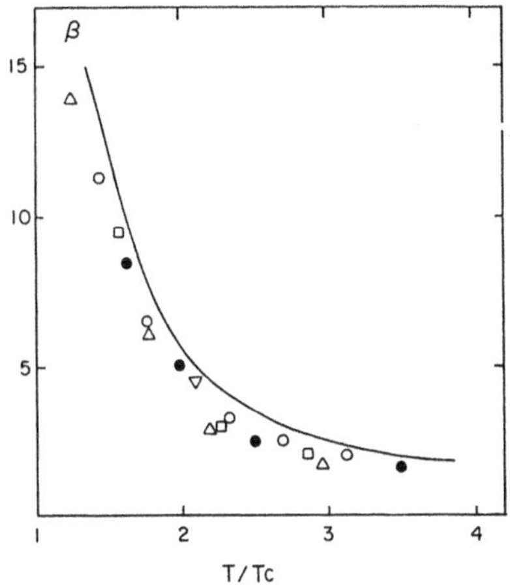

Fig.2: Amplitude factor of the anomalous magnetoresistance in Aℓ films (after Ref.18). Full line is the β function of LARKIN [17]

verification was first performed on thin Aℓ films [18] and gave complete agreement with Larkin's theory; similar agreement was reported by GERSHENSON, GUBANKOV and ZHURAVLEV [19] also on Aℓ films, and on other materials by different authors. (For instance on Pb-Cu proximity effect samples [20] and on In-Ge mixture films [12]). The effect is quite spectacular in Aℓ films where the spin orbit effect is small enough for the anomalous MR to have its usual negative sign far above T_c; then, as the temperature is reduced towards T_c, the MR becomes positive and very large near T_c. (Fig.2). Although the sign reversal of the MR at low temperatures may in general be due to the spin orbit interaction, its divergence near T_c as well as its detailed field and temperature-dependence constitute a clear verification of Larkin's theory.

As first noticed by KAPITULNIK et al. [12], the quality of the fit between experiment and theory allows one to identify for the first time the field at which the function β becomes field-dependent, i.e. the "ghost" critical field H_{c2} necessary to quench superconducting fluctuations. It has thus become possible to measure this ghost field as a function of $R_\square$ [12] and of (T/T_c) [20]. The data is in agreement with the proposed interpretation that $H_{c2} = \phi/2\pi\, \xi_s^{*\,2}(T)$ where $\xi_s^*(T)$ differs from the value $\xi_s(T)$ below T_c (for equal values of $|T_c-T|/T_c$) by a factor $\sqrt{2}$. In particular, H_{c2} goes linearly to zero as $T \to T_c$ (Fig.3).

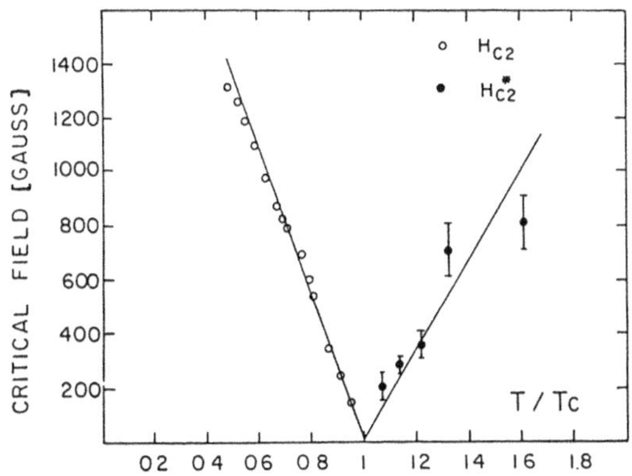

Fig. 3: Behaviour of the upper critical field H_{c2} $(T<T_c)$ and of the ghost critical field H^*_{c2} $(T>T_c)$ in Pb-Cu films (after [20])

Thus, in addition to the verification of Larkin's theory, it has been possible to extract from the same set of MR data the values of L_{in} and of $\xi_s(T)$. Because both of these lengths are proportional to $\sqrt{D}$, it is possible to eliminate this parameter and to obtain directly τ_{in} [12]. Such measurements and analysis outline the power of current theories of weak localization in 2D to obtain the values of fundamental solid state parameters.

3. Practical Materials for the Study of 3D Localization and its Influence on Superconductivity

As mentioned in the introduction, metals used for the study of 3D localization, and its influence on superconductivity, must in practice be metal-insulator mixtures [21]. We review in this section some of the relevant properties of these materials, including their normal state behaviour near the M-I transition.

3a. Structures

In order to approach this transition, it is in general necessary to use mixtures with insulator volume fractions of the order of several tens of percent. Since the constituents used are in general immiscible in each other (such as Aℓ-Ge, Aℓ-$A\ell_2O_3$, In-Ge, Hg-Xe), the mixtures may by no means be considered as homogeneous. Their structure then becomes an important factor in the understanding of their physical properties. Direct analysis through electron microscopy has revealed the existence of random structures (In-Ge, Pb-Ge) where both constituents are crystalline with similar grain sizes. Such mixtures have clearly a percolative structure, and their transport and superconducting properties follow the predictions of percolation theory (see below). The structure of the mixture can be characterized by a correlation length $\xi_p = L(p-p_c)^{-\nu}$ where L is of the order of the elementary grain size and $\nu_p = 0.88$ in 3D. ξ_p is the length scale beyond which the physical properties of the mixture (metal concentration, conductivity, etc.) become scale-independent. In granular structures - composed of metallic grains with a thin, amorphous, insulating coating - the main element of disorder is provided by variations in the thickness of the thin insulating barriers.

It seems now well established that the metal forms grains of a finite size in most metallic metal-insulator mixtures, including mixture films such as Hg-Xe quenched condensed at very low temperatures (∿ 10 K), with the exception of mixtures where the metal *itself* is amorphous when quench-condensed (such as Bi or Sn-Cu) [22]. Hence, the elementary length L that sets the scale for ξ_p varies in practice from a few 100 Å (In-Ge, Pb-Ge) to a few 10 Å (Aℓ-$A\ell_2O_3$, quench condensed Hg-Xe films) to a few Å (amorphous films). It will turn out that the length L plays a crucial role in the discussion of localization and superconductivity in metal-insulator mixtures.

3b. Conductivity and the Percolation-Localization Cross-Over

In random structures with well defined metallic grains (In-Ge, Pb-Ge and most probably Hg-Xe), there exists clear experimental evidence that the critical metal volume fraction X_c at which the M-I transition occurs is very close to the geometric percolation threshold p_c = 15% in 3D. This immediately suggests that localization effects are not very important in these mixtures, except of course in the immediate vicinity of X_c. Indeed, the percolation law $\sigma = \sigma_0$

$(p-p_c)^\mu$, with $\mu \simeq 2$, is followed essentially up to X_c in In-Ge and Pb-Ge [21], and close to it in Hg-Xe [23]. In granular mixtures, X_c is much larger than p_c (typically $X_c > 50\%$), but this is because of the asymmetry of the structure (thin insulating coating around metallic grains), and percolation critical laws still apply [24][25]. The situation is remarkably different in *amorphous* mixtures. In Bi-Kr films for instance [26], $X_c = 43\% \gg p_c$ *and* the index of the conductivity power law is close to unity, very much lower than the value of the percolation index, but in agreement with the value expected for a localization transition. Hence, there exists empirical evidence that a percolation-localization transition takes place when the scale L is reduced from a few 100 Å to a few Å.

A scaling argument first proposed by KHMELNITSKII [27] has been used [22] to explain this transition. In a percolative structure, the importance of localization effects is determined by the resistance of a cube of size ξ_p measured in units of $\hbar/e^2$. Once this normalized resistance is equal or larger than unity, it must be a decreasing function of length scale at larger length scales, according to the scaling theory of localization [7]. The sample must then be an insulator. In other terms, the Anderson transition in a percolative network is predicted to occur at a metal volume fraction X_c given by:

$$\frac{e^2}{\hbar}[\rho_0 (X_c - p_c)^{-\mu}] \;/\; \xi_p \;\simeq\; 1 \tag{3}$$

where ρ_0 is the resistivity on the length scale L. Equation (3) can be rewritten as:

$$X_c = p_c + y^{[1/(\mu-\nu_p)]} \tag{4}$$

where we have defined $y = \rho_0\, e^2/L\,\hbar$. The parameter y may be understood as the microscopic disorder parameter in the sample. When $y \ll 1$, $X_c \simeq p_c$ and we have in practice a percolation transition. On the other hand, when y approaches unity X_c may be significantly larger than p_c. The importance of the length scale L is immediately understood as we notice that the resistivity on the scale L is, in general, a decreasing function of L. If one makes the rough assumption that the mean free path on the scale of the elementary grain is of the order of the grain's size, it follows that $y \propto L^{\,2}$. We estimate [22] that $y \simeq 10^{-4}$ in the large-grained random mixtures and approaches unity in the amorphous ones. Equation (4) then predicts $X_c \simeq p_c$ in the first case and $X_c > p_c$ in the second one, in agreement with experiments.

One may go one step further, and make the assumption that localization effects (on the behaviour of the conductivity for instance) become important (as compared to the classical percolation effects) at the concentration where the localization correlation length ξ_ℓ becomes larger than the value of ξ_p (at X_c). In other terms, localization effects above X_c (weak localization) are observable in a concentration range $(X-X_c)$ which grows with (X_c-p_c), and turns out to be roughly proportional to it if one uses the accepted values of μ, ν_p and ν_ℓ (the index for ξ_ℓ). This scaling assumption satisfactorily explains the conductivity behaviour above X_c in all known structures [22].

Irrespective of the detailed nature of the M-I transition, the conductivity always decreases exponentially with temperature for $X < X_c$, $\sigma = \sigma_m \exp - (T_0/T)^n$, with values of n ranging from 1 [28] to $\frac{1}{4}$ [29]. Very often, $n=\frac{1}{2}$ in granular structures with a fairly broad range of grain sizes [1].

3c. Magnetoresistance near the M-I Transition

While we have now detailed measurements of the threshold and of the conductivity behaviour in zero-field in many systems, only recently have systematic magnetoresistance measurements been carried out on metal-insulator mixtures near the M-I transition. In granular Aluminium ($A\ell$-$A\ell_2O_3$), there is evidence for a large negative magnetoresistance near X_c, on both sides of the transition [30]. The amplitude of this magnetoresistance decreases clearly on both sides, hence it is a natural assumption to make [30] that it has its maximum at the transition itself. Its amplitude also decreases at high temperature [31] a clear indication that it is a quantum phenomenon (Fig.4). For a more complete discussion of the MR of granular $A\ell$, we refer the reader to the paper of the Rutgers group in the Proceedings of this Conference [30].

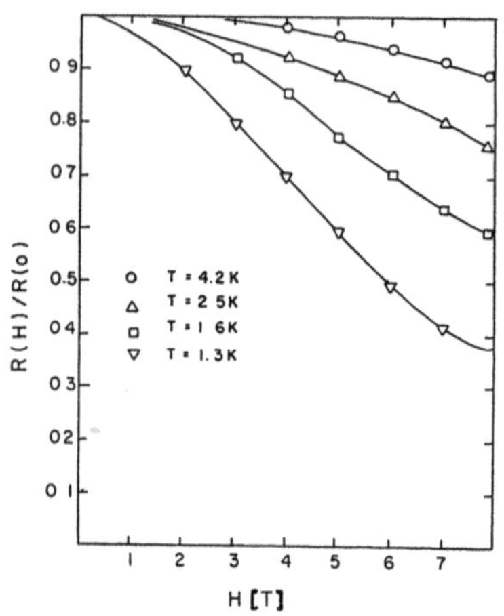

Fig.4: Anomalous magnetoresistance of $A\ell$-$A\ell_2O_3$ films (after [31]). Note the weakening of the MR at high temperatures

The behaviour of the MR in $A\ell$-Ge films is also quite interesting. In contrast with $A\ell$-$A\ell_2O_3$, the MR is in general positive [32] except very close to the transition [33]. The positive magnetoresistance seems to have two different origins: i) at high temperatures and at high fields, it is a classical magnetoresistance, with a field dependence $\propto H^2$ and an amplitude modified by the percolation effects, ii) at low fields and low temperatures it is a weak localization effect, where the sign of the MR is most probably due to a combination of strong spin-orbit interaction at the $A\ell$-Ge interface (Ge being a heavy element [16]) and the proximity of the superconducting transition. An example of a typical behaviour observed on the metallic side of the transition is given in Fig.5.

It has very recently been observed that the sign of the MR is *reversed* on the insulating side of the transition in $A\ell$-Ge [33]. This appears to be a very interesting phenomenon, possibly related to the phase diagram recently proposed by SHAPIRO [34] for the Anderson transition in a magnetic field. The experimental observations may be summarized in the following way:
i) for a sample very close to the transition ($\rho \simeq 3 \times 10^{-2}$ Ωcm at 4.2K, and about double that value at 1.6K), the magnetoresistance is still positive at 4.2K, but becomes negative above 1T at 1.6K with a weak positive part up to 1T. The trend of the data indicates that the MR would probably become negative at all fields at lower temperature. Such a behaviour cannot be explained by the weak localization theory in the presence of spin-orbit

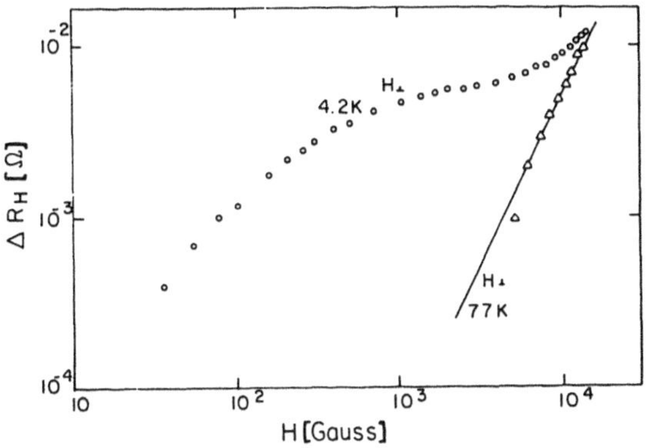

Fig.5: Magnetoresistance of Aℓ-Ge films at liquid nitrogen and liquid He temperatures (after [32]). The low field, low temperature part is due to localization while the high field, high temperature behaviour is classical

interaction. In that case, it is known that the MR turns from negative at high temperatures to positive at low temperatures (as the inelastic scattering time becomes longer than the spin orbit one). ii) A more strongly insulating sample (ρ = 1.5 Ωcm at 4.2K, ρ = 20 Ωcm at 1.6K) exhibits a negative MR at all fields already at 4.2K, and stays that way at lower temperatures. iii) A crossover from positive to negative MR at low fields is observed for an intermediate sample. Moreover, PALEVSKI et al. [33] notice that the excess conductivity $\Delta\sigma(H)$ varies approximately as H^2, and that at a given field *it tends to become temperature-independent* at high fields and low temperatures (Fig.6). This observation is of crucial importance, as it indicates that the insulating samples have become metallic in the presence of a magnetic field. If a form $\Delta\sigma = \sigma_g\ (H/H_0)^2$ is used to fit the data, with σ_g some characteristic (sample independent) conductivity, it also appears that the characteristic field H_0 becomes small as the transition is approached.

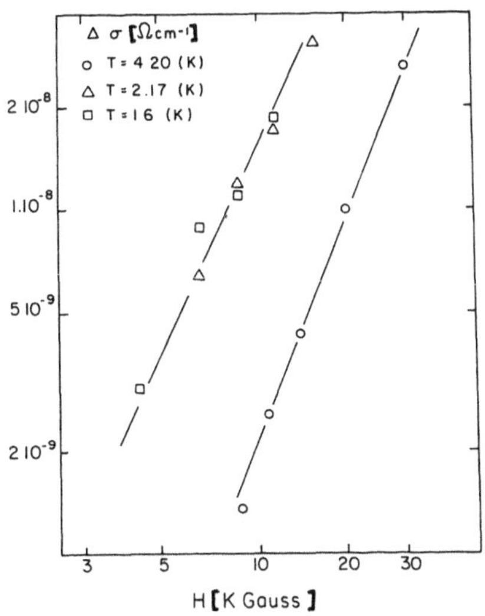

Fig.6: Magnetoconductivity $\Delta\sigma(H) = \sigma(H) - \sigma(0)$ in an Aℓ-Ge sample slightly below the M-I transition. Note the saturation of $\Delta\sigma$ at low temperatures, indicating field-induced metallic behaviour

The above observations are in agreement with the theoretical arguments of SHAPIRO [34], who has proposed, on the basis of general scaling arguments, that there exists a characteristic field above which an insulator becomes metallic close to the M-I transition, this field going to zero at the transition. We propose that, in the presence of strong spin-orbit interactions, the sign reversal of the MR signals the M-I transition itself. We note that, if this statement is correct at arbitrarily small fields, it implies the existence of a finite metallic conductivity at the transition. (We also note that no such conclusion can be drawn when the MR is negative on both sides of the transition).

4. Superconductivity near the M-I Transition

Although the question of the disappearance of superconductivity at or near the M-I transition has received attention for quite some time [35], the subject seems still to be far from being completely understood.

The feature that is generally observed is that near the M-I transition the superconducting critical temperature becomes lower, and that there is an increasing broadening of the transition. It was already noticed several years ago [4] that the decrease of T_c with ρ_n is *not* universal: superconductivity persisted up to higher values of ρ_n in samples prepared at room temperature, compared to samples deposited on liquid nitrogen cooled substrates (Fig.7). Electron micrographs showed that the former had larger grains, and the empirical conclusion was that superconductivity was less affected in large than in small grain structures near the M-I transition. The proposed interpretation relied on the observation that bulk-like superconductivity cannot persist in very fine-grained structures once the grains are effectively decoupled, i.e. when the effective superconducting coherence length becomes of the order of the grain size: in that limit there are very large thermodynamic fluctuations of the order parameter in the grains [6], which will necessarily broaden and lower the transition.

Recent experimental data obtained on quench condensed mixture films [23][26] have provided further evidence that ρ_n is not a universal parameter for the depression of T_c. In films known to be amorphous (such as Bi-Kr) the depression of T_c is much more pronounced than in films where the pure metal is crystalline when quench-condensed, and probably remains so in the

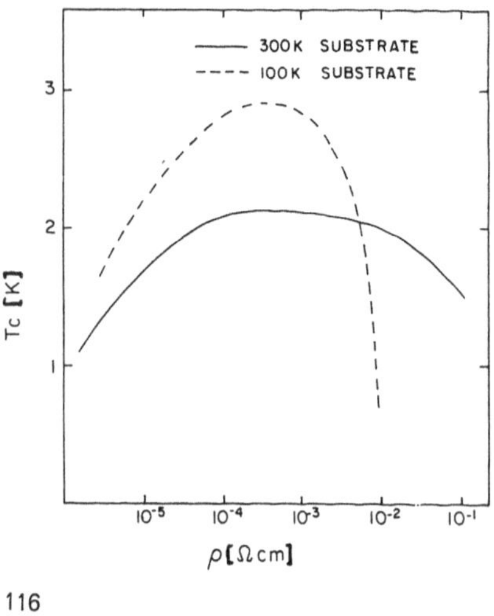

Fig.7: T_c of $A\ell\text{-}A\ell_2O_3$ films deposited at room and liquid nitrogen temperatures (after [4])

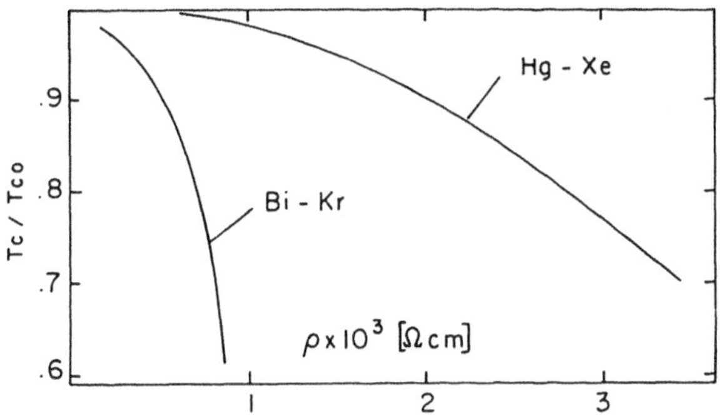

Fig.8. T_c depression for amorphous and crystalline quench condensed films

mixture film (such as Hg-Xe) (Fig.8). Although the grain size has not yet been measured in these films, it is probably on the order of 20-30 Å.

Thus, it is quite clear that the elementary scale of the disorder is an important parameter in the problem of the destruction of superconductivity near the M-I transition. An interpretation of the above experiments has been proposed [22] in terms of the percolation-localization crossover discussed above. The first remark is that percolation, being a purely classical phenomenon, cannot by itself affect the superconducting transition. Hence, in the region $\xi_p \gg \xi_\ell$, where all transport properties are governed by percolation, we do not expect superconductivity to be affected - in that limit, T_c should remain that of the bulk metal. If we remember that in large-grained structures the condition $\xi_p \simeq \xi_\ell$ is met only very close to X_c (which itself is very close to p_c), it is now clear why T_c remains unchanged in these structures up to very close to X_c where ρ_n is already large. By contrast, in amorphous films where $\xi_p \simeq \xi_\ell$ already far from X_c (which is itself significantly larger than p_c), T_c is depressed already far from the transition, at relatively low values of ρ_n.

Although a detailed theory clearly remains to be worked out, our conclusion is, therefore, that there is a close relationship between the appearance of localization effects - marked by a displacement of X_c towards values larger than p_c, by a conductivity index smaller than that predicted by percolation and by anomalous magnetoresistance effects - and the depression of T_c.

A fine point which has been raised already several times [36][28][35] is that of the exact value of the concentration at which superconductivity is quenched, compared to X_c. It is, of course, well known empirically that insulators are not superconductors, hence one would a priori expect superconductivity to be quenched at $X \geqslant X_c$. However, there exists as far as we know no general proof that an insulator cannot be a superconductor, and it has been argued [6][37] and reported [28] that in the case of a granular metal with large enough grains, superconductivity may persist in the insulating phase in a small concentration range below X_c. As pointed out in [28], a large grain size plays two distinct roles for the existence of a semiconductor-superconductor transition: i) it allows the Josephson coupling energy to be larger than the regular tunneling coupling energy, a situation that arises as soon as there exists on the average more than one Cooper pair per grain; ii) it reduces the grain's electrostatic charging energy; taking into account the divergence of the effective dielectric constant at the M-I transition [38], this opens up the range of concentrations below X_c where the effective charging energy is smaller than the Josephson coupling energy, and thus allows superconductivity to be observed. (We note, however, that these theoretical arguments do not take into account the effect of disorder.) For Aℓ, the critical grain diameter is 75 Å, and indeed a semiconductor-superconductor transition has been clearly observed for Aℓ-Ge with grain size 120 Å, but not for $A\ell$-$A\ell_2O_3$ with grain size of about 30 Å [2][36][39]. As a matter of fact, the Rutgers group has now come to the conclusion that in $A\ell$-$A\ell_2O_3$ superconductivity disappears exactly at the M-I transition, assuming that it occurs at the point of maximum magnetoresistance [30]. More work is certainly needed for a complete understanding of the intriguing semiconductor-superconductor [40].

As a last conjecture, one may wonder at the possibility of a magnetic field turning an insulator into a superconductor near the M-I transition [30]. This may not be impossible since - if the conjecture of SHAPIRO [34] is correct, and the first experimental indications are that it is [33] - the critical field required to turn an insulator into a metal goes to zero at the transition, while the value of the upper superconducting critical field H_{c2} (at T=0) might remain high (although not infinite [14]) in this very dirty limit.

5. Conclusions

In contrast with the situation in 2D, where the detailed interplay between (weak) localization and superconductivity is now fairly well understood (may be with the exception of the small T_c depression predicted in that limit, which is difficult to check experimentally), our knowledge of what happens in 3D, in mixture films, near the M-I transition is far more limited. One point that has now been made clear is the strong influence of the elementary length scale of the mixture on the percolation-localization crossover. For scales (grain sizes) of a few 100 Å, or more, the M-I transition is essentially a percolative one: the transition occurs practically at the percolation threshold, the conductivity index is the percolative one and T_c remains equal to the bulk value up to very close to the transition. When the disorder is on the atomic scale (amorphous mixtures), the transition takes place well above the percolation threshold, the conductivity index is the localization one, and T_c is strongly depressed already far from the transition. Simple scaling arguments give a satisfactory explanation to these observations, but we still need a detailed theory for a quantitative description of the quenching of superconductivity in these inhomogeneous media. Another point which is emerging is the critical behaviour of the magnetoresistance at the M-I transition. There is experimental evidence that it is negative and goes through a maximum amplitude in the case of weak spin-orbit interaction ($A\ell$-$A\ell_2O_3$) while a sign reversal is observed in the weak spin-orbit case (Aℓ-Ge). In the latter case, the excess conductivity produced by the applied field in the insulating phase appears to be temperature-independent at low temperatures and sufficiently high fields, indicating field-induced metallic behaviour as first suggested by SHAPIRO [34]. The implications for superconductivity remain to be investigated. Finally, there is some evidence that the disappearance of superconductivity does not necessarily coincide with the M-I transition, at least in large grained structures, but we still lack a general understanding of this question.

One of us (G.D.), has greatly benefited from illuminating discussions with A. Goldman and H. Micklitz. This work has been partially supported by an endowment from the Oren Family and the U.S.-Israel Binational Science Foundation.

References

1. B. Abeles in Applied Solid State Science, edited by R. Wolfe (Academic Press, New York, 1976); Vol.6, p.1
2. G. Deutscher, H. Fenichel, M. Gershenson, E. Grunbaum and Z. Ovadyahu, J. Low Temp. Phys. 10, 231 (1973)
3. M. Strongin, R.S. Thompson, O.F. Kammerer and J.E. Crow, Phys. Rev. B1, 1978 (1970)
4. G. Deutscher, M. Gershenson, E. Grunbaum and Y. Imry, J. Vac. Sci. Technol. 10, 697 (1973)
5. L.G. Aslamazov and A.I. Larkin, Phys. Letters 26A, 238 (1968)
6. G. Deutscher, Y. Imry and L. Gunther, Phys. Rev. B10, 4598 (1974)
7. E. Abrahams, P.W. Anderson, D.C. Licciardello and T.V. Ramakrishnan, Phys. Rev. Lett. 42, 673 (1979)
8. L. Van den Dries, C. Van Haesendonck, Y. Bruynseraede and G. Deutscher, Phys. Rev. Lett. 46, 565 (1981) and Physica (Utrecht) 107B, 7 (1981)
9. S. Maekawa, H. Ebisawa and H. Fukuyama, Technical Report of the Institute for Solid State Physics of the University of Tokyo, Ser.A. No.1426 April 1984, and to be published
10. See for instance, B.Shinozakij, T. Kawaguti and Y. Fujimori,J.Phys.Soc. Jpn.52, 2297 (1983)
11. A. Palevski M.L. Rappaport, A. Kapitulnik, A. Fried and G. Deutscher, J. Physique Lett. 45, L-367 (1984)
12. A. Kapitulnik, A. Palevski and G. Deutscher, to appear in J. of Phys.C
13. For a review, see for instance, J.H. Mooij in the Proceedings of the NATO Advanced Summer Institute on "Percolation, Localization and Superconductivity" (Les Arcs, France, 1983), edited by A.M. Goldman and S. Wolf
14. G. Deutscher, I. Grave and S. Alexander, Phys. Rev. Lett. 48, 1497 (1982)
15. B.L. Altshuler, D. Khmel'nitskii, A.I. Larkin and P.A. Lee, Phys. Rev. B22, 5142 (1980)
16. G. Bergmann, to be published in the Proceedings of LT17, Karlsruhe 1984
17. A.I. Larkin, Pis'ma Zh. Eksp. Teor. Fiz. 31, 239 (1980) [J.E.T.P. Lett. 31, 219 (1980)]
18. Y. Bruynseraede, M. Gijs, C. Van Haesendonck and G. Deutscher, Phys. Rev. Lett. 50, 277 (1983)
19. M.E. Gershenson, V.N. Gubankov and Yu. E. Zhuravlev, Solid State Commun. 45, 87 (1983)
20. D. Avraham, Ph.D. Thesis, Tel Aviv University (1984)
21. For a review of the properties of metal-insulator mixtures, see for instance, G. Deutscher, A. Kapitulnik and A. Palevski in "Percolation, Structures and Processes", edited by G. Deutscher, R. Zallen and J. Adler, published by Adam Hilger, Bristol, and the Israel Physical Society, Jerusalem (1983), p.207
22. G. Deutscher, A.M. Goldman and H. Micklitz, submitted to Phys. Rev. Lett
23. K. Epstein, A.M. Goldman and A.M. Kadin, Phys. Rev. B27, 6685 (1983)
24. B. Abeles, H. L. Pinch and J. Gittleman, Phys. Rev. Lett. 35, 247 (1976)
25. G. Deutscher, M. Rappaport and Z. Ovadyahu, Solid State Commun. 28, 593 (1978)
26. R. Ludwig and H. Micklitz, Solid State Commun. 49, 519 (1984)
27. D.E. Khmelnitskii, Pis'ma Zh. Eksp. Teor. Fiz. 32, 248 (1980) [JETP Letters 32, 229 (1980)]
28. Y. Shapira and G. Deutscher, Phys. Rev. B27, 4463 (1983)
29. J.J. Hauser, Phys. Rev. B3, 1611 (1971)
30. See H.K. Sin, P. Lindenfeld and W.L. McLean, these Proceedings (Suppl.)

31. W.L. McLean, T. Chui, B. Bandyopadhyay and P.Lindenfeld, in Inhomogeneous Superconductors -- 1979, edited by D.W. Gubser, T.L. Francavilla, J.R. Leibowitz and S.A. Wolf, AIP Conf. Proc. No.58 (AIP, New York, 1980), p.42
32. A. Kapitulnik, Ph.D. Thesis, Tel Aviv (1983) and A. Kapitulnik, A. Palevski and G. Deutscher, to be published
33. A. Palevski, R. Rosenbaum and G. Deutscher, to be published
34. B. Shapiro, in "Percolation, Structures and Processes", ibid p.367
35. For a review, see for instance, Y. Imry in the Proceedings of the NATO Advanced Summer Institute on "Percolation, Localization and Superconductivity", (Les Arcs, France, 1983) edited by A.M. Goldman and S. Wolf
36. G. Deutscher, B. Bandyopadhyay, T. Chui, P. Lindenfeld, W.L. McLean and T. Worthington, Phys. Rev. Lett. 44, 1150 (1980)
37. Y. Imry and M. Strongin, Phys. Rev. B24, 6353 (1981)
38. D. J. Bergman and Y. Imry, Phys. Rev. Lett. 39, 1222 (1977)
39. T. Chui, G. Deutscher, P. Lindenfeld and W.L. McLean, Phys. Rev. B23, 6172 (1981)
40. E. Simanek, Solid State Commun. 31, 419 (1979)

Aperiodic Structure in the Magnetoresistance of Very Narrow Metallic Rings and Lines

R.A. Webb, S. Washburn, C.P. Umbach, and R.B. Laibowitz

IBM T.J. Watson Research Center, PO Box 218, Yorktown Hts., NY 10598

ABSTRACT We report the low-temperature magnetoresistance of very small normal metal (Au and $Au_{60}Pd_{40}$) rings and lines. The wires composing all of the devices were typically 50 nm in diameter, and the ring diameters were about 350 nm. The resistances of the rings were measured in perpendicular magnetic field at temperatures down to T = 5 mK. For T $\lesssim$1 K, a reproducible structure appeared in the magnetoresistance, and it grew as the temperature decreased. Similar structure was seen in several Au and $Au_{60}Pd_{40}$ rings and lines. Although there is some evidence for flux quantization effects in one device, the structure was sample specific, and it is apparently the result of the extremely small device size.

Over thirty years ago, DINGLE [1] analyzed the magnetoresistance (neglecting scattering) of very small devices, and concluded that it should contain dense, complicated structure. The new features would arise from the finite spacing between the energy levels; this spacing is negligible for large devices, but as the size of the device becomes small, the level spacing becomes important. More recently, considerable interest has developed around several experiments which discovered periodic oscillations in the magnetoresistance of very small normal metal cylinders [2,3,4,5]. These oscillations arise from the the interference of the carrier waves after the phases have been tuned by the magnetic flux passing through the cylinder [6,7]. The observed resistance oscillations were periodic with respect to a flux quantum of h/2e, and diminished in amplitude as the temperature or magnetic field was increased. For rings, there is a competing theory which directly calculates the transmission coefficient of the ring from the Landauer formula [8]. This theory predicts magnetoresistance oscillations with period h/e and weaker oscillations with period h/2e superimposed [9,10]. To date, no one has observed clear evidence for flux periodic effects in a single ring. As discussed below, experiments on arrays of rings are probably equivalent to experiments on long cylinders.

In this paper, we will describe measurements of the magnetoresistance of small Au and $Au_{60}Pd_{40}$ rings and lines. The samples were patterned with a scanning transmission electron microscope [11] from 38 nm thick, polycrystalline Au and $Au_{60}Pd_{40}$ films. Aluminum wires were ultrasonically bonded to the connection pads about 0.5 mm from the current and voltage probes. The samples were placed in the mixing chamber of a ^3He-^4He dilution refrigerator which reached temperatures as low as 5 mK. The resistance was measured with an AC bridge consisting of two PAR124 lockins, a room temperature standard resistor, and an HP3456a digital voltmeter. Figure 1a is a transmission micrograph of one of the Au rings measured in this experiment (sample 1). The average diameter of the ring was 320 nm and the widths of the lines forming the ring were approximately 45 nm. The resistance of the gold samples indicated that the mean free path was limited by the smallest dimension of the sample, the thickness of the wires in the ring (38 nm). (Note that this implies that at low temperatures, the inelastic diffusion length L_{in} will be larger than the width of the wires in the sample: the wires will be quasi one-dimensional.) For the films used in the gold samples, the grain size was $\simeq$ 100 nm. For the $Au_{60}Pd_{40}$ devices (Fig. 1b), the grain size was $\simeq$ 3 nm, and the mean free paths were limited by bulk scattering.

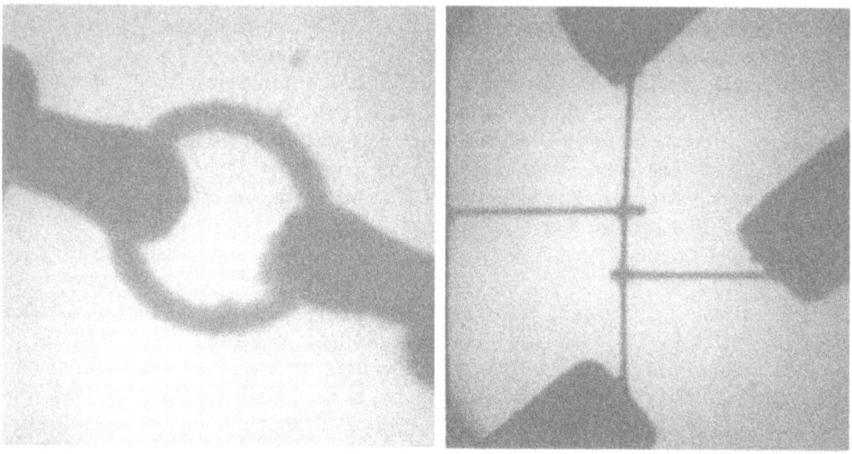

Fig. 1. (a) A transmission electron micrograph of sample 1, the Au ring which yielded the data in Fig. 2a; the average diameter of the ring was 320 nm, and the width of the wires forming the arms of the ring was 45 nm. (b) A micrograph of a $Au_{60}Pd_{40}$ line 50 nm wide and 790 nm long

Representative magnetoresistance data from sample 1 are shown in Fig. 2a. Below T $\simeq$ 1K, a complex, reproducible structure appeared in the magnetoresistance. The amplitude of this structure increased to a few parts in 10^3 of the sample resistance for temperatures less than T $\simeq$ 0.1 K. The positions of the peaks on the magnetic field axis were temperature independent. Figure 2b contains the magnetoresistance of a $Au_{60}Pd_{40}$ line. The resistances of the $Au_{60}Pd_{40}$ samples were 10 to 15 times that of the Au rings. However, the structure in the magnetoresistance was qualitatively the same (although less densely spaced on the magnetic field axis) as for the pure Au samples. In all samples (Au and $Au_{60}Pd_{40}$, rings and lines), the magnetoresistance structure persisted with no sign of attenuation out to magnetic fields of more than 8 T. The sharp drop in resistance near H = 0 in Fig. 2a is a superconducting proximity effect from the aluminum wires bonded to the pads. This proximity effect disappeared abruptly at H = 0.009 T, and it was seen in all of the Au samples. The $Au_{60}Pd_{40}$ samples with their shorter mean free paths had no such feature in the magnetoresistance.

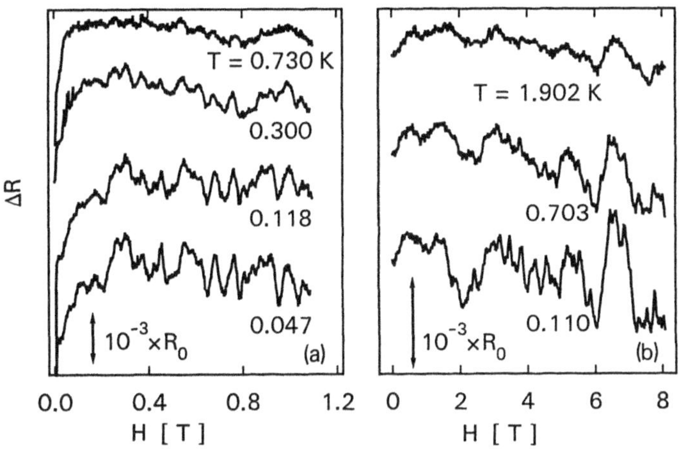

Fig. 2. (a) Representative magnetoresistance data from the sample 1. The resistance of the sample is $R_0 = 7.6\ \Omega$ at $H = 0$. (b) The magnetoresistance of the $Au_{60}Pd_{40}$ line in Fig. 1b. $R_0 = 102\ \Omega$

The positive magnetoresistance between 0.009 T and ≃ 0.2 T is attributed to the spin-orbit scattering from the heavy gold atoms [14]. On the scale of several tesla, the background magnetoresistance was generally temperature independent and sometimes negative. This behavior is probably associated with the magnetic impurities. We emphasize that the magnetoresistance of the line samples had structure with roughly the same amplitude and density along the field axis as the rings made from the same material; temperature dependence of the structure from the lines was also comparable to that in the rings.

The flux periodic effects require that the inelastic diffusion length $L_{in} = (D\tau_{in})^{1/2}$ be comparable to the circumference of the device else the phase of the carrier is destroyed before it can reach the point where it interferes with other carriers. (When present, other mechanisms, such as paramagnetic scattering, which destroy the phase of the electron must also be accounted for. The phase coherent diffusion length L_ϕ would then be, roughly speaking, the shortest of these scattering lengths.) In previous experiments [3,5], L_{in} was determined from weak localization properties. At $H = 0$, the resistances of our samples generally rose logarithmically as the temperature decreased. This implies that either the samples are effectively two-dimensional [12] or that this temperature dependence arises from the presence of magnetic impurities [13]. Considering that the cross section of the wires in all of the samples is roughly square, it is very unlikely that the samples are ever two-dimensional as far as the carrier diffusion is concerned. We are left with localized magnetic impurities (*e.g.* Kondo effect) as a mechanism for the logarithmic temperature dependence. (Materials analysis revealed traces of Fe and Cr which are known to cause such behavior in Au.) No information about L_{in} could be obtained from the magnetoresistance either. The expected background magnetoresistance from the spin-orbit interaction and weak localization [14] was not present in the samples. It may have been masked by the magnetic impurities or it may have been reduced in amplitude by the narrow conduction channels [15]. This leaves us without a method of measuring the inelastic diffusion length L_{in}. We can only rely on the estimate that the in the gold samples, L_{in} must be larger than the thickness of the wires. We can make a crude estimate of L_{in} using previous measurements of inelastic scattering time τ_{in} in thin gold films [16]. In quenched-condensed Au films with resistivities two orders of magnitude greater than our samples, $\tau_{in} \simeq 10^{11}$ s at T = 4.5 K. The mean free path in sample 1 was $\ell \simeq 10$ nm, and therefore the carrier diffusion constant is $D = \frac{1}{3} v_F \ell \simeq 0.008\ m^2/s$. If we assume that τ_{in} scales inversely with resistivity (since the mean free time does), in our Au samples, we set $\tau_{in} \sim 10^9$ s which leads to $L_{in} \sim 1\ \mu m$. (This is also consistent with experimental results from Mg films [5].) Typically for one-dimensional samples, L_{in} is expected increase as $T^{-1/2}$ implying that at least in the Au rings, L_{in} is more than long enough to permit the flux periodic effects to be seen. Empirically, the structure appears for $T \lesssim 1$ K so a safer estimate might be that $L_{in} <$ diameter for temperatures greater than 1 K. However, we emphasize that this is speculative until the origin of the structure is ascertained. We do not know how much paramagnetic scattering occurs in our samples so we cannot estimate its effects on L_ϕ.

Contrary to the predictions of [7] and [9], these data appear not to have one or two periods of oscillation; they are either random or have a very complicated oscillation spectrum. In order to analyze the periodic components of the magnetoresistance, Fourier power spectra were calculated for each of the curves in Fig. 2a; these are displayed in Fig. 3a. There are a number of periods as expected

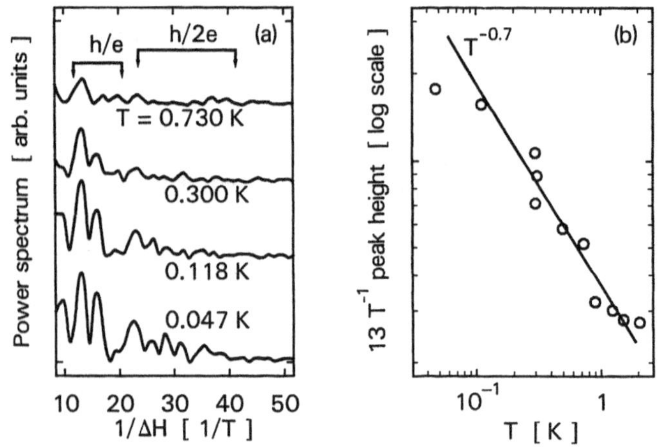

Fig. 3. (a) The power spectra of the data in Fig. 2a. The marks at the top of the figure illustrate the size of the flux quanta that were expected for this ring; the left-most of each pair of marks is the value for flux quantization within the inside area of the ring, and the right-most is the expected value for the average area. The uncertainty in the measurement of the diameter of the ring is about 15%. (b) The temperature dependence of the 13 T^{-1} peak in the power spectra; the straight line illustrates that the temperature dependence is roughly a power law between 0.1 K and 2 K

from the complexity of the resistance curves. At the top of the graph are marks indicating the expected locations of components arising from h/e oscillations for flux quantized entirely inside (left-most) the ring and by the average area (right-most) of the ring. Similar marks indicate the locations expected from h/2e oscillations. There are at least half of a dozen peaks within the range of periods that could result from flux quantization. If the inside area of the ring is the relevant dimension for normal metal flux quantization, there are two tantalizing peaks in the data. One, $1/\Delta H = 13\ T^{-1}$, is near the value expected for h/e, and the weaker peak near $1/\Delta H = 23\ T^{-1}$ is near h/2e. A strong h/e oscillation with a weaker h/2e component is predicted by [9]. However, this scheme of interpreting the data does not provide an explanation for one of the largest features in the spectra: the peak at $16\ T^{-1}$. According to theory [7], the flux periodic oscillations should decrease in amplitude as the magnetic field increases. However, Fourier transforms of our data suggest that the two peaks mentioned above do not depend on the size or location of the "window" in H over which the data were transformed. For example, transforming the data for T = 0.047 K in the range 0 to 1.1 T yields the same height $13\ T^{-1}$ peak as a transform over the range 0 - 0.6 T. If the peaks were the result of the mechanism in [7], then we would expect that the 0 - 0.6 T result would be larger. To illustrate the temperature dependence in the data, we have plotted the height of the $13\ T^{-1}$ peak as a function of temperature in Fig. 3b. From T = 1.5 K down to 0.15 K, the peak height grows as $T^{-0.7}$. (This is consistent with predictions of the the amplitude of flux periodic oscillations [7].) Below 0.1 K, the $13\ T^{-1}$ peak is temperature independent, but other features ($16\ T^{-1}$ and $23\ T^{-1}$) continue to grow as the temperature decreases.

It was also found that the structure in the magnetoresistance and the positions of the Fourier peaks could be altered by carelessly disconnecting and reconnecting the leads to the samples. We believe that the changes in the structure reflected real physical changes in the delicate samples caused by small current transients. If such electrical shocks were avoided, however, the structure was reproducible upon temperature cycling above T = 1 K, and its amplitude had a smooth temperature dependence (Fig. 3b). These two features of the data indicate that the anomalous structure reflects the transport properties of the sample and not spurious instrumental effects. The $Au_{60}Pd_{40}$ samples were more resilient, perhaps because the measurement circuit for these samples included low-pass RC filters in the leads. Such filtering softens the transients and protects the samples. Another possible explanation for the difference between the two classes of sample is that the structure results from grain boundary scattering. The grains in the Au samples were larger than the width of the wire in the samples. Any electromigration resulting from the transients when the measurement circuit is connected could cause significant changes in the array of grain boundaries and therefore in the structure. The $Au_{60}Pd_{40}$ samples with very small grains and bulk scattering would be less disturbed by the transients because any migration would be a smaller perturbation of the scattering potentials.

We were particularly startled by the discovery that when the magnetic field direction was reversed under carefully maintained experimental conditions, both the complex structure in the magnetoresistance and the power spectra of the Au samples were found to be asymmetric with respect to zero field. This can be seen

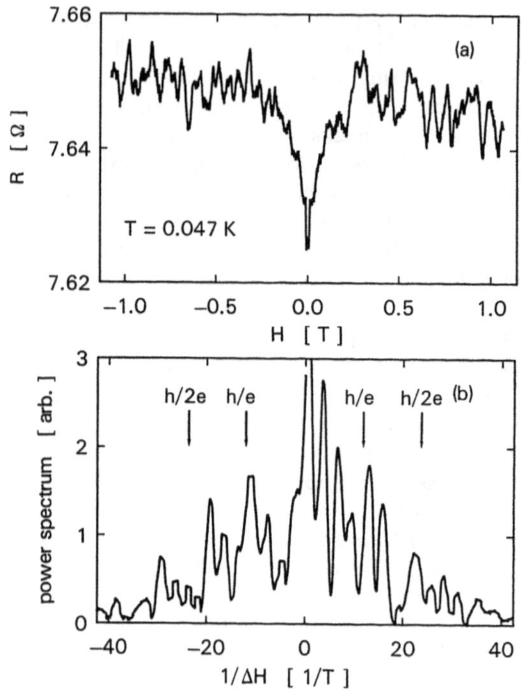

Fig. 4. (a) The magnetoresistance of sample 1 from -1.1 T to 1.1 T illustrating the asymmetry of the data about the point H = 0. (b) The power spectra of the data on the two sides of H = 0. (The negative axis is artificial, of course, and is used for convenience of presentation.) The arrows at the top of the graph illustrate the size of the flux quanta based on the inside area of the ring

from Fig. 4a which contains magnetoresistance data (T = 0.047 K) for $-1.1\ T < H < 1.1\ T$. On a rough scale, the data are symmetric about H = 0, but in detail, the two sides differ substantially. The anomalous structure for $H < 0$ is of the same average amplitude and density along the magnetic field as for $H > 0$, but the positions of the individual peaks are different from one side to the other. The two sides of the H = 0 also have different sets of frequency components as can be seen from the power spectra in Fig. 4b. As expected, the background magnetoresistance, including the "spin-orbit" and superconducting proximity effects, was symmetric about H = 0. The asymmetry with respect to zero field, and the variability of the complex structure mentioned above, indicate that the data displayed in Fig. 2a are not clear evidence for flux quantization in the Au rings. The $Au_{60}Pd_{40}$ devices were more conventional in their behavior. The structure in the data from these samples was symmetric with respect to H = 0.

One conclusion to be drawn from the above observations is that the structure arises from effects inside the wires which make up the arms of the ring, and that these mechanisms dominate (or perhaps even prevent) flux periodic effects. The major reason for assuming this is the equivalence of structure from the rings and the lines: any oscillations from the flux enclosed inside the arms of the rings would be eliminated from measurements on the single line. The similarity between the behavior of the rings and the lines leads us to believe that magnetoresistance structure is dominated by the small sample effects like those described by DINGLE [1]. The quantitative differences in peak size and location are thought to be due to the detailed differences among the scattering potentials in the various samples. Dingle's calculations apply to systems that are completely "size-quantized", *i.e.* even the elastic scattering is ignored. Such effects should be observed in small systems with diffusing carriers, and they may be related to structure seen in the resistance (as a function of gate voltage and magnetic field) of pinched accumulation layers [17]. These "1-D MOSFET" experiments have been interpreted as exhibiting transmission resonances [18] or statistical fluctuations [19] in this very small system. In our samples, each arm of the ring is a more or less linear maze of weak potential barriers, and the Lorentz force on the electron in each of the available conduction paths through the maze is different. Transmission through each path will vary as the as the relative attractiveness of the path changes with the strength of the magnetic field [10]. We believe that the apparently random, sample specific structure in the magnetoresistance is the result of effects similar to those seen in the "1-D MOSFET". A sample in which flux does not pierce the wires themselves, but is enclosed entirely within the ring, would be free of these effects, and then the flux periodic oscillations should be observed.

The choice of rings over cylinders seems to have one subtle consequence [10]. The cylinders used in previous experiments [2,3,4,5] are self averaging in the sense that they contain $\sim 10^4$ independent samples. The cylinders had diameters comparable to the inelastic lengths, but axial lengths several thousand times greater. This huge aspect ratio serves to experimentally average the transmission of electrons through the cylinder in a way which is equivalent to the ensemble impurity averaging in the diagram calculations of [7]. Any pair of perpendicular slices of the cylinder which are more than L_{in} apart are, in the sense of interference effects, separate samples which are averaged incoherently against each other. Any sample specific contributions arising from the locations of the impurities in the slice are averaged out automatically by the sample geometry. The h/e oscillation has an inherent *random* term of π in its phase relative to $H = 0$; the sum of large number of rings (or slices of a cylinder) should therefore average the h/e oscillations to zero. In contrast, the h/2e component is "phase-locked" to the origin (apart from an overall offset of π resulting from any spin-orbit contribution) and therefore adds constructively across the whole sample. Furthermore, both h/e and h/2e oscillations may be phase shifted randomly around the point $H = 0$ by a time-reversal symmetry breaking mechanism [10]. The amount of phase shift will vary from ring to ring (or section to section of the cylinder) and average to zero in samples containing many rings or sections. Neither of these averaging mechanisms will be operative in single rings where there is only one sample in the sense of coherent interference of the carriers (so long as $L_{in} > r$ where r is the width of the wires in the ring). We emphasize that arrays of rings (linear or planar) [20] should accomplish the same averaging as the long cylinders and yield only h/2e oscillations. All experiments reported to date on single ring samples [21,22] have observed complex magnetoresistance structure like that in Fig. 2.

In summary, a careful study of the magnetoresistance of small normal metal devices has discovered an aperiodic, reproducible structure for $T \lesssim 1$ K. The structure is clearly the result of the transport physics in the devices since it is reproducible for each sample. However, each sample had a different characteristic structure, and we believe that the structure for a particular sample was representative of the specific disordered potential of that sample. Although Fig. 3a and 4b contain some evidence of h/e flux periodic oscillations, the variation from sample to sample leads us to believe that the structure has its origin in the small sample size. The asymmetry of the structure about $H = 0$ is thought to result from the presence of magnetic impurities in the samples.

Acknowledgement We are grateful to Y. Imry, M. Buettiker, S. Maekawa, P.A. Lee, P. Chaudhari, C.C. Chi, R. Bruinsma, and R. Landauer for useful conversations, and to J. Bucchignano, K. Nichols and J. Speidell for technical assistance.

References

1. R. Dingle, *Proc. Roy. Soc.* 212A, 47 (1952)
2. D.Yu. Sharvin and Yu.V. Sharvin, *JETP Lett.* 34, 272 (1981)
3. B.L. Al'tshuler, A.G. Aronov, B.Z. Spivak, D.Yu. Sharvin and Yu.V. Sharvin, *JETP Lett.* 35, 588 (1982)
4. F. Ladan and J. Maurer, *Comptes Rendus de l'Academie des Sciences,* (Paris, 1983)
5. M. Gijs, C. van Haesendonck, and Y. Bruynseraede, *Phys. Rev. Lett.* 52, 2069 (1984)
6. Y. Aharonov and D. Bohm, *Phys. Rev.* 115, 485 (1959)
7. B.L. Al'tshuler, A.G. Aronov, B.Z. Spivak, *JETP Lett.* 33, 94 (1981)
8. R. Landauer, *IBM J. Res. Dev.* 1, 223 (1957)
9. M. Buettiker, Y. Imry and R. Landauer, *Phys. Lett.* 96A, 365 (1983); Y. Gefen, Y. Imry, and M.Ya. Azbel, *Proceedings of the 5th International Conference on the Electronic Properties of Two-Dimensional Systems*, Oxford, U.K. (1983), p.294; Y.Gefen, Y. Imry and M.Ya. Azbel, *Phys. Rev. Lett.* 52, 129 (1984)
10. Y. Gefen, Y. Imry and M. Buettiker, private communication.
11. R.B. Laibowitz and A.N. Broers, *Treatise on Materials Science*, 24, ed. K.N. Tu and R. Rosenberg, (Academic Press, New York, 1982) p.285; W.W. Molzen, A.N. Broers, J.J. Cuomo, J.M.E. Harper and R.B. Laibowitz, *J. Vac. Sci. Tech.* 16, 269 (1979).
12. E. Abrahams, P.W. Anderson, D.C. Licciardello, and T.V. Ramakrishnan, *Phys. Rev. Lett.* 42, 673 (1979); B.L. Al'tshuler, A.G. Aronov, and P.A. Lee *Phys. Rev. Lett.* 44, 1288 (1980)
13. F.J. Ohkawa, H. Fukuyama, and K. Yosida, *J. Phys. Soc. Jpn.* 52, 1701 (1983), and references therein.
14. S. Hikami, A.I. Larkin, and Y. Nagaoka, *Prog. Theor. Phys.* 63, 707 (1980)
15. B.L. Al'tshuler and A.G. Aronov, *JETP Lett.* 33, 499 (1980)
16. G. Bergmann, *Z. Phys. B* 48, 5 (1982)

17. A.B. fowler, A. Hartstein, and R.A. Webb, *Phys. Rev. Lett.* 48, 196 (1982); R.A. Webb, A.B. Fowler, A. Hartstein, J.J. Wainer, to be published
18. M. Ya. Azbel and P. Soven, *Phys. Rev. B*, 27, 831 (1983)
19. P.A. Lee, to be published
20. B. Pannetier, J. Chaussy, R. Rammal, and P. Gandit, to be published
21. G. Blonder, *Bull. Am. Phys. Soc.* 29, 535 (1984)
22. C.P. Umbach, S. Washburn, R.B. Laibowitz, and R.A. Webb, submitted to *Phys. Rev. B*

Theory of Weak Localization and Superconducting Fluctuations

S. Maekawa

The Research Institute for Iron, Steel and Other Metals, Tohoku University
Sendai 980, Japan*, and

Institut für Festkörperforschung der Kernforschungsanlage Jülich, Postfach 1913
D-5170 Jülich, Fed. Rep. of Germany

We discuss the superconducting transition temperature T_c, the upper critical field, and the conductivity near T_c in the weakly localized regime of dirty superconductors. It is shown that T_c decreases with increasing atomic disorder, since the Coulomb repulsive interaction is enhanced and the density of states of electrons is depressed in dirty metals. In addition, disorder induces the depairing effect due to the retardation of the Coulomb interaction. The conductivity σ induced by the superconducting fluctuations depends on disorder through the diffusion constant and the pair-breaking parameter δ. The parameter δ is related to the phase coherence time τ_ϕ. We propose that the magnetoresistance near T_c may be useful for studying τ_ϕ.

1. Introduction

Atomic disorder introduces various quantum corrections in the transport properties in dirty metals at low temperatures. Such quantum corrections are called weak localization effects, and are characterized by an expansion parameter $(k_F l)^{-1}$, where k_F and l are the Fermi wave number and the mean free path of an electron, respectively. The precursor of Anderson localization is one of the origins for the corrections. The electron-electron interaction is modified by disorder, and results in another origin for the corrections.

The purpose of this paper is to discuss the weak localization effects on dirty superconductors, especially on thin superconducting films. Recently, it has been revealed that the weak localization competes with superconductivity [1-4]. The superconducting transition temperature T_c in thin films decreases in proportion to the sheet resistance R. Such universal phenomenon is understood as an effect of weak localization. This idea was also extended to bulk superconductors [5-7]. We review the effect of weak localization T_c and the upper critical field H_{c2} in Sec. 2.

---

* Permanent address

Enhanced conductivity in dirty superconductors above T_c (paraconductivity) may be considered as quantum corrections with regard to the parameter $(k_F l)^{-1}$. This is the same parameter as the weak localization, and therefore both effects have to be considered in parallel. The paraconductivity results from two terms; one of which is the excess current carried by fluctuating Cooper pairs and is called Aslamazov-Larkin (AL) term [8]. The other is the forward scattering effect on normal electrons due to Cooper pair fluctuations and is called Maki-Thompson (MT) term [9,10]. The latter term is sensitive to the pair-breaking parameter δ. Since δ is related to the phase coherence time of an electron τ_ϕ [11,12], which is a characteristic time for Anderson localization, the paraconductivity may be used to study τ_ϕ. In Sec. 3 we discuss the magnetoresistance near T_c. In Sec. 4, the correction in the superconducting fluctuations due to the localization is examined, with special emphasis on the AL term of paraconductivity in thin films. Finally, the results are summarized in Sec. 5. We take units $\hbar = k_B = 1$.

2. Superconducting Transition

It is well known that the superconducting transition temperature T_c is not affected by static and non-magnetic disorder [13,14]. We note that in this argument the interplay between disorder and the electron-electron interaction is neglected. However, in thin films the interplay, which is quantum corrections, gives essential contribution and modifies T_c drastically. To examine the quantum corrections, let us start with the BCS Hamiltonian,

$$H = H_o + g_{BCS} \int d\vec{r} \{\Delta^+(r)\psi_\uparrow(r)\psi_\downarrow(r) + \Delta(r)\psi_\downarrow^+(r)\psi_\uparrow^+(r)\}, \tag{1}$$

$$H_0 = H_{KE} + H_{imp} + H_{int} . \tag{2}$$

where $\psi_s(r)$ is the electron field operator with spin s and g_{BCS} is the BCS attractive interaction ($g_{BCS} < 0$). The first term H_{KE} in (2) denotes the kinetic energy and H_{imp} denotes impurity potentials. The Hamiltonian H_{int} includes all kinds of electron-electron interaction except the BCS interaction term given in (1). The superconducting order parameter $\Delta(r)$ is given by

$$\Delta(r) = \mathrm{Tr} e^{-H/T} \psi_\uparrow(r)\psi_\downarrow(r)/\mathrm{Tr} e^{-H/T} . \tag{3}$$

Introducing (1) into (3), we have the equation for the condition of the second order transition;

$$\Delta(r) = |g_{BCS}| \int d\vec{r}' \int_0^{T^{-1}} du \langle T_u \psi_\downarrow^+(r',u)\psi_\uparrow^+(r',u) \times \psi_\uparrow(r,o)\psi_\downarrow(r,o)\rangle_{H_0} \Delta(r'), \tag{4}$$

where u is the imaginary time, Tu the time-ordering operator, and the bracket means the thermal average in the space of H_0. It is useful to introduce the eigenfunction in the disordered system which are diagonal in the space of the non-interacting Hamiltonian [13],

$$\psi_s(r) = \sum_\alpha \phi_\alpha(r) c_{\alpha s}, \tag{5}$$

where $\phi_\alpha(r)$ is the eigenfunction of state α and $c_{\alpha s}$ is the annihilation operator of the state with spin s. Then, inserting (5) into (4), we have

$$\begin{aligned} |g_{BCS}|^{-1} &= \sum_{\alpha\sim\delta} \int d\vec{r} \int d\vec{r}' \phi_\alpha^+(r')\phi_\beta^+(r')\phi_\gamma(r)\phi_\delta(r) \\ &\times \int_0^{T^{-1}} du \langle T u c_{\alpha\downarrow}^+(u) c_{\beta\uparrow}^+(u) c_{\gamma\uparrow}(o) c_{\delta\downarrow}(o)\rangle_{H_0} \\ &= \sum_{\alpha,\beta} \int_0^{T^{-1}} du \langle T u c_{\alpha\downarrow}^+(u) c_{\bar{\alpha}\uparrow}^+(u) c_{\bar{\beta}\uparrow}(o) c_{\beta\downarrow}(o)\rangle_{H_0}, \end{aligned} \tag{6}$$

where $\bar{\alpha}$ denotes the time-reversal state of α. When the interaction H_{int} in (2) is neglected, it is easy to take the thermal average in (6).
We obtain

$$\begin{aligned} |g_{BCS}|^{-1} &= T\sum_{n\alpha}\sum G_\alpha(\varepsilon_n) G_{\bar{\alpha}}(-\varepsilon_n) \\ &= T\sum_n \pi N(o)/|\varepsilon_n|, \end{aligned} \tag{7}$$

where $G_\alpha(\varepsilon)$ is the thermodynamic Green's function, $\varepsilon_n = 2\pi T(n+\frac{1}{2})$, n being integer, and N(o) is the density of states of electrons per spin. If N(o) is assumed to be independent of disorder, we find that T_c is not affected by disorder. This result is called the Anderson's theorem of dirty superconductors [13,14].

Let us next consider the interplay between disorder and the electron-electron interaction. In disordered systems, the electron density fluctuations diffuse slowly so that the Coulomb interaction becomes dependent on disorder [15,16]. The electron-phonon interaction may also be modified by disorder [17,18]. However, this interaction will not depend on the thickness (d) of films, since we are interested in the films with d >> 1. Therefore, we neglect the modification of the electron-phonon interaction, and take into account the interplay between disorder and the Coulomb interaction in the effective interaction between Cooper pair electrons and the self-energy of electrons. Then, introducing the magnetic field (H), we obtain the equation for the superconducting transition in the mean field theory,

$$\ln \frac{T}{T_{co}} = \psi(\tfrac{1}{2}) - \psi(\tfrac{1}{2} + \frac{eDH_{c2}}{2\pi T_c}) + R_V + R_{HF} , \qquad (8)$$

where T_{co} is the transition temperature in the pure system, D is the diffusion constant, H_{c2} is the upper critical field, and $\psi(z)$ is the digamma function. Here, R_V and R_{HF} indicate the effects of the Coulomb repulsion and the self-energy due to the interplay, respectively. Mathematically, (8) is calculated from (6) retaining the linear term of $\mathcal{H}_{in}$. In thin films with $\sqrt{D/T} \gg d \gg l$, d and l being the thickness of film and the mean free path, respectively, we have, in the lowest order of $(k_F l)^{-1}$,

$$R_V = R_{HF} = -\lambda^*\{AB^2 - \frac{2}{3} B^3 + (2AB - B^2)(\psi(\frac{1}{2}) - \psi(\frac{1}{2} + \frac{eDH_{c2}}{2\pi T_c}))\}, \qquad (9)$$

$$A = \ln \frac{D\kappa_3^2}{2\pi T} \quad \text{and} \quad B = \ln \frac{1}{2\pi T\tau} , \qquad (10)$$

where we used the dynamically screened Coulomb interaction in the random phase approximation [4] and retained only the leading contributions with regard to lnT, by assuming $\omega_D > \tau^{-1}$, ω_D and τ being the Debye frequency and the relaxation time, respectively. Here, κ_3 and λ^* are given by $\kappa_3^2 = 8\pi e^2 N(o)$ and $\lambda^* = (\pi k_F l(2k_F d/3\pi))^{-1}$, respectively. Since λ^* is proportional to the sheet resistance as $R = 2\pi^2\lambda^*/e^2$, we find that T_c decreases in proportion to R. This is because (i) the Coulomb repulsion is enhanced (R_V) and (ii) the density of states of electrons is depressed in the disordered systems (the half of R_{HF}). In addition, disorder induces the depairing effect due to the retardation of the Coulomb interaction. This effect gives the other half of R_{HF}. Experimentally, it is known that T_c decreases in proportion to R and becomes the half of the bulk value when R reaches 1 kΩ in many thin films. This fact is consistent with the present theory. The upper critical field is also given in (9). Here, we must note the additional contribution in H_{c2} due to the localization: As seen in (9). H_{c2} depends on the diffusion constant D, which is affected by the localization. Mathematically, D in (9) is derived by inserting the maximally crossed diagrams in Fig. 1 in the Cooper pair propagator in Fig. 2. Then, D is given by

$$D = \frac{1}{3} v_F l(1 - \lambda^* \ln \frac{1}{2\pi T\tau}) . \qquad (11)$$

with v_F being the Fermi velocity. The upward curvature in an H_{c2} vs. T curve is observed in thin films as well as dirty bulk systems and layered materials. This curvature may be explained as a localization effect by (8).

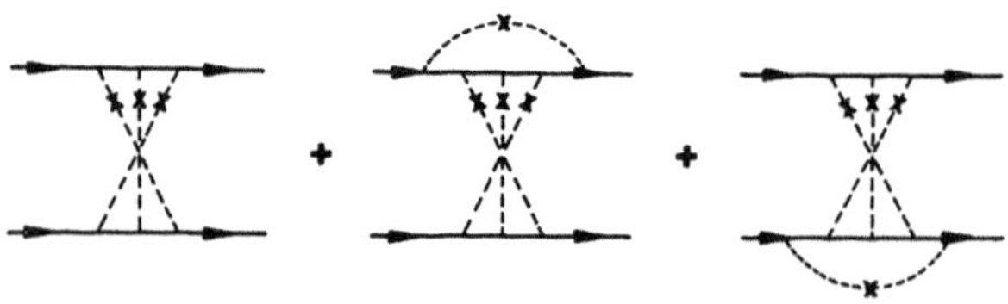

Fig. 1: Diagrams which contribute to the diffusion constant in (8)

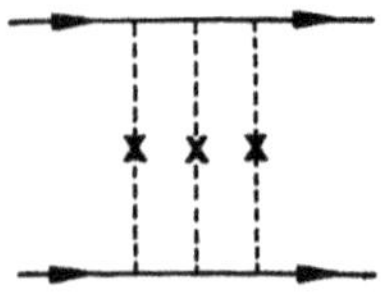

Fig. 2: Pair propagator in the lowest order of $(k_F l)^{-1}$

3. Magnetoresistance near T_c

The AL and MT terms in paraconductivity are the lowest order effects of superconducting fluctuations. As for the parameter $(k_F l)^{-1}$, we observe that they are of the same order as the weak localization effects. The AL and MT conductance in dirty films is given [8-10], respectively, by

$$\sigma_{AL} = \frac{e^2}{16\eta} \quad , \tag{12}$$

$$\sigma_{MT} = \frac{e^2}{8}\frac{1}{\eta-\delta} \ln\frac{\eta}{\delta} \quad , \tag{13}$$

where $\eta = \ln(T/T_c)$ and δ is the pair breaking parameter. These equations may be compared with the conductance due to the Anderson localization [19];

$$\sigma_L = -\frac{e^2}{2\pi^2} \ln\frac{\tau_\phi}{\tau} \quad , \tag{14}$$

where τ_ϕ is the phase coherence time.
Since δ is related to τ_ϕ as [11]

$$\delta = \frac{\pi}{8T\tau_\phi} \quad , \tag{15}$$

both the MT process and the Anderson localization are characterized by τ_ϕ. The magnetoresistance due to the weak localization has extensively been used to study τ_ϕ [20]. Since the MT term is also sensitive to τ_ϕ, the paraconductivity may be used to study τ_ϕ [21]. LARKIN [22] has proposed the magnetoresistance of the MT term in the temperature range;

$$\ln\frac{T}{T_c} >> \frac{1}{T\tau_\phi} \quad . \tag{16}$$

However, when we consider that the MT term can be applied in the range that $\ln(T/T_c) << 1$, the condition (16) in Larkin's theory restricts its validity. We calculate the magnetoresistance in the MT term in the temperature range

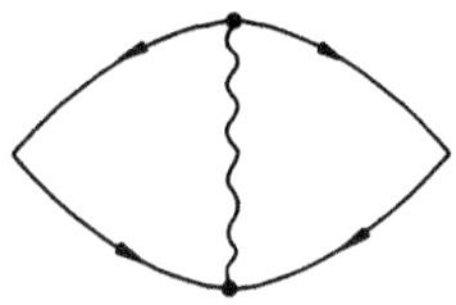

Fig. 3: Diagram of the MT term

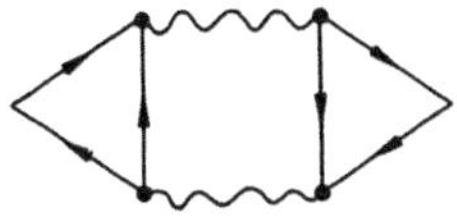

Fig. 4: Diagram of the AL term

that $\ln(T/T_c) \ll 1$ without the restriction of (16). The diagram of the MT process is given in Fig. 3, where the wavy line denotes the fluctuation and the dots denote the vertices. Here, the fluctuation and the vertices are functions of magnetic field. Replacing the summation over momentum states with Landau states, we obtain

$$\sigma_{MT}(H) = \frac{e^2}{8(\eta-\delta)} \{\psi(\frac{1}{2} + \frac{\eta}{\lambda_o a}) - \psi(\frac{1}{2} + \frac{1}{a\tau_\phi})\} , \tag{17}$$

where $a = 4DeH/c$ and $\lambda = \pi/8T$. LOPES DOS SANTOS and ABRAHAMS [23] have also obtained (17) independently. We calculate the magnetoresistance due to the AL term in Fig. 4 in the temperature range that $\ln(T/T_c) \ll 1$ in the same way and have [24]

$$\sigma_{AL}(H) = \frac{e^2}{8a\lambda_o} \{ 1 - \frac{2\eta}{\lambda_o a}(\psi(1 + \frac{\eta}{\lambda_o a}) - \psi(\frac{1}{2} + \frac{\eta}{\lambda_o a}))\} . \tag{18}$$

Therefore, the positive magnetoresistance near T_c is given by the sum of (17) and (18), and provides an opportunity for studying τ_ϕ.

4. Localization Effect on Superconducting Fluctuations

In the previous section, the lowest order effects on the conductivity near T_c were examined with regard to both the fluctuation and the parameter $(k_F l)^{-1}$. In this section, we propose a phenomenological argument for the higher order effect of the localization within the lowest order of the fluctuation. The superconducting fluctuation propagator is written as

$$\Pi(q,\omega_1) = N(o)^{-1}(\eta + \lambda_o Dq^2 + \lambda_o|\omega_1|)^{-1} . \tag{19}$$

Therefore, the effects of the localization will appear in the parameters T_c and D. As discussed in Sec. 2, T_c decreases with increasing disorder. The diffusion constant D was studied in connection with H_{c2} in Sec. 2: When the maximally crossed diagrams (Fig. 1) are introduced in the pair propagator (Fig. 2), D is given by (11).

We assume that the MT term is already suppressed by large δ, and only the AL term determines the paraconductivity. This situation is realized in many amorphous systems [25]. Using (19), we recalculate the AL term given in Fig. 4, and have

$$\sigma_{AL} = \sum_q \left(\frac{4e}{3m}\lambda_o k_F l\right)^2 \frac{q^2 \lambda_o T}{(\eta + \lambda_o D q^2)^3} \quad . \tag{20}$$

Taking the summation over q, we obtain

$$\sigma_{AL} = \frac{e^2}{16\eta}\left(\frac{D_o}{D}\right)^2 , \tag{21}$$

where $D_o = \frac{1}{3} v_F l$. If we neglect the correction of the localization in D, we have $D = D_o$ and, thus (12) instead of (21). Therefore, when D is depressed by the localization, the paraconductivity is enhanced. Here, we note that T_c will also be depressed strongly in this case. In very dirty films, the superconducting transition is observed to be broad. It is interesting to examine this fact in the light of (21).

5. Summary

We have discussed the effects of weak localization on T_c, H_{c2}, and the fluctuations. It was shown that T_c decreases with increasing atomic disorder, since the Coulomb repulsive interaction is enhanced and the density of states is depressed by disorder. In addition, the depairing effect induced by the retardation of the Coulomb interaction gives an important contribution to T_c in dirty metals.

The pair-breaking parameter in the paraconductivity is related to the phase coherence time τ_ϕ. We have derived the expression of the magnetoresistance may be used to study τ_ϕ.

In Sec. 4, we have proposed a phenomenological argument to study the higher order effect of the localization on the fluctuation phenomena. Finally, superconductivity and localization exhibit strong contrasts in various aspects. Therefore, the interplay between them will provide further interesting quantum effects.

Acknowledgements

I would like to thank H. Fukuyama and H. Ebisawa for collaborations in various stages of the work. I am greatly indebted to G. Bergmann for valuable discussions and the critical reading of the manuscript.

References

1 S. Maekawa and H. Fukuyama: J. Phys. Soc. Jpn. 51, 1380 (1982)

2 S. Maekawa: "Anderson Localization", ed. Y. Nagaoka and H. Fukuyama (Springer, 1982), pp. 103

3 S. Maekawa, H. Ebisawa, and H. Fukuyama: J. Phys. Soc. Jpn. 52, 1352 (1984)

4 S. Maekawa, H. Ebisawa, and H. Fukuyama: J. Phys. Jpn. 53, No. 8 (1984)

5 P.W. Anderson, K.A. Muttalib, and T.V. Ramakrishnan: Phys. Rev. B 28, 117 (1983)

6 L. Coffey, K.A. Muttalib, and K. Levin: Phys. Rev. Lett. 52, 783 (1984)

7 J. Fukuyama, H. Ebisawa, and S. Maekawa: to be published in J. Phys. Soc. Jpn.

8 L.G. Aslamazov and A.I. Larkin: Phys. Lett. 26A, 238 (1968) and Sov. Phys. Solid State 10, 875 (1968)

9 K. Maki: Prog. Theor. Phys. 40, 193 (1968)

10 R.S. Thompson: Phys. Rev. B 1, 327 (1970)

11 H. Ebisawa, S. Maekawa, and H. Fukuyama: Solid State Comm. 45, 75 (1983)

12 B.L. Altshuler, A.G. Aronov, and D.E. Khmel'nitskii: Solid State Comm. 39, 619 (1981)

13 P.W. Anderson: J. Phys. Chem. Solids 11, 26 (1959)

14 L.P. Gorkov: Sov. Phys. JETP 10, 998 (1960)

15 B.L. Altshuler, A.G. Aronov, and P.A. Lee: Phys. Rev. Lett. 44, 1288 (1980)

16 H. Fukuyama: J. Phys. Soc. Jpn. 48, 2169 (1980) and 50, 3407 (1981)

17 A. Schmid: Z. Physik 259, 42 (1973)

18 B. Keck and A. Schmid: J. Low Temp. Phys. 24, 611 (1976)

19 P.W. Anderson, E. Abrahams, and T.V. Ramakrishnan: Phys. Rev. Lett. 43, 718 (1978)

20 See, for example, G. Bergmann: Phys. Reports 107, 1 (1984)

21 G. Bergmann; Phys. Rev. B 29, 6114 (1984)

22 A.I. Larkin: JETP Lett. 31, 219 (1980)

23 J.M.B. Lopes dos Santos and E. Abrahams: preprint

24 E. Abrahams, R.E. Prange, and M.J. Stephen: Physica 55, 230 (1971)

25 See, for example, S. Maekawa, S. Takahashi, and M. Tachiki: J. Phys. Soc. Jpn. 53, 702 (1984)

Preparation and Physics of Novel Microstructures for Localization Studies

M.R. Beasley and S. Bending

Department of Applied Physics, Stanford University, Stanford, CA 94305, USA

J. Graybeal

Department of Physics, Stanford University, Stanford, CA 94305, USA

1. Introduction

It is now well established that disorder and reduced dimensionality have both interesting and profound effects on the behavior of conducting systems.[1-4] Anderson localization and modified electron-electron interactions are now familiar consequences of disorder. Interesting questions remain, however, as the various papers being presented at this conference attest. But, like many areas of contemporary condensed matter physics, further progress depends on more than good theory and experiment. One needs good materials and experimental systems - not simply systems that manifest the phenomena of interest, but so-called model systems, i.e., systems that exhibit the desired phenomena particularly clearly, and in which the relevant material parameters, structure, etc., can be controlled and varied. It is with such model systems that one generally gains the clearest insight and obtains the best quantitative tests of theory. And it is on the basis of this understanding that more complicated (and probably typical) systems can then be understood. This approach should be contrasted with exploratory systems, in which one attempts to find new phenomena or enter new parameter regimes by any means possible.

In some cases model systems can be found among naturally occurring materials and systems. However, increasingly, particularly when reduced dimensionality plays a crucial role, advanced vapor deposition and micropatterning are being brought to bear to produce the desired systems. The organizers of this conference have asked us to review this approach in the context of localization phenomena. However, such a task is too large for the time and space permitted. The arsenal of useful vapor deposition and micropatterning tools and techniques is growing rapidly, as is the range of materials to which they can be applied. Application of these tools has played a role in localization studies right from the beginning. In fact the demands of the field have led to some of the most important new developments in this area, particularly as regards micropatterning. Fortunately, some good reviews already exist [5-8].

In light of the above, we would like to take a more limited, admittedly parochial point of view. We would like to discuss some recent examples of our own work that serve to illustrate this "high tech" approach to model systems, but which are not intended to be comprehensive. Specifically we want to illustrate how multilayer vapor deposition can be used to produce interesting structures for localization physics studies. Here, clearly, the emphasis is on using vertical structures to advantage rather than the more common (and admittedly spectacular) applications of state-of-the-art microlithography. Of course, these two approaches are by no means mutually exclusive. At the same time the examples we should discuss serve to illustrate how the advancing understanding of localiza-

SOME CHARACTERISTIC LENGTH SCALES OF INTEREST

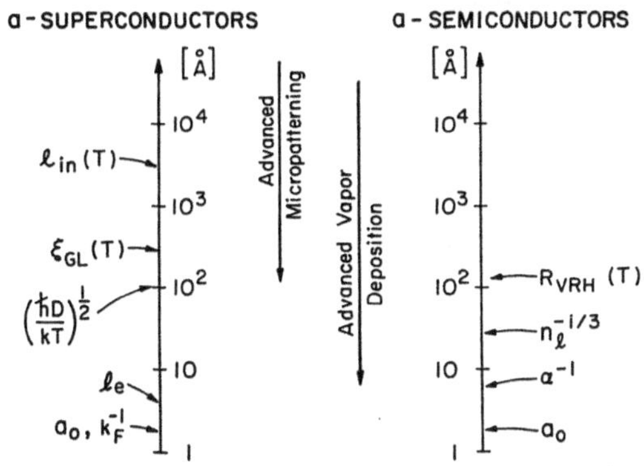

Fig. 1: Characteristic length scales relevant to 2D superconductivity in disordered systems and tunneling through amorphous semiconductors. Here ℓ_{in} is the inelastic scattering length, ξ_{GL} is the Ginzburg-Landau coherence length, $(\hbar D/kT)^{1/2}$ is the characteristic decay length of electron density fluctuations, ℓ_e is the elastic scattering length, k_F is the Fermi wave number, and a_0 is the mean interatomic distance. Also, R_{VRM} is the Mott variable range hopping length, n_ℓ is the number density of localized states, and α^{-1} is the decay length of the localized slots.

tion is beginning to be applied in other interesting problems in condensed matter physics. There are two topics. They deal, respectively, with amorphous (i.e., disordered) superconductors and amorphous semiconductors, specifically the competition between localization/interaction effects and superconductivity in two dimensions, and resonant tunneling via localized states in amorphous semiconducting tunnel barriers. Both problems are of increasing importance in superconductivity research, and are also of broader interest as well. It is the connection to superconductivity research that makes them appropriate to this conference, which is focused on metallic systems.

Figures 1 and 2 serve to illustrate the approach. In Fig. 1 we show the length scales of interest in the two examples at hand. The symbols are defined in the figure caption. Also shown are the length scales over which advanced deposition and micropatterning tools can be used to shape or structure (e.g. layer) materials. Note that advanced vapor deposition techniques permit exploration of length scales not accessible by micropatterning (see Ref. 8). Figure 2 shows a vertical section of the two illustrative ultra-thin-film multilayered structures of interest to us here. As seen from Fig. 1, it is relatively easy to make superconducting films that are 2-D with respect to Anderson localization ($d < \ell_{in}$), superconductivity ($d < \xi_{GL}$) and electron-electron interaction effects ($d < \hbar D/kT$), but not the various microscopic processes ($d > \ell_e, k_F^{-1}$) or the atomic structure of the material ($d > a_0$). To produce homogeneous ultra-thin films approaching thickness ≃ 10 Å in order to go deeply into the 2-D region is another matter, as we shall see. (We exclude the use of monolayer epitaxial growth as a general approach, although its use where appropriate is certainly very effective and on the increase.) In the case of amorphous semiconductor tunnel barriers, it is

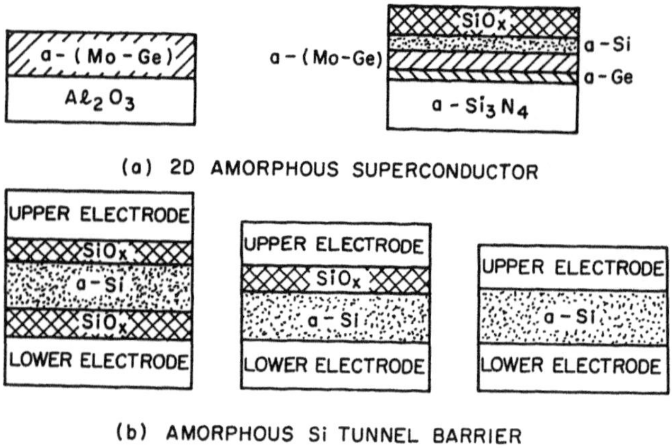

Fig. 2: Two examples of multilayered thin film structures suitable for producing model systems for (a) 2D superconductivity and (b) tunneling through amorphous semiconductors.

possible to make effectively pinhole-free layers that are thin compared with inelastic processes $(d < R_{VRH})$, the average distance between localized states $(d < n_o^{-1/3})$, and comparable to the decay lengths of the individuals localized states $(d \simeq \alpha^{-1})$. Using the most advanced microlithography to narrow such thin film structures, one can also begin to imagine being able to approach the 1-D limit in these cases as well. In the remainder of this paper we discuss these two examples in greater detail, from the points of view of how they were made and also what interesting results they are yielding.

2. Effects of Disorder in 2-D Superconductors

As we know from the theory of localization and interaction effects in disordered systems, disorder leads to anomolies in the diffusion constant D, the density of states N(E) near the Fermi energy, and in the electron-electron interaction. It is not hard to imagine that such changes in the normal state properties of a metal would also affect superconductivity. In particular, since for a 2-D system localization/interaction theory predicts $\sigma \to 0$ as $T \to 0$, there clearly is a competition between these effects and superconductivity for which $\sigma \to \infty$ as $T \to 0$. The competition exists also in 3-D and should be even more pronounced in 1-D.

In order to study this competition we have been preparing ultra-thin films of amorphous superconducting Mo-Ge alloys.[9,10] Other researchers have looked at W-Re alloy films finding similar results.[11] Several groups have treated the question theoretically, beginning first with the work of MAEKAWA and FUKUYAMA [12]. Up-to-date review articles are also available [13,14]. The theoretical predictions are best defined in the weakly-localized regime where perturbation theory can be applied. As these theories show, the initial change in the superconducting transition temperature due to disorder is such that universally one expects

$$\frac{\Delta T_c}{T_{co}} \propto \left[\frac{1}{E_F \tau}\right]^{D-1} \tag{1}$$

where here D is the dimensionality and $(E_F\tau)^{-1}$ the expansion parameter of the perturbation theory. This equation implies that to a first approximation ΔT_c is proportional to $R_\square$ and ρ^2 in 2 and 3 dimensions, respectively. These predictions are shown schematically in Fig. 3. Here the T_c - $R_\square$ and T_c - ρ planes correspond to the pure 2-D and 3-D limits, respectively. In practice, however, thin films generally lie somewhere in the full ρ, $R_\square$, T_c space. Thus, presumably, if in a film ρ changes as $R_\square$ is varied, the equivalent bulk (3D) T_c also changes, making clean tests of the 2D theoretical predictions very difficult. Consequently, in order to have a model system with which to study the crossover to 2-D unambiguously one needs a system in which $R_\square$ can be increased by reducing d but for which ρ remains constant, i.e. a system that follows the dashed, not dotted, line in Fig. 3. Technically speaking, one is then looking at the 1/d dependence of T_c.

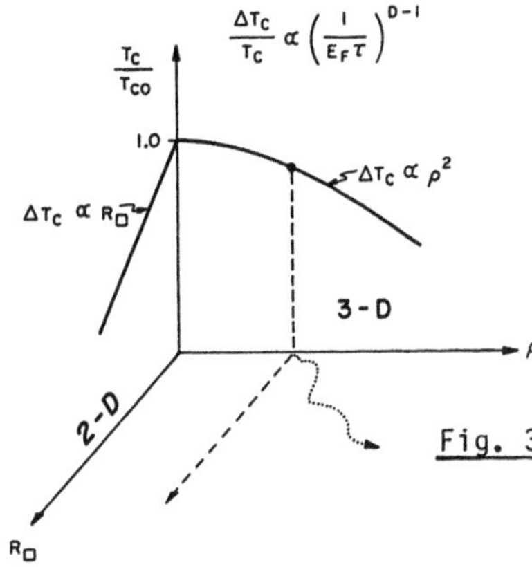

Fig. 3: Schematic representation of the suppression of T_c due to disorder in 2 and 3 dimensions.

This challenge has been met recently by Graybeal and coworkers using advanced multilayer sputtering techniques to produce homogeneous amorphous films with thicknesses down to ~10 Å that have effectively constant bulk resistivity. Clearly an amorphous film is desirable if one wants to produce successively thinner films with no charge in ρ. Amorphous films have the added advantage that all order-sensitive band structure effects have already taken place and are not further altered by making the film thin. For low $R_\square$ straightforward single-film depositions are (top picture of Fig. 2(a)) sufficient. However, as found by Graybeal, the thinner films tend to break up, leading to large $R_\square$ but also increased ρ, and almost certainly resulting in an inhomogeneous structure. Inhomogeneous structures may also arise if an inappropriate alloy is used. For example, Mo-Ge alloys appear to work well, while even relatively thick films of Al-Si made using the same techniques show evidence of inhomogeneities. In any event, to produce ultra-thin homogeneous amorphous films (d < 50 Å), a more complex vertical structure is required, including a suitable substrate, rapid rotation during deposition, and the use of appropriate over and under layers (see Fig. 2). For the a-(Mo-Ge) system specifically, Graybeal used on amorphous substrate (a-Si_3N_4), a 10 Å thick amorphous underlayer of Ge, and a protective overlayer (25 Å of a-Si, subsequently oxidized).

The desirability of an amorphous substrate is perhaps obvious. The need for the a-Ge underlayer may reflect the desirability of having a chemically compatible interface for the particular amorphous film in question (Mo-Ge). The overlayer in this case chemically protects the thin film and provides a tunnel barrier for tunneling studies of these ultra thin films. The multilayer nature of these samples is thus not only essential in order to form the desired model system but also can be used to build into the structure a very important physical probe. The whole system is stable at ambient temperature and atmosphere, which means it can be characterized readily by various relevant materials characterization techniques (e.g., X-rays, TEM, XPS, etc).

The physical characteristics of these films have been reported by GRAYBEAL and coworkers [9,10,15]. The most important results, the transport properties and the dependence of T_c and the critical field slope dHc_2/dT at T_c on $R_\square$ are shown in Figs. 4 and 5 along with the latest theoretical predictions using the best independent estimates or direct measurement of the relevant material parameters. Figure 4 shows the normalized excess conductance $\Delta g_\square = \Delta G_\square e^2/\hbar$ which according to weak localization theory should be constant and of order unity for a homogeneous system as $R_\square$ increases providing the basic bulk resistivity of the material (i.e., the resistivity excluding localization effects themselves) remains constant. This is exactly what is observed in films with thicknesses down to ~10Å. Thus, any change in the resistivity of these films is consistent with localization itself, and breakup apparently

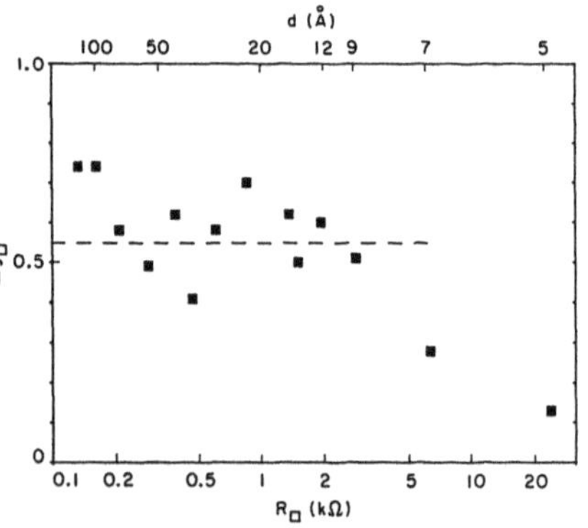

Fig. 4: Normalized reduction in the sheet conductance $-\Delta g_\square = \Delta G_\square (\hbar/e^2)$ low temperatures as a function of sheet resistance in ultrathin amorphous Mo-Ge films. In the weakly localized regime ($R_\square < 81$ kΩ), theory predicts $-\Delta g_\square$ to be a constant independent of $R_\square$. The dotted line shows the average constant value.

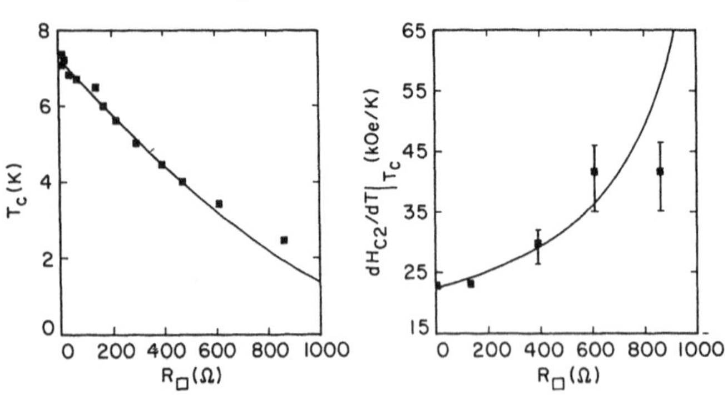

Fig. 5: Comparison between theory and experiment (Ref. 9) for the depression of T_c and increase in critical field slope dH_{c2}/dT due to disorder as measured by $R_\square$.

does not occur until d < 10 Å. The universal prediction that $\Delta T_c \propto R_{\Box}$ is also borne out, and even the rate of decrease $(dT_c/dR_{\Box})$ and the sense of the curvature in the data are in good accord with theory. The theory is still evolving, however, so one should not seek quantitative agreement yet.

Nonetheless, the agreement can be considered satisfactory at the present stage. Thus there seems little doubt that disorder has a profound effect on superconductivity in 2-D. The possible broader implications of these results have yet to be adequately addressed. Present theory predicts a sizeable effect in 3-D. Of course in 1-D dimension perturbation theory breaks down, and the situation should become even more dramatic and interesting. Recalling Fig. 1 we see that to produce such samples (d and w < ℓ_{in}, ξ_{GL} and $(\hbar D/kT)^{1/2}$) is perhaps just possible with advanced microlithography. Certainly one could look for the leading corrections in w^{-1}.

3. Tunneling via Localized States Through Amorphous Semiconductor Tunnel Barriers

The second example we want to discuss relates to transport through amorphous semiconductors, not in the usual context of variable-range hopping in bulk materials, but rather the related process of tunneling through an amorphous semiconductor tunnel barrier. The second example differs from our first in several regards. First, it deals with semiconductors, not metals, i.e., it is on the insulator side of the metal insulator transition. Second, we are here interested in transport through a vertical structure, not along a layer. The important length scales are very short (~ 10 - 100 Å). Finally, unlike localization effects in superconductors, this is not a new subject. Tunneling through amorphous semiconductor barriers has been discovered independently by various workers, all with rather different motivations [15-19]. A proper understanding of the phenomenon is not yet at hand. Moreover, the processes involved appear to relate to some of the most interesting current questions in the theory of localization, specifically resonant tunneling and the distinction between elastic and inelastic processes in electronic transport.

The structures of interest are shown schematically in Fig. 2b. They differ only in the presence or absence of conventional tunnel barriers (SiO_x) between the electrodes and the amorphous semiconductor (a-Si) layer. The function of these barriers is both practical (to block pinholes, if necessary) and scientific (to decouple the localized states from the continuum in the metal electrodes). The SiO_x layer also hinders any interdiffusion between the Si layer and the two metal electrodes. Each of these structures has been made and studied by one group or another. Both sputtering and electron beam evaporation have been successfully employed. Pinhole free barriers without the SiO_x layers have been achieved in some cases [18]. Amorphous germanium has also been studied to a lesser extent [20]. The bottom SiO_x layer is obtained by depositing a thin (~ 20 Å) layer of a-Si which is subsequently oxidized. Then a thicker layer of a-Si is deposited and oxidized to form an a-Si/SiO_x bilayer. Obvious variations are used in the other cases. Various electrode materials have been used. XPS studies show that a uniform oxide layer is formed with a well-defined thickness [21].

The physical interest in these structures lies in the question of how the tunneling process takes place and how it is affected by the existence of the high density of localized states n_ℓ present in the barrier. For typical densities (10^{18}-10^{20} states/eV-cm^3) there are on average only one or perhaps two localized states across the thickness of the tunnel barrier. Thus, the situation is different from that normally considered in transport phenomena in amorphous semiconductors. Specificially, only a few localized states are involved and the total length of the sample is short compared to the variable range-hopping length ($d < R_{VRH}$, see Fig. 1).

Some of the most characteristic physical properties of the SiO_x/a-Si/SiO_x barriers are summarized in Figs. 6 to 8 from the work of BENDING [22]. At high bias voltages (V > 10 meV) all the various types of barriers appear to show qualitatively similar behavior. However, when one or more super-conducting electrodes are used, only the SiO_x/a-Si/SiO_x barriers show unmodified superconductive tunneling characteristics ($V \approx 2\Delta \sim$ meV). The behavior appears to be related to the presence of a

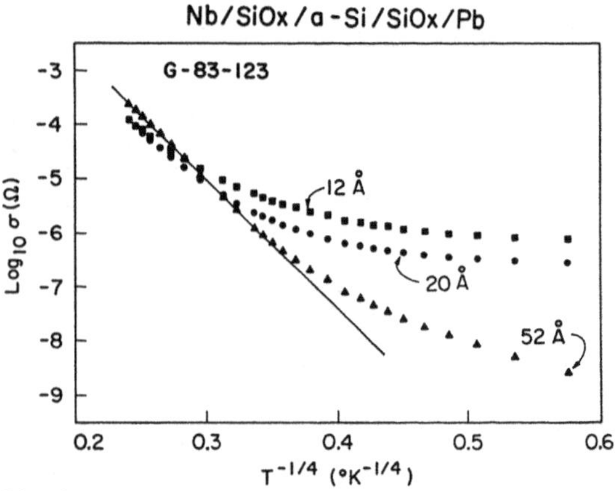

Fig.6: Temperature-dependence of the junction conductance for SiO_x/ a-Si/SiO_x tunnel barriers. Recall that variable hopping in bulk amorphous semiconductors exhibits an $\exp[-(1/T)^{1/4}]$ dependence on temperature.

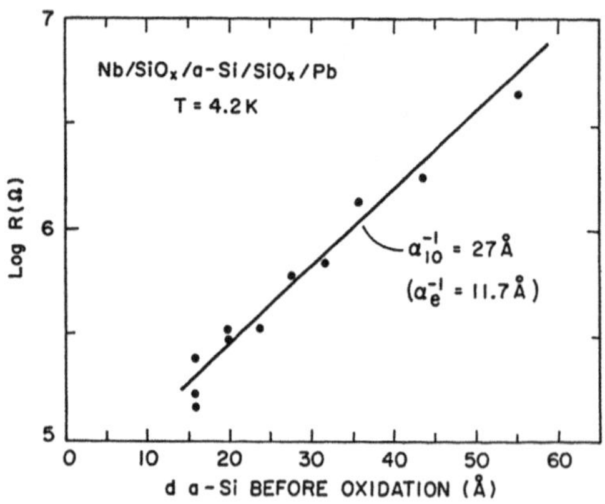

Fig. 7: Dependence of junction resistance on the thickness of the a-Si layer at low temperatures. The observed exponential dependence demonstrates transport is by tunneling.

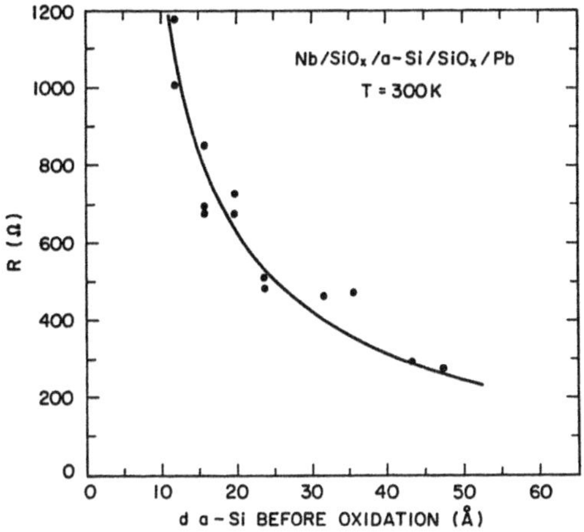

Fig. 8: Dependence of junction resistance on the thickness of a-Si layer at room temperature. Note that the resistance is a decreasing function of thickness over the range of thicknesses shown.

direct a-Si/metal contact, and we avoid any such complications (however interesting) in the data shown. In any event, this situation illustrates the importance and effectiveness of using more complex vertical structures to avoid unwanted complications.

A proper understanding of these results is not yet available, and may require original theoretical input. Some seemingly relevant pieces of the theory can be found in the literature, although in most cases they were developed for other purposes [23-26]. The broad outline of what is going on can be deduced directly from Figs. 6 to 8, however. As seen in Fig. 6, the conductance of these junctions goes from being very temperature dependent at high temperatures (in a fashion not unlike bulk amorphous semiconductors in which conduction is via variable range hopping) to temperature-independent at low temperatures. The crossover temperature depends on the thickness of a-Si barrier. In the low temperature regime transport is by tunneling, as demonstrated by the exponential increase of the resistance with barrier thickness shown in Fig. 7. Because of large density of localized states, resonant tunneling is almost certainly involved. The decay length of the localized state implied by the data is α_e^{-1} = 12 Å. (Since the localized states decay as $e^{-\alpha x}$ for resonant tunneling, the tunneling current is given by $(e^{-\alpha d/2})^2 = e^{-\alpha d}$.)

At room temperature the resistance of these junctions actually decreases as the barrier thickness increases (Fig. 8). Clearly coherent quantum mechanical tunneling all the way across the composite SiO_x/a-Si/SiO_x barrier is no longer taking place. Rather, it appears that, at least in part, inelastic processes intervene (akin to those in variable range hopping) so as to permit the transport to take place by means of two incoherent tunneling processes through the two conventional SiO_x barriers as well as by direct and/or resonant tunneling. Evidently the probability of such inelastic processes is increasing with barrier thickness. A proper understanding of this interplay between quantum mechanically coherent and incoherent tunneling processes is not at hand. Nevertheless, the physical observation of a decreasing resistance with barrier thickness surely signals interesting physics.

In conclusion, we hope that with these two examples we have been able to demonstrate the power of modern vapor deposition tools to fabricate multilayer structures that permit new length scales of localization physics to be explored in a controlled way. Yet to be agressively exploited is the combination of multilayered structures and advance micropatterning. The combination of these two approaches will almost certainly make possible entirely new classes of model systems.

Acknowledgements

This paper was prepared under the support of the United States National Science Foundation and the Office of Naval Research. Also we would like to express our thanks to T.W. Barbee, S. Ruggiero, D. Rudman, R. Hammond, and T.H. Geballe for helping us to see the potential of modern vapor phase deposition for science.

References

1. N.F. Mott and E.A. Davis, Electronic Processes in non-crystalline materials. Oxford University Press (1979).
2. Y. Imry, in the Proceedings of the NATO Advanced Study Institute on Localization, Percolation and Superconductivity, A.M. Goldman and S. Wolf, Eds., (1984).
3. H. Fukuyama, in the Proceedings of the NATO Advanced Study Institute on Localization, Percolation and Superconductivity, A.M. Goldman and S. Wolf, Eds., (1984).
4. G. Bergmann, PHysics Reports 107, 1 (1984).
5. R.E. Howard and D.E. Prober, VLSI Electronics: Microstructure Science, Vol. 5, N.G. Einsprush, Ed., Academic Press (1984).
6. D.E. Prober, in the Proceedings of the NATO Advanced Study Institute on Localization, Percolation and Superconductivity, A.M. Goldman and S. Wolf, Eds., (1984).
7. R. Laibowitz, in the Proceedings of the NATO Advanced Study Institute on Localization, Percolation and Superconductivity, A.M. Goldman and S. Wolf, Eds., (1984).
8. M.R. Beasley, in the Proceedings of the NATO Advanced Study Institute on Localization, Percolation and Superconductivity, A.M. Goldman and S. Wolf, Eds., (1984).
9. J.M. Graybeal and M.R. Beasley, Phys. Rev. B. 29, 4167 (1984).
10. J.M. Graybeal, M.R. Beasley and R.L. Greene, in the Proceedings of the 17th International Conference on Low Temperature Physics, to appear in Physica.
11. H. Raffy, R.B. Laibowitz, P. Chaudhari, and S. Maekawa, Phys. Rev. B 28, 6607 (1983).
12. S. Maekawa and H. Fukuyama, J. Phys. Soc. Japan 51, 1380 (1981).
13. H. Fukayama, in the Proceedings of the 17th International Conference on Low Temperature Physics, to appear in Physica.
14. S. Maekawa, this conference proceedings.
15. J.M. Graybeal, Ph.D. Thesis, Department of Physics, Stanford University (1984).
16. J.A. Sauvage, C.J. Mogals, and D. Adler, Phil. Mag. 25, 1305 (1972).
17. D.A. Rudman, and M.R. Beasley, Appl. Phys. Lett. 36, 1010 (1980).
18. R. Meservey, P.M. Tedrow and J.S. Brooks, J. Appl. Phys. 53, 1563 (1982).
19. E.S. Yang, Q.H. Hua, D.K. Yang, G.S. Yang, and P.S. Ho, Appl. Phys. Letters 44, 1144 (1984).
20. J.W. Osmun and H. Fritzshe, AIP Conference Proceedings #20, p. 333. "Tetrahedrally Bonded Amorphous Semiconductors" (Yorktown

Heights 1974). Edited by M.H. Brodsky, S. Kirkpatrick and D. Weaire; also see paper by J.J. Hauser in this same proceedings (p. 338).

21. K. Takei, T. Iwata and M. Igarashi, Proceedings of the Sixth Symposium on Ion Sources and Ion-Assisted Technology (Japan) Edited by T. Takagi (1982).
22. S.J. Bending, Ph.D. Thesis and to be published.
23. I.M. Lifshitz and V. Ya. Kirpichenkov, Sov. Phys. JETP 50, 499 (1979).
24. J. Halbritter, Surface Sci. 122, 80 (1982).
25. M. Ya Azbel, Solid State Commun. 45, 527 (1983); M.Ya. Athel and P. Soven, Phys. Rev. B 27, 831 (1983).
26 V. Kresin and J. Halbritter, Proceedings of LT-17 Edited by U. Eckern, A. Schmid, W. Weber and H. Wuhl (1984).

Fluctuation and Localization Effects in Quasi-One-Dimensional Metallic Structures

D.E. Prober, S. Wind, and P. Santhanam

Section of Applied Physics, Yale University, P.O. Box 2157
New Haven, CT 06520, USA

1. Introduction

In recent years the understanding of electron transport in metallic systems with random impurities has advanced dramatically. Since the original predictions of one-dimensional (1D) localization by THOULESS [1], localization theory has been developed more formally, and has been justified by microscopic calculations [2]. Some of the predictions for two-dimensional (2D) systems have been verified by experiments on metal films or MOSFETs [3]. Many experiments in recent years have attempted to verify the 1D predictions by Thouless; however, despite significant experimental effort, there has not been a clear verification of the predictions for 1D metallic wires. In the earliest experiments [4,5], it appeared that the results could be interpreted in terms of 1D localization theory. However, very large inelastic scattering rates, much larger than usual electron-phonon rates, were required. These large inelastic scattering rates remain unexplained. In subsequent experiments [6-8] 1D localization effects appeared to be absent. Thus, significant issues on electron transport in 1D systems remained open - as to whether real one-dimensional systems differ subtly (but critically) from the model understood by theory, and whether new pathways for electron energy loss (inelastic scattering) occur in such lower-dimensional metals.

Recently we have reported new experiments on Al wires [9]. These experiments confirm unambiguously, and for the first time, the localization predictions of Thouless for metallic wires, using magnetoresistance measurements. We have extended [10] the 1D theory to include spin-orbit scattering. The inelastic scattering rates inferred in our 1D experiments are in excellent quantitative agreement with rates determined independently for 2D metal films with the same film properties. The inelastic mechanisms are well understood from previous 2D studies [11,12] as being due to electron-phonon and electron-electron scattering. Thus, there appear to be no new inelastic mechanisms, at least for the wires studied. The theoretical understanding of such 1D systems does, in fact, appear to be in good order.

Using the newly formulated 1D theoretical result, we have reanalyzed the results of previous experiments on narrow wires. We find that most of those wires were not in the fully-1D regime (discussed below). Moreover, probably none of the samples had material parameters such that the original formulas derived by Thouless would be appropriate. The more complete theoretical results presented here do apply.

In this paper we review the work on aluminum films and wires at Yale, and then discuss the interpretation of previous experiments on wires. We start by reviewing the theory necessary for understanding the 1D experiments. In this paper we focus on metal wires, though studies of narrow MOSFETs have also yielded significant results [13,18]. The theoretical results we treat are those of weak localization theory, as this is appropriate for all experiments on metal wires to date. An excellent review of the general (weak) localization theory from an experimental viewpoint, and a full review of 2D localization experiments has been given by BERGMANN [3].

2. Theoretical Background

A. Introduction: Effects Contributing to R(T,H)

There are various effects which contribute to the resistance change as a function of temperature or magnetic field. The details of the theoretical results for 2D systems have been given in [2,3, and 14]. In addition to the term due to classical (Drude) elastic scattering, R_0, the main effects which contribute to the total resistance R(T,H) are

i. electron localization - ΔR^{Loc}
ii. Coulomb interactions - ΔR^{Int}
iii. Maki-Thompson fluctuations - ΔR^{MT}
iv. Aslamozov-Larkin superconducting fluctuations - ΔR^{AL}
v. "classical" magnetoresistance - ΔR^{cl}
vi. "classical" electron-phonon scattering - ΔR^{ph}.

For the samples studied at Yale, the terms which make the dominant contributions at low fields to the magnetoresistance (MR), $\delta R \equiv R(H,T)-R(H=0,T)$, are electron localization and Maki-Thompson fluctuations. For the change of resistance with temperature, $R(H,T_2)-R(H,T_1)$, the dominant contributions are electron localization, MT fluctuations, and "classical" electron-phonon scattering. In general, measurements of the magnetoresistance allow more direct access to the localization effects and inelastic times, and thereby allow tests of the inelastic mechanisms. Below, we shall treat in detail the terms which are relevant for the recent 1D studies. We note that the length scale(s) determining the effective system dimensionality for any given effect (e.g., localization) can be different from the length scale setting system dimensionality for some other effect (e.g., Coulomb interactions). This is of particular importance for treating the behavior of narrow wires, which in practice cannot be made arbitrarily narrow.

B. Electron Localization

For narrow wires of width W and thickness d such that W, d << $\ell_i (=\sqrt{D\tau_i}$, where τ_i is the inelastic scattering time and D is the diffusion constant), ALTSHULER AND ARONOV [15] calculated the localization contribution to the MR for small magnetic fields in the <u>absence</u> of spin-orbit scattering. We have recently reported the extension of the existing theory to include spin-orbit scattering, and have also presented experimental confirmation of these new predictions [10]. We ignore magnetic scattering in our discussions as magnetic scattering is claimed to be negligible in Al [11]. (Our results can, in any case, be generalized to include magnetic scattering). We also assume the elastic scattering time, $\tau << (\tau_i, \tau_{so})$; τ_{so} is the spin-orbit scattering time.

1. Case of no spin-orbit scattering - results of Altshuler and Aronov

The quantum correction to the dc conductivity due to localization is given by [15] $\Delta\sigma(\bar{r},\bar{r}') = -(2e^2D/\pi)C(\bar{r},\bar{r}')$. Altshuler and Aronov found by using perturbation theory, for a wire with the length along the x axis and width W along the y axis in a field $\vec{H} = H\hat{z}$, that the correlator is $C(y,y') = \sum_{q_x} C_{q_x}\phi_{q_x}(y)\phi_{q_x}(y')$ where for small H

$$C_{q_x} = \frac{1}{\hbar[Dq_x^2 + D(\frac{2eH}{\hbar c})^2 \frac{W^2}{12} + \frac{1}{\tau_i}]} \tag{1}$$

and $\phi_{q_x}(y)$ are the eigenfunctions of the operator in (2) of Ref. [10]. Equation (1) allows the evaluation of conductivity. One obtains for the extra resistance due to localization at a given temperature and magnetic field,

$$\frac{\Delta R}{R}^{Loc}(T,H) = \frac{R_\square}{(\pi\hbar/e^2)} \frac{\sqrt{\hbar c/4e}}{W\sqrt{H_i}} [1 + \frac{H^2}{48H_iH_W}]^{-1/2} = f_1(H,H_i) \tag{2}$$

in terms of sheet resistance $R_\square$ and characteristic fields $H_i = \hbar c/4eD\tau_i$ and $H_W = \hbar c/4eW^2$. For H = 0, (2) is just the result originally derived by Thouless. The range of validity of the small field approximation which gives (2) is $H < 12.5\ H_W$ [10]. Thus, for a sample of W = 0.4μm, a limit of ∿80G for using (2) applies.

2. Effects of spin-orbit scattering

When spin-orbit scattering is not negligible the correlator sum has two terms, corresponding to triplet and singlet wave functions respectively:

$$C_{q_x} = \frac{3}{2}\frac{1}{\hbar[Dq_x^2 + D(\frac{2eH}{\hbar c})^2 \frac{W^2}{12} + \frac{1}{\tau_2}]} - \frac{1}{2}\frac{1}{\hbar[Dq_x^2 + D(\frac{2eH}{\hbar c})^2 \frac{W^2}{12} + \frac{1}{\tau_i}]} \tag{3}$$

with $\tau_2^{-1} = (4/3)\tau_{so}^{-1} + \tau_i^{-1}$ and τ_{so} the spin-orbit scattering time. The second term (singlet term) has a length scale of $\ell_i = (D\tau_i)^{1/2}$ determining system dimensionality. The first (triplet) term, however, has as its relevant time τ_2, and thus has a different dimensional length scale $\ell_2 \equiv (D\tau_2)^{1/2}$, which is always $<\ell_i$. If $\tau_{so} << \tau_i$, then $\ell_2 << \ell_i$.

The fully-one-dimensional case

The fully-1D (narrow) case requires $W < \ell_2$ and ℓ_i, so that both sums for C(y,y') are one-dimensional. The change in fractional resistance is then

$$\frac{\Delta R}{R}^{Loc}(T,H) = \frac{3}{2} f_1(H,H_2) - \frac{1}{2} f_1(H,H_i) \tag{4}$$

where $H_2 = \hbar c/4eD\tau_2$ and $f_1(H,H_i)$ is the 1-D form, (2). We may define $H_{so} = \hbar c/4eD\tau_{so}$, so that $H_2 = (4/3)H_{so} + H_i$. Note that for negligible spin-orbit scattering ($\tau_{so}^{-1} \to 0$), (4) correctly reduces to (2).

The Mixed-Dimensional Case

For intermediate widths such that $\ell_{so} << W << \ell_i$ the correlator sum for the triplet term will be two-dimensional as in the case of wide films, since $\ell_2 << W$.

The singlet term will be one-dimensional. (Note that d is still $<\ell_{so}$.) Thus, for such wires of intermediate width we find that

$$\frac{\Delta R^{Loc}}{R}(T,H) = \frac{3}{2} f_2(H,H_2) - \frac{1}{2} f_1(H,H_i) \tag{5}$$

where $f_2(H,H_2)$ is the 2D form, given in Ref. 14a, (13). The localization behavior for wires of intermediate width is thus of a mixed-dimensional nature.

C. Maki-Thompson Fluctuations

The Maki-Thompson contribution is significant for superconductors [16]. It is given for 1D samples by [10]

$$\frac{\Delta R^{MT}(T,H)}{R} = -\beta(\frac{T}{T_c})\, f_1(H,H_i). \tag{6}$$

β is the electron interaction parameter introduced by Larkin; β is independent of sample dimensionality [2], and diverges as $T \to T_c$. Equation (6) applies for small fields, and for $k_B(T-T_c) > \hbar/\tau_i$. The length scale determining sample dimensionality for the MT term is ℓ_i; the MT contribution is not affected by spin-orbit scattering. At high fields the prediction of (6) is depressed [17].

D. Interaction Effects and Aslamazov-Larkin Fluctuations

Other terms which can contribute to the temperature-dependence are those describing the diffusion (particle-hole) channel and the Cooper (particle-particle) channel [2,3]. For 1D systems, one of these terms has been derived for H=0 [7]. This term has a form very similar to (2) with H=0. The major difference is that in ΔR^{Int}, τ_i of (2) is replaced by the quantum diffusion time, $(\hbar/k_BT)$. The magnetoresistance of this interaction term is typically much smaller than that of the localization and MT terms. The length scale for these electron interaction terms is $\ell_{int} \equiv (\hbar D/k_BT)^{1/2}$

The AL fluctuation contribution, derived ~15 years ago, is due simply to fluctuation-induced conductivity above T_c. For clean superconductors with weak electron-phonon coupling (e.g., Al) the AL term is much smaller than the MT term (except for $T \to T_c$) and the field-dependence is negligible.

E. "Classical" Contributions

In addition to the above quantum corrections, there are two "classical" corrections, familiar from standard electron transport theory. These contributions are significant for cleaner metals ($\rho \lesssim 10\,\mu\Omega$-cm) with low resistance ($R_\square \sim 1$ ohm). The classical MR is given by $\sim(\omega_c\tau)^2$, with ω_c the cyclotron frequency and τ the elastic scattering time.

For the temperature-dependence of the resistance, we find that in our 1D Al wires and in our 2D Al films at $T \leq 20$K the contribution due to electron-phonon scattering is given experimentally by

$$\Delta R^{ph}/R = C_{ph} T^3$$

with C_{ph} the same for wires and films if $R_\square$ is the same. A similar term is found for Cu wires [7]. ABRAHAM AND ROSENBAUM [24] find for Cu films a term $\propto T^4$. The usual theory of low-temperature electron-phonon scattering in simple metals predicts a temperature-dependence of $\Delta R^{ph} \propto T^{p+2}$, with p the exponent of the electron-phonon inelastic rate, $\tau_i^{-1} \propto T^p$;p=3 for 3D metals. The extra power of 2 in the formula for ΔR^{ph} is due to the requirement of small-angle scattering at low temperatures. It appears that either the small sample size or the fairly rapid elastic scattering relieves the requirement for small-angle scattering. Indeed, the magnitude of the experimentally determined coefficient C_{ph} is in excellent quantitative agreement with that expected from the electron-phonon scattering rate without a limit on scattering angle.

To summarize the theoretical results for clean Al wires and films ($R_\square \sim 1\Omega$): the dominant terms contributing to the low field magnetoresistance are the localization term and the MT term. Thus, the magnetoresistance is given as

$$\delta R(T,H) = \delta R^{Loc}(T,H) + \delta R^{MT}(T,H) \tag{7}$$

with $\delta R^{Loc}(T,H)=[\Delta R^{Loc}(T,H)-\Delta R^{Loc}(T,H=0)]$. $\delta R^{MT}(T,H)$ is defined similarly. For fully-1D systems, the ΔR terms (the extra resistance at a fixed T and H) are given by (4) and (6) respectively; for wires of mixed dimensionality, (5) and (6) are appropriate. For 2D systems the localization and MT terms are given, respectively, in Ref. 14a, (13) and in [16] for ΔR.

3. Experimental Results - Aluminum Films and Wires

In this section we shall treat results of studies at Yale on aluminum wires and 2D aluminum films. Aluminum represents a good choice for such studies since it has a long intrinsic inelastic time (for bulk samples), moderate (i.e., not large) spin-orbit scattering rate, negligible magnetic scattering [11], and can be produced in stable, thin films of thickness $d \sim 100$ Å on room-temperature substrates. Clean films, with elastic mean free paths $\ell = v_F\tau \sim d$ can be achieved, yielding low values of sheet resistance $R_\square \sim 1$ ohm. With low $R_\square$ films the inelastic times are long, and the low-field MR is fully determined by the localization and MT terms.

In general, fabrication of narrow wires can be accomplished by microlithographic techniques [18] or by some solid state phenomenon [6,8]. Microlithographic techniques include step-edge methods, where the wire is built directly on the edge of a substrate step [4,7], direct electron-beam lithography [5], and x-ray lithography. We selected x-ray lithography for the present studies instead of a step-edge method [4] because we wished to produce a 2D film with thickness and properties identical to those of the film from which the 1D wire was formed. The contact mask for x-ray lithography was made by etching a step with reactive ion etching in a 1-μm-thick-polyimide film, and depositing a Au film (the x-ray absorber) over the edge of the etched step, so that a gap is formed. The electrical contact pad areas in this Au mask are defined by a metal mask during deposition of the Au film. After exposure and development, Al is deposited by thermal evaporation and resist liftoff is used. The Appendix shows these processes.

A. Results from 2D Al Films: Inelastic Mechanisms and Spin-Orbit Times

To establish length scales ℓ_i and $\ell_2=(D\tau_2)^{1/2}$, and to quantitatively test the known 2D theory for localization and MT fluctuations, we first studied

2D films. The details are given in [11]. The inelastic-scattering rates determined from our 2D experiments are due to a combination of electron-electron and electron-phonon processes, such that

$$\tau_i^{-1} = \tau_{ee}^{-1} + \tau_{ep}^{-1} . \qquad (8)$$

The electron-phonon rate we measure agrees quantitatively with that of LAWRENCE AND MEADOR [19] who calculated for Al that

$$\tau_{ep}^{-1} = 0.91 \times 10^7 \ \mathrm{sec}^{-1}\mathrm{K}^{-3}T^3.$$

For electron-electron scattering, ABRAHAMS et al. [20] have calculated for a dirty ($\hbar/\tau > k_BT$), 2D system ($\hbar D/k_BT > d^2$)

$$\tau_{ee}^{-1} = (e^2R_\square/2\pi\hbar^2)k_BT \ \ln(T_1/T) \approx 13.4 \times 10^7 R_\square T.$$

Here $T_1 = 9\times10^5(k_F\ell)^3 \sim 10^{12}$. The observed inelastic rates are in excellent quantitative agreement with these two predictions, for each of the six films studied, with $R_\square$ =0.8-8 ohms.

We thus find that the inelastic mechanisms operative in our clean Al films are dirty-limit, 2D electron-electron scattering, and clean-limit 3D electron-phonon scattering. These results provide at least a partial resolution of questions raised by superconducting charge-relaxation experiments for $R_\square$ >1 ohm [11,21]. A recent study of microwave gap enhancement comes to a similar conclusion. This is discussed elsewhere in this volume by Mooij.

The spin-orbit scattering rates determined in our studies of Al films are described by $(\tau_{so}^{-1})/\tau^{-1} \approx 2\times10^{-4}$ with τ^{-1} the elastic scattering rate. ABRIKOSOV AND GORKOV [29] originally estimated that the spin-orbit rate should be a fraction $(\alpha Z)^4$ of the elastic rate. α is the fine-structure constant. For Al with Z = 13, one obtains a value of 0.8×10^{-4}. This agreement with the experimental result is sufficient to demonstrate that spin-orbit scattering results from the elastic scattering process. For granular Al films [22], the fraction is almost 10^3 smaller, using values of τ derived from resistivity. Weak tunneling between grains determines ρ for granular Al, and the mechanism(s) contributing to spin-orbit scattering in granular Al must differ from the mechanism in "pure" Al.

B. Experimental Results - Al Wires

The Al wires studied [9,10] had widths $0.2\mu m \lesssim W \lesssim 0.6\mu m$, lengths $\sim 200\mu m$, and thicknesses 150-250 Å. Sheet resistances were $R_\square \sim 1\Omega$, comparable to the 2D films discussed in the last section. Wide (2D) films were also deposited at the same time, to allow comparison of their properties. Properties of two of the wires are given in Table I. Data for other similar samples are given in [9]. The measurements were done as in [11].

Table I Sample Parameters for Al Wires. $R_\square$ at 4.5K

Sample	$R_\square$ (Ω)	d (Å)	W (μm)	ℓ_{so} (μm)	ℓ_i(4.5K) (μm)	Magnetoresistance Dimensionality	
A	0.9	250	0.20	0.48	1.47	1D	$T\lesssim15K$
C	2.8	150	0.60	0.32	0.98	Mixed dim.	$T\lesssim6K$
						2D	$T\gtrsim15K$

Fully-one-dimensional case (Wire A)

The theory for fully-1D systems requires $[d,W]<[\ell_i,\ell_{so}]$. This fully-1D theory uses (4) and (6) for the localization and MT terms. To test this theory we show in Fig. 1 the normalized MR for sample A, the narrowest wire studied. The theoretical expression for the MR is shown by solid lines.

Fitting was done as in the 2D studies. We show data only up to 300G, as (6) is not valid in large fields, and (4) is valid only for $H<12.5H_W$. We find that the MR obeys the one-dimensional form over the entire temperature range 1.8K to 15K. Fits to the 2D or mixed-dimensional form were not satisfactory. H_{so} is found to be independent of temperature, as for 2D films.

The inelastic scattering length $\ell_i(T)$ determined from this experiment is plotted in Fig. 2. For comparison we show $\ell_i(T)$ for the co-deposited wide (2D) film. The agreement is excellent, further confirming our use of the fully-1D analysis. We find that for $T \lesssim 15K$, $\ell_i(T)$ and ℓ_2 are indeed larger than W, so that use of the 1D theory is self-consistent. Similar agreement is seen for another fully-1D wire [9].

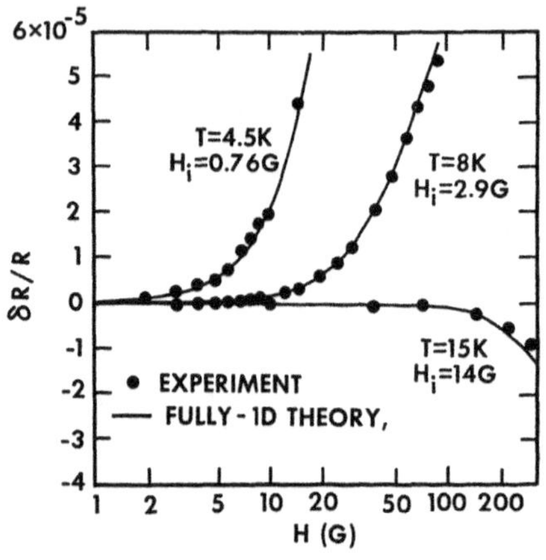

Fig. 1: Normalized Magnetoresistance, wire A. H_{so}=7G for theory curves.

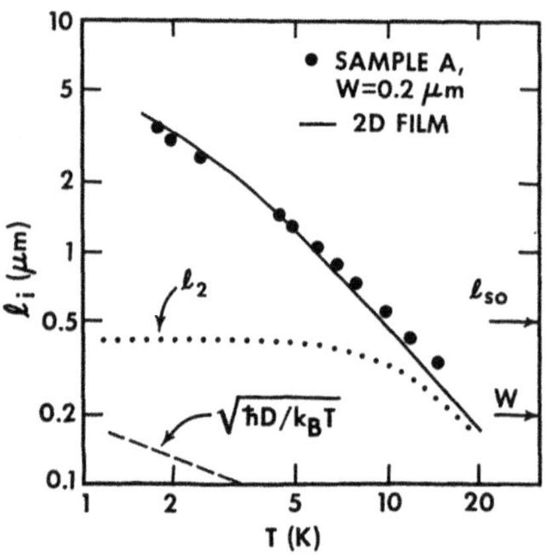

Fig. 2: Inelastic diffusion length vs. temperature for wire A which is fully one-dimensional to 15K. The solid line is the experimentally determined ℓ_i for the codeposited wide film.

The agreement of inelastic scattering rates in Fig. 2 is the first reported agreement of inelastic rates for samples of different localization dimensionality. This result is in fact sensible. The dimensional size scales for electron-phonon inelastic scattering and for dirty-limit electron-electron scattering are, respectively, the phonon wavelength and the quantum diffusion length $(\hbar D/k_BT)^{1/2}$. These are both less than the wire width, so that inelastic mechanisms for our wires and for 2D films should be the same. As seen in Fig. 2, wires of width $\lesssim 0.1\mu m$ would be required to study the 1D electron-electron inelastic mechanism.

Most previous experiments on wires measured only the resistance change with temperature. For our samples, MR data and resulting τ_i values provide a more direct test of localization theory. Still, to verify the R(T) pre-

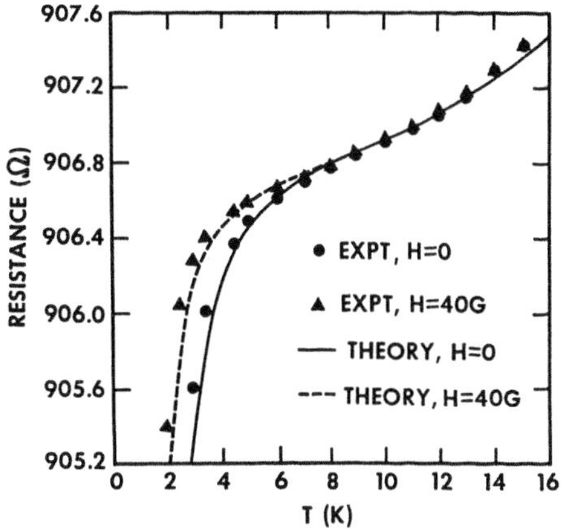

Fig. 3: Resistance as a function of temperature, wire A. Theory curves were matched to the data at 6K. 1D theory for localization and MT terms are used over the entire temperature range.

diction, in Fig. 3 we plot R(T) for sample A. We expect that the localization and MT contributions to R(T) will be 1D up to 15K. The electron-phonon contribution is well fitted by $\Delta R^{ph}/R = C_{ph}T^3$ for both the wires and our wide 2D samples, independent of H. Theoretical plots of R(T) are also given in Fig. 3. The parameters of the theoretical curves, H_i and H_{so}, are taken from the 1D MR studies. The values are not adjustable here. C_{ph} was determined from the 2D film. In Fig. 3 there is very good quantitative agreement over the full temperature range, further confirming the 1D theoretical prediction. At low temperatures the MT fluctuation term is dominant. The agreement of theory and experiment at low temperatures confirms that ℓ_i is the relevant length scale for the MT term.

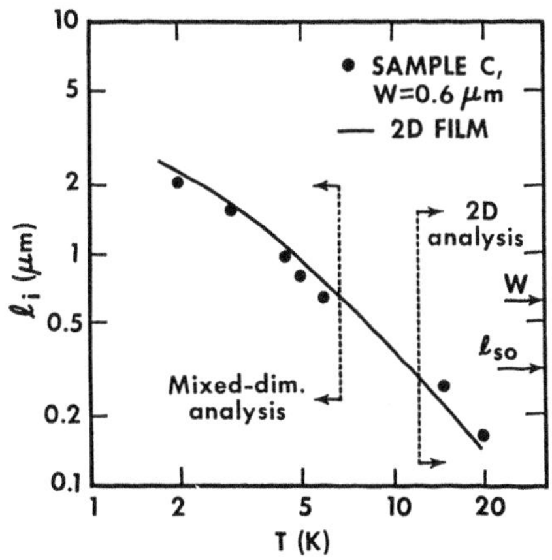

Fig. 4: Inelastic diffusion length vs. temperature for a wire of intermediate width; at low temperature (T≤6K), the localization behavior is of mixed dimensionality, as $\ell_{so}<W<\ell_i$. For T≥15K, the localization behavior is that of a 2D film, since $\ell_{so},\ell_i<W$. In each temperature range appropriate theoretical analysis was used to extract values of ℓ_i.

Mixed-dimensional case (Wire C)

Now we discuss a wider wire sample C, of width 0.6μm. The MR data at high temperatures, 15K and 20K, can be satisfactorily fit to the 2D theory with reasonable values of H_i, and H_{so}=16G (independent of temperature). This implies $\ell_{so}\equiv(D\tau_{so})^{1/2}\approx0.3$μm. In contrast, for T$\lesssim$12K the experimental data cannot be fit to the 2D theory for any set of parameters H_i and H_{so}. However, when T$\lesssim$7K the behavior of this sample is described well by the MR formula corresponding to the mixed dimensional behavior, with the same H_{so} as at the highest temperatures. (The fully-1D theory does not provide as good fits.) This can be explained as due to the fact that at low temperatures, $\ell_2 < W < \ell_i$, so that the singlet and triplet terms have different dimensionalities (1D and 2D, respectively). The inelastic diffusion length inferred from the fits are shown as a function of temperature in Fig. 4 along with those for a codeposited wide 2D film.

The agreement seen for the inferred inelastic rates confirms the mixed-dimensional behavior. Experimentally, when $\ell_i \gtrsim 1.3W$, the mixed dimensional theory is applicable, and when $\ell_i \lesssim 0.5W$ the 2D theory is applicable. Thus, for the singlet and MT terms, there is a crossover of dimensionality when $\ell_i \sim W$. This crossover of localization dimensionality is in accord with the original ideas of Thouless.

4. Previous Experiments on Metal Wires

The first experiments testing the predictions of Thouless were carried out by GIORDANO et al. [4]. They made narrow wires of $Au_{60}Pd_{40}$ with $(Area)^{1/2} \sim$ 300 Å to 2000 Å using a novel step-edge shadowing technique. The low temperature resistivity of the films used ranged from 100μΩcm to 400μΩcm. An increase in resistance with decreasing temperature was observed; the samples of smaller cross-sectional area showed proportionately larger effects such that $\Delta R/R \propto A^{-1}$, in agreement with the theoretical prediction for one-dimensional behavior. The interpretation of the R(T) data was in terms of Thouless' 1D localization prediction, (2) with H=0. This yielded an inelastic scattering rate $\tau_i^{-1} = 9\times10^{12}(T/\rho)$ with ρ the resistivity in μΩcm. This inelastic rate far exceeds that due to clean-limit electron-phonon scattering, which of course also does not depend on ρ. The dependence $\tau_i \propto \rho$ is also hard to understand within any simple model of disorder-related inelastic scattering. No satisfactory theoretical explanation of the inferred inelastic rate has emerged to date.

Since these experiments on Au-Pd wires were performed, the theoretical understanding of the effects of spin-orbit scattering and of the MR have advanced significantly. Recent MR measurements on 2D Au-Pd films [23] infer an inelastic rate of 2.5×10^{10} at 2K. The inelastic rate increased faster than T between 2K and 5K. These results are for a film of ρ=112μΩcm and d=145 Å. It was also found that $\tau_{so}^{-1} \gtrsim 5\times10^{12}$, so that for this Au-Pd alloy, $\ell_{so} \lesssim 80$ Å.

These quantitative results for the scattering rates call into question the original interpretation of the results for Au-Pd wires. Assuming that the 2D value of τ_{so} applies also for the wires, the narrowest wires of Giordano et al. could only be in the mixed dimensional regime ($\ell_{so} < W$), and (2) should not be applied. In fact, considering the scattering parameters of the 2D films, R(T) for the wires should decrease as T is reduced. This definitely was not the experimental result. If significant magnetic scattering was present, the localization effects would be decreased in magnitude and the temperature-dependence "flattened." In any case we do not know of any plausible set of parameters τ_i, τ_{so}, or τ_s (magnetic) which could give the observed rise of resistance from localization effects.

WHITE et al. have suggested [7] that the resistance rise observed for the Au-Pd wires was in fact due to interaction effects, and not localization effects. They could not then explain why localization effects were absent. As seen above, we can now provide a plausible explanation for the absence of localization effects, as being due to spin-orbit and magnetic scattering. We find the identification by White of electron-electron interaction effects as the cause of the observed resistance rise to be fully plausible. The absence of localization effects for the clean Cu wires of White et al. may have been due to magnetic scattering, which "flattens" the temperature-dependence of ΔR^{Loc}. Indeed, MR studies of 2D Cu films [24] do yield a large magnetic scattering rate. Furthermore, additional phase relaxation mechanisms due to electromagnetic fluctuations [25] might become important in such one-dimensional systems where $W,d \ll (\hbar D/k_BT)^{1/2}$. The predicted magnitude of this rate is large, so that ΔR^{Loc} would be quite small.

One problem which is not addressed by White's explanation of the Au-Pd wire results is that for the AuPd wires, $\Delta R/R$ is experimentally proportional to ρ/A. Theoretically, it is predicted that $\Delta R^{Int}/R \propto \rho^{1/2}/A$. Over the range of ρ studied for the Au-Pd wires, these forms differ by a factor of 2. We speculate that this extra factor results from the mean free path being very short, so that the theoretical result acquires an extra prefactor when $k_F \ell \sim 1$.

Another early study of narrow wires used W-Re wires of width 700 Å to 5000 Å fabricated by electron-beam lithography [5]. Resistivity was $\sim 400 \mu\Omega$cm and films were ~ 50 Å thick. Measurements on superconducting phase-slip centers in 250 Å-wide wires were also performed, to obtain an independent measure of τ_i [26]. The inelastic scattering rate found from the resistance rise, using (2) with H=0, was in reasonable agreement with $\tau_i^{-1}(T_c)$ for the phase-slip experiments. However, a more recent experiment on 2D W-Re films [27] has concluded that the inelastic scattering length ℓ_i was <200 Å for $T \lesssim 5$K, leaving the 1D behavior of W-Re wires in question. That study also found that spin-orbit scattering is strong in W-Re, so that use of (2), or even (4), is not justified. We note that White et al. have also reanalyzed the data for W-Re wires, and found there also that the resistance rise was consistent with the electron-electron interaction theory.

We last discuss two experiments where the wires were defined without lithographic processes. In the case of platinum wires of ~ 1000 Å diameter produced by a mechanical drawing process [8], the authors found an unusually large fractional change of resistance with temperature and also an anomalous positive magnetoresistance. Neither of these could be explained by 1D-localization theory or electron-electron interaction theory, nor by any combination of them, as was noted by those authors. The unusually large disorder produced in the drawing process was suggested as a reason, but the mechanism is not completely understood.

In a second study, single-crystal whiskers of bismuth less than 2000 Å in diameter were made by stress-accelerated growth [6]. The mean free path of electrons at low temperatures was estimated to be very large. Resistance measurements from 20K to 0.5K showed only a decrease in resistance with decreasing temperature. The authors concluded that there was no clear evidence for 1D localization. We conjecture that this apparently null result may in fact be a case of weak localization, but one where only the (negative) singlet term in (4) is significant. The authors estimated the inelastic length to be $\sim 100 \mu$m, and given the large Z of Bi, ℓ_{so} is likely much shorter. MR measurements on these samples would be of significant interest.

To summarize our interpretation of the results on narrow metallic wires other than aluminum: i) For the alloys Au-Pd and W-Re, it appears that magnetic scattering and/or large spin-orbit scattering rates may give very small localization effects, smaller than ΔR^{Int} due to interaction effects. ii) For Cu the resistance rise is also due to ΔR^{Int}, as shown in [7]. Localization contributions may be absent because of magnetic scattering, or because the phase-relaxation rate is very large due to new mechanisms. iii)

For drawn Pt wires the observed MR is not due to 1D localization effects. iv) In Bi whiskers, it is possible that the observed resistance decrease is explained by 1D localization theory, but in the limit of strong spin-orbit scattering. Without systematic magnetoresistance data for i, ii, and iv, further interpretation of those data is difficult.

5. Conclusions

A. Localization and System Dimensionality

In our studies of 2D aluminum films and aluminum wires we have confirmed in detail the predictions of modern localization theory. While the 2D localization theory has previously been verified in dirty metal films [3], our studies are the first to validate the 1D theory which had originally started this whole field. With our clean metal films and wires we also find understandable inelastic times (and mechanisms), forming a full and self-consistent picture of how inelastic scattering limits the localization effects. In contrast, the inelastic mechanisms in the dirty films have remained, in general, unexplained.

In our studies of wires we find inelastic times which are the same as those in the 2D films. This is the first instance of systems with different localization and Maki-Thompson dimensionality having the same inelastic times. This agreement of inelastic times is sensible, and further strengthens the interpretation of the data and our conclusions.

B. Inelastic Mechanisms

The studies of Al wires and films have, in addition to developing and testing predictions of localization theory, produced a clear, quantitative understanding of the mechanisms determining inelastic scattering in such clean films. The two operative mechanisms have been determined to be clean-limit electron-phonon scattering, $\propto T^3$, and dirty-limit electron-electron scattering, $\propto T$.

Recent studies of other 2D metal films have yielded a richness of experimental results for the inelastic scattering rate (BERGMANN [3]). Evidence for dirty-limit electron-electron inelastic scattering, $\propto T$, has been provided by numerous studies. For systems of other dimensionality, such electron-electron scattering has a temperature-dependence $\tau_{ee}^{-1} \propto T^{d/2}$. An apparently different inelastic mechanism is related to coherent effects of electromagnetic fluctuations [25]. The rate due to this Nyquist mechanism in 2D goes as $\tau_N^{-1} \propto T$, with a magnitude for clean films very similar to that of the previously noted electron-electron scattering, $\tau_{ee}^{-1} \propto T$. (The $\ell n(T_1/T)$ term has only very weak dependence on T). It appears that 1D wires $[W<(\hbar D/k_B T)^{1/2}]$ provide the only systems in which τ_N processes may be observable. No experiments on this exist to date.

Electron-phonon scattering is the other mechanism expected to be important at low temperatures. We have seen in the 2D Al film studies that $\tau_{e-p}^{-1} \approx 10^7 T^3$, in full accord with the microscopic theory calculation [19] employing a realistic band structure for aluminum. This is for clean aluminum. For dirty metals, for which $q_{ph}\ell<\pi$, there are apparently two different theoretical predictions for electron-phonon scattering in 3D. These are, first [1], $\tau_{e-p}^{-1} \propto T^4$, and second [3], $\tau_{e-p}^{-1} \propto T^2$. (Most films on solid substrates should be in the 3D regime.) The magnitude of the T^2 inelastic rate observed by BERGMANN [14] and others is, however, much larger than this theoretical prediction. An additional candidate for an inelastic rate is electron-impurity vibration mechanism, $\propto T^2$ [28]. These issues are not yet resolved, and future experiments can still profitably explicate inelastic scattering in such high-resistance films.

We acknowledge useful conversations with numerous colleagues, especially J. M. Gordon, Y. Imry, W. E. Lawrence, W. L. McLean and J. W. Serene. This research was supported by U.S.-NSF grant DMR-8207443.

Appendix A - Fabrication of Narrow Aluminum Wires by X-Ray Lithography

Since x-ray lithography is at present limited to 1:1 contact printing, the x-ray absorber must be patterned at the appropriate small dimensions. We have accomplished this by reactive ion etching a 1-μm-thick polyimide film (the mask membrane) so that a step is formed in the polyimide. The Au absorber is deposited at an angle over the edge of the step, leaving a gap at the base of the step (Fig. 5a). Electrical contact pads are defined by a metal mask during the Au deposition. PMMA resist is exposed by C_K or Cu_L line radiation (Fig. 5b) followed by resist development. Aluminum is then deposited by thermal evaporation. Because resist liftoff is used, a directional deposition technique, such as thermal evaporation, must be employed. The final lifted off pattern is shown in Fig. 5c.

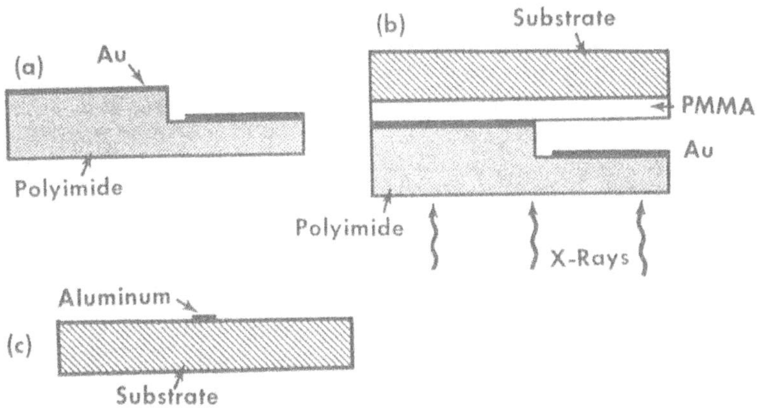

Fig.5: a. Au x-ray mask absorber pattern formed by evaporating Au over a RIE-formed step in polyimide. (side view); b. PMMA is exposed by x-ray photons. Mask and substrate are held in contact electrostatically. c. Aluminum is evaporated onto developed PMMA pattern and "lifted off", leaving a narrow wire whose width is determined by the gap in the x-ray mask.

References

1. D.J. Thouless: Phys. Rev. Lett. 39, 1167 (1977); Solid State Comm. 34, 683 (1980). The prefactor in (2) can be written as $\ell_i/(\pi\hbar/e^2)\times[\rho/A]$ (with A the cross sectional area of the wire). This is larger than Thouless' result by a factor of $\sqrt{2}$ due to differences in the definition of the diffusion constant we use.

2. B.L. Altshuler, A.G. Aronov, A.I. Larkin, and D.E. Khmel'nitskii: Zh. Eksp. Teor. Fiz. 81, 768 (1981) [Sov. Phys. JETP 54, 411 (1981)]; H. Fukuyama: Surf. Sci. 113, 489 (1982).

3. G. Bergmann: Phys. Reports 107, 1 (1984); R. C. Dynes: Physica 109 & 110B, 1857 (1982).

4. N. Giordano, W. Gilson and D.E. Prober: Phys. Rev. Lett. 34, 725 (1979); N. Giordano: Phys. Rev. B22, 5635 (1980); N. Giordano, Proceedings of the International Conference on Physics in One Dimension, J. Bernasconi and T. Schneider, editors (Springer, 1980).

5. P. Chaudhari and H. U. Habermeier: Phys. Rev. Lett. 44, 40 (1980); Solid State Comm. 34, 687 (1980).

6. D.R. Overcash, B.A. Ratnam, M.J. Skove and E.P. Stillwell: Phys. Rev. Lett. 44, 1348 (1980).

7. A.E. White, M. Tinkham, W.J. Skocpol and D.C. Flanders: Phys. Rev. Lett. 48, 1752 (1982).

8. A.C.Sacharoff, R.M. Westervelt, and J. Bevk: Phys. Rev. B29, 1647 (1984).

9. P. Santhanam, S. Wind and D.E. Prober: Phys. Rev. Lett., to be published.

10. P. Santhanam, S. Wind and D.E. Prober: Physica, to appear (Proceedings of the LT-17 Conference).

11. P. Santhanam and D.E. Prober: Phys. Rev. B29, 3733 (1984).

12. J.M. Gordon, C.J. Lobb and M. Tinkham: Phys. Rev. B29, 5232 (1984).

13. R.G. Wheeler, K.K. Choi, A. Goel, R. Wisnieff, and D.E. Prober: Phys. Rev. Lett. 49, 1674 (1982); M. Pepper, this volume.

14a G. Bergmann: Phys. Rev. B29, 6114 (1981); b. Z. Phys. B48, 5 (1982).

15. B.L. Al'tshuler and A.G. Aronov: Pis'ma Zh. Eskp. Teor. Fiz. 33, 515 (1981) [JETP Lett. 33, 499 (1981)].

16. A.I. Larkin: Pis'ma Zh. Eksp. Teor. Fiz. 31, 239 (1980) [JETP Lett. 31, 219 (1980)].

17. J.M.B. Lopes dos Santos and E. Abrahams: to be published.

18. D.E. Prober: chapter in Percolation, Localization and Superconductivity, A.M. Goldman and S. Wolf, editors (Plenum, New York, 1984), p.231.

19. W.E. Lawrence and A.B. Meador: Phys. Rev. B18, 1154 (1978).

20. E. Abrahams, P.W. Anderson, P.A. Lee and T.V. Ramakrishnan: Phys. Rev. B 24, 6783 (1981).

21. C. C. Chi and J. Clarke, Phys. Rev. B19, 4495 (1979).

22. K.C. Mui, P. Lindenfeld and W.L. McLean, this volume.

23. W. McGinnis, Ph.D. Thesis, U.C.L.A., 1983 (unpublished).

24. D. Abraham and R. Rosenbaum: Phys. Rev. B27, 1409 (1983); Phys. Rev. B27, 1413 (1983).

25. B.L. Altshuler, A.G. Aronov and D.E. Khmel'nitskii, J. Phys. C15, 7367 (1982); B.L. Altshuler, A.G. Aronov, D.E. Khmel'nitskii and A.I. Larkin, in Quantum Theory of Solids, I.M. Lifshitz, editor (Mir Publishers, Moscow, 1982), p. 146.

26. P. Chaudhari, A.N. Broers, C.C. Chi, R. Laibowitz, E. Spiller and J. Viggiano: Phys. Rev. Lett. 45, 930 (1980).

27. H. Raffy, R. Laibowitz, P. Chaudhari and S. Maekawa: Phys. Rev. B28, 6607 (1983).

28. Yu. Kagan and A.P. Zhernov: Zh. Eksp. Teor. Fiz. 50, 1107 (1966) [Sov. Phys. JETP 23, 737 (1966)].

29. A.A. Abrikosov and L.P. Gor'kov: Zh. Eksp. Teor. Fiz. 42, 1088 (1962) [Sov. Phys. JETP 15, 752 (1962)].

Theory of Fluctuation Phenomena in Kinetics

M. Ya Azbel'
IBM T.J. Watson Research Center, P.O. Box 218, Yorktown Heights, NY 10598, USA
and
Department of Physics, Tel-Aviv University, Tel-Aviv, Israel

Localization leads to an unusual kind of kinetics. At sufficiently low temperatures transport in a macroscopic system is dominated by a single quenched fluctuation ("flucton"). Kinetic coefficients are unique for each system and exhibit huge oscillations. Even the effective dimensionality of the system may fluctuate. Some manifestations of flucton kinetics are discussed in the paper.

1. Mesoscopic Systems and Fluctons

Until recently the reproducibility in physics was taken for granted. In fact reproducibility was related only to the fact that the systems studied were either microscopic or "conventional" macroscopic ones. "Conventional" macroscopic systems have relatively small fluctuations (which $\rightarrow 0$ when the size of the system $\rightarrow \infty$). Their behavior depends only on their average macroscopic characteristics. The "individual" differences are negligible, and the same behavior can be reproduced at systems different microscopically but identical macroscopically. Microscopic systems exhibit an individual behavior, which allows one to identify the system by its spectroscopy. Such systems (atoms, molecules) are exactly "reproducible" because there are so many kinds of them. Very low temperatures lead to a new type of physical system, which may be denoted, following N. Van Kampen, as mesoscopic systems. These systems contain millions of atoms. It is doubtful that their microscopic structure may be exactly reproduced. Meanwhile, their behavior, just as the behavior of microscopic systems, is determined by the exact microscopic structure. It is unique and extraordinarily rich. Moreover, it is super-sensitive to microscopic changes in the structure. For instance, a few extra electrons or a shift of a single impurity atom may increase or decrease the resistance by orders of magnitude. This may be true for a system of arbitrarily large size at sufficiently low temperatures $T < T_l$. Remarkably, T_l only logarithmically decreases with the increasing size.

The sensitivity to microscopic changes is related to a very special feature of the considered systems. Their kinetics reduces to a certain quenched microscopic fluctuation, whose impact is exponentially (with T^{-1}) blown up. Such quenched fluctuations are later denoted as "fluctons". (The word was suggested by A.D. Stone). Fluctons may even change the effective dimensionality of the system. Microscopic fluctons are very sensitive to any change of system parameters. As a result, such usually "stable" characteristics, as the temperature dependence of conductance, may become individual and fluctuating.

Experimentally, fluctuation kinetics is manifested in high oscillations of, e.g., conductance in low-dimensional devices. These oscillations look chaotic and are irreproducible from sample to sample or upon significant temperature cycling. However, they are perfectly reproducible in the same sample and upon moderate temperature cycling. Such oscillations were observed since mid-sixties [1] and discussed [1a] in 1979. However, their detailed study started only recently, in independent and almost simultaneous experimental [2] and theoretical [3] papers. It was followed by new experimental [4,4a] and theoretical [5-5d] developments and ideas [6,7]. Still, the study is very far from being completed, and is related almost exclusively to quasi-one-dimensional (1D) systems. Even then a lot of open problems remain, and the purpose of this paper is to draw attention to them, as well as to the fluctuation kinetics, fluctons and mesoscopic systems in general.

2. Fluctons and Strong Localization

Fluctuation kinetics is related to the localization of eigenstates. Eigenstates are believed to be always localized in 1D and 2D and to be localized below a certain (mobility edge) energy in 3D [8]. To understand the impact of localization on kinetics, consider first zero temperature: $T = 0$. Then the conductance $G \propto \exp(-L/L_0)$, where L is the length of the system and L_0 is the probability density localization length. Localization length (and therefore ℓn G) is a " well-behaved" quantity. Its fluctuations yield [8-10] the Gaussian distribution, and large fluctuations are exponentially (with L) little probable. However, in virtue of $\ell n\, G \propto -L$, large fluctuations may lead to exponentially higher conductance. So, the average conductance $<G>$ is exponential different [8,9] from G, and the probability distribution of G has a large tail. At first glance this result looks somewhat academic, since exponentially rare samples have no chances to be observed in real life when L is large. Rather, one should consider [8,9] only a "representative" average over a reasonable number of samples, thus cutting off unreasonably rare fluctuations. However, one may take exceptions to this general statement on three points. First, in small systems L/L_0 may be large, but "reasonable". Then , highly "inordinary" samples may indeed be observed. Second, even in "representative" samples the fluctuations ΔL_0 of L_0 are $\Delta L_0/L_0 \sim \sqrt{L_0/L}$. So, their conductances differ exponentially [11], by a factor $\sim \exp(a\sqrt{L/L_0})$, where $a \sim 1$. And, most important, third - the physical meaning of the conductance maxima. They are not incidental, but rather related to the resonance tunneling through localized states [3,5]. Their number equals the number of localized states and the distance between them is $\propto 1/L$. Their conductance is related to the situation of the eigenstates, i.e. of the corresponding fluctuations in the impurity concentration. In agreement with the exponentially small probability of a large fluctuation, the energy width of the conductance maxima is exponentially small (and related to the position of the corresponding eigenstates). So, a randomly chosen Fermi energy always yields a representative conductance. However, an eigenenergy is one of the major intrinsic characteristics of a system rather than just an arbitrary energy. It manifests itself at finite temperatures.

Of course, sufficiently high temperature may wipe out all these eigenstate fluctuations [10]. But a sufficiently low temperature actually allows for their experimental observation. The role of temperature T is twofold. First, it leads to the population of various energies and to their contribution into the conductivity. The conductance G(T) is the "thermally weighted energy average"[10a]. This makes the "theoretical average" meaningful (although amended by the thermal weight). The competition between the thermal population (which decreases with the distance $\Delta\epsilon$ from the Fermi energy) and the best conducting eigenstate (the larger is the energy interval $\Delta\epsilon_F$, the larger is the number of contributing eigenstates and the better can be the choice) selects the resonance eigenenergy, which dominates G(T). Naturally, G(T) depends on the eigenstates in the immediate vicinity of the Fermi energy ϵ_F. This leads [6] to the uniqueness of G(T) for every system and each ϵ_F.

Another implication of finite temperature is the possibility of (inelastic) Mott hops from one localized eigenstate to another. A finite lifetime in the eigenstate leads to its finite "inelastic" width and thus to the broadening of tunneling resonances.

A. Fowler was probably the first to emphasize that at low temperatures the number of important Mott hops in the whole system may not be too large, and that this may lead to large fluctuations in the conductance. P. Lee demonstrated by the direct computer experiment [7] that the Mott hopping in a finite 1D system yields huge irregular conductance oscillations. Since successive hops form the "series of resistances", the total resistance R(T) revives the idea of the "average resistance". Since the 1D Mott resistances are exponentially large with 1/T and exponentially different from each other, at low enough T one of them dominates the total resistance of the system. This quenched hop is the flucton. Its change with ε_F leads to giant oscillations of $R(\varepsilon_F)$. (The Fermi energy ε_F may be changed by the gate voltage, magnetic field ect.). Resonance tunneling is the elastic coherent transmission through a specific localized eigenstate. The Lee oscillation is related to the inelastic incoherent transition between specific localized eigenstates; its resonance nature is explicit in the R resonances at the corresponding r.f. frequencies or magnetic fields. In the next section I demonstrate the features of resonance tunneling and transition and speculate that these features are common for flucton kinetics whatever is its specific origin.

3. Localized Eigenstates, Resonances and the Lee Oscillations

Two cases allow for the accurate analytical calculation: resonance tunneling with no inelastic scattering [12,5c] and the Mott inelastic hopping [7,17] in the absence of resonance tunneling. Their refinements (inelastic scattering in resonance tunneling, resonance tunneling in the Mott hopping) pose considerable and mostly still unresolved problems. So, to suggest a general approach, I present the exact results [12,5] in a kind of a general "uncertainty principle". The resonance transmission coefficient t for a monochromatic electron beam and the eigenstate width $\delta\varepsilon$ satisfy the relation

$$t\delta\varepsilon^2 \sim t_{nr}\Delta\varepsilon_0^2 . \tag{1}$$

So, $t\delta\varepsilon^2 \sim$ const. Outside the resonance t and $\delta\varepsilon$ reduce to non-resonant transmission coefficient t_{nr} and to the interlevel spacing $\Delta\varepsilon_0$ in a given localization well (whose size is the wave function localization length $2L_0$) . The simplest case of Eq. (1) is related to the resonant tunneling through a potential well between two nearly opaque potential barriers.

A non-monochromatic ("nm") beam with the characteristic energy width ε_{nm} yields [5,12] the "uncertainty relation" between the resonant total transmission coefficient t_{nm} and the total eigenstate width $\delta\varepsilon$ (inelastic width included):

$$(t_{nm}\delta\varepsilon)\varepsilon_{nm} \sim (t_{nr}\Delta\varepsilon_0)\Delta\varepsilon_0 . \tag{2}$$

By Eq. (1), this reduces to

$$t_{nm}\varepsilon_{nm} \sim t^{1/2} t_{nr}^{1/2}\Delta\varepsilon_0 . \tag{3}$$

Resonances exponentially die out when the eigenstate width $\delta\varepsilon$ smears out the interlevel spacing $\Delta\varepsilon$ in the whole system and makes eigenstates indistinguishable, i.e. when

$$\delta\varepsilon > \Delta\varepsilon \sim \frac{1}{\rho_D V_D} . \tag{4}$$

Here ρ_D and V_D are correspondingly the volume of the D-dimensional system and its density of states.

The eigenstate width is related to the hopping or tunneling from one localized state to another. In both cases $\delta\varepsilon$ may be exponentially small, and $<< \Delta\varepsilon$, although $\Delta\varepsilon << \varepsilon_{nm}$ . . This is a characteristic feature of localized eigenstates; "conventional" resonances yield

$\delta\varepsilon \sim \varepsilon_{nm}$ and, thus, by Eq. (4), die out when $\varepsilon_{nm} \sim \Delta\varepsilon$.

The strongest resonances may have t ~ 1. Then, by Eqs. (1,3)

$$\delta\varepsilon \sim t_{nr}^{1/2}\Delta\varepsilon_0;\ t_{nm} \sim t_{nr}^{1/2}\Delta\varepsilon_0/\varepsilon_{nm}\,. \tag{5}$$

The strongest resonances die out, by Eqs. (4,5) when

$$t_{nr}^{1/2} \sim \Delta\varepsilon/\Delta\varepsilon_0\,. \tag{6}$$

All the above relations look very general; they are also consistent with the Breit-Wigner formula [5a]. So one may believe that they are always valid.

The dimensionless (in the units of e^2/h) conductance G is

$$G \sim t\,. \tag{7}$$

This Landauer formula [13] is applicable [5a] to low-temperature conductance and (qualitatively) to the Mott hopping. In all cases, when inelastic diffusion between localized states is related to the Mott variable range hopping [14], t_{nr} is described by the formula

$$t_{nr} \sim \exp\left[-(T_0/T)^{\frac{1}{D+1}}\right], \tag{8}$$

where T_0 is the Mott temperature and D is the dimensionality. An exception to this formula is [15,16] a 1D system at relatively high temperatures $T > T_t$, where T_t may be presented in the form [17]:

$$T_t \sim T_0\,[\ell n\,(L/L_0)]^{-2}\,. \tag{9}$$

Then $\ell n\, G \propto -T^{-1}$. At temperatures lower than T_t the Mott Eq. (8) is valid, but with the renormalized Mott temperature. Whereas $T_0 \sim \varepsilon_F/(k_F L_0)^{-D}$ in $D \geq 2$, (ε_F and k_F are the Fermi energy and wave vector correspondingly), in 1D $T_0 \sim (\varepsilon_F/k_F L_0)\ell n\,(L/L_0)$ in a one channel case and $T_0 \sim (\varepsilon_F/k_F^2 L_0 w)\ell n\,(L/L_0)$ in a polychannel case.
Equations (5,7,8) yield:

$$\ell n G \propto -\tfrac{1}{2}(T_0/T)^{\frac{1}{D+1}}\,. \tag{10}$$

By Eqs. (6,8), resonances vanish at $T > T_t$, where

$$T_t \sim T_0[\ell n(2\Delta\varepsilon_0/\Delta\varepsilon)]^{-(D+1)} \tag{11}$$

Note that T_t (here and on temperature is measured in the energy units) may be larger than $\Delta\varepsilon$, in agreement with the general reasoning. In a 1D case $\Delta\varepsilon_0 \propto \frac{1}{L_0}$, $\Delta\varepsilon \propto 1/L$, and Eq. (11) is equivalent to Eq. (9).

The energy distance between resonances is determined similar to the variable Mott hopping and is [5c] $\sim \sqrt{T_0 T}$. Various (not the largest) resonances have [5a] $\ell n\, G \propto -(T^*/T_0)^{1/2}$, where $(T^*/T_0)^{1/2}$ are randomly distributed between 0.5 and 1.

When the Fermi energy ε_F moves away from the resonant eigenstate energy ε_r, then the resonant eigenstate population decreases $\propto \exp\,(-\,|\,\varepsilon_F - \varepsilon_r|\,/T)$. So, the slope of the resonant $G(\varepsilon_F)$ is $1/T$.

All these results, derived from the suggested "uncertainty relations", are valid for the Lee's oscillations. The energy distance between their maxima is related to the average energy hop, i.e. $\sim\sqrt{T_0T}$ The slope, in agreement with the Lee's numerical calculations [7], is [17] $1/T$. When $T < T_1$, the major contribution into G comes from a single (highest resistance) Mott hop [7,17].

The hop changes with ϵ_F. This leads to the giant oscillations. Thus, the predictions of the uncertainty relations do look very general. (Simultaneously, this complicates the verification of a specific theory). Not surprising, they agree with experiments [2,4a]. The difference between the resonance tunneling and the Lee oscillations lies outside the "uncertainty" relations. First, the energy of the resonance tunneling equals the eigenenergy and therefore is independent of temperature T, as long as the "best" eigenstate does not change with T. The maximum in the Lee oscillations is related to the change in the "decisive hop" and as a result is continuously T-dependent. On the average, it shifts by $\sim\frac{1}{4}\sqrt{T_0T}$. Second, certain hops lead [7] to T-independent plateaus in G.

Highly unusual features of the fluctuation kinetics are well demonstrated by the fluctuation change in the effective dimensionality of a system. The "conventional" Mott hoping changes its dimensionality when the average hopping distance becomes less than the width w of the system. However, a fluctuation may make the hopping distance somewhere in the system larger than w. If such a flucton dominates the resistance, the situation returns to effectively 1D, with $\ell n\, G \sim -(T_0/T)^{1/2} \propto -[\ell n\,(L/L_0)/wT]^{1/2}$. (Note the unusual G dependence on L and w). The calculation [17] demonstrates that this flacton-induced dimensionality change typically occurs in a certain temperature interval. For instance, 2D → 1D corresponds to

$$\epsilon_F L_0/k_F^2 w^3 < T < (\epsilon_F L_0/k_F^3 w^3)\ell n(L/L_0),\ (\epsilon_F/k_F^2 w L_0)/\ell n(L/L_0)\,. \tag{12}$$

A flucton, which changes the dimensionality of a specific system, may appear also in a certain region of ϵ_F Then one will observe the fluctuation change in the dimensionality in this region. Naturally the dependence of a flucton on ϵ_F and/or T (i.e. on the average energy hop) leads [6] to non-thermodynamic fluctuations in the T-dependence of G.

4. Fluctons and other Phenomena

Previous sections demonstrate that fluctons and fluctuation kinetics are a direct implication of localization. Therefore, everything that affects the localization and/or the Fermi energy is important for fluctuation kinetics. Eigenstate resonances and the Lee oscillations may show up in, e.g., the G dependence on the gate voltage or magnetic field. Highly monochromatic a.c. field may lead to the transitions between localized eigenstates and to the corresponding impedance resonances at certain frequencies. Exponentially large hopping times may lead to significant difference between d.c. and a.c. behavior at very low frequencies. Possibly, 1/f noise and its oscillations [18] are related to those frequencies. High sensitivity to the energy change implies non-linearity in very weak electric fields [18a,7].

Localization is characteristic in low-dimensional and/or sufficiently random media of any waves: acoustic [12], electromagnetic [12,19], hydrodynamic [20]. For instance, in 2D a dielectric constant $\epsilon = \epsilon(x,z)$ allows for the TE wave $E_x = E_z = 0$, $E_y = E_y(x,z)\exp(i\omega t)$. The equation for E_y reads:

$$\left(\frac{\partial^2}{\partial x^2} + \frac{\partial^2}{\partial z^2}\right)E_y + \frac{\omega^2 \epsilon_d}{c^2}E_y = 0\,; \tag{13}$$

with $E_y = 0$ at the edges. It is equivalent to the 2D Schroedinger equation with the energy $W = \hbar^2\omega^2 \max_{x,z} \varepsilon_d(x,z)/2m^*c^2$, in the potential $U = W - (\hbar^2\omega^2\varepsilon_d/2m^*c^2)$. Thus, it yields localized solutions. Localization is strong if energy losses are low.

5. Summary

Fluctuation kinetics and fluctons are characteristic of strong localization, i.e. of low dissipation phenomena in random media. In particular, fluctons determine the conductance at temperatures

$$T < T_t, T_t \sim T_0[\ell n(V_D/L_0^D)]^{-(D+1)}, \quad (14)$$

where V_D is the volume of D-dimensional system with the localization length L_0, and T_0 is the Mott temperature. Temperatures $T < T_t$ may be much larger than the interlevel spacing in the system.

At such temperatures the conductance G as a function of the Fermi energy has extraordinary high and narrow maxima, whose relative height is exponential with $(T_0/T)^{\frac{1}{D+1}}$. The average height of the highest maxima is $\propto \exp\left[0.5(T_0/T)^{\frac{1}{D+1}}\right]$. The G is supersensitive to the number of electrons in the system, because an extra electron, which goes to the eigenstate, localized at a different place, yields a different effective Mott temperature. The energy distance between the G maxima is $\sim T_0(T/T_0)^{\frac{1}{D+1}}$. The slope of $G(\varepsilon_F)$ is $1/T$.

References

1. W.E. Howard, F.F. Fang, Solid State Electronics 8, 82 (1965), and J.A. Pals, W.J.J.A. Van Heck, Appl. Phys. Letters 23, 550 (1973), and J.M. Voschenko, J.N. Zemel, Phys Rev. 9B, 4410 (1974), and R.J. Tidey, R.A. Stradling, M.Pepper, J. of Phys. C7, L353 (1974), and G. Voland, K. Pagnia, Appl. Phys 8, 211 (1979), and M. Pepper, J. Phys. C12, L617 (1979), and Phil Mag. B42, 947 (1980), and Surf. Phys. 98, L218 (1980), and M. Pepper, M.J. Wren, R.E. Oakley, J. Phys. C12, L897 (1979), and M. Pepper, M.J. Wren, J. Phys. C15, L617 (1982).

1a. I.M. Lifshitz, V. Ya Kirpichenkov, Sov. Phys. JETP 50, 499 (1979).

2. A.B. Fowler, A. Hartstein, R.A. Webb, Phys. Rev. Lett. 48, 196 (1982).

3. M. Ya. Azbel', P. Soven, Phys. Rev. B27, 831 (1983).

4. W.J. Skocpol, L.D. Jackel, E.L. Hu, R.E. Howard, L.A. Fetter, Phys. Rev. Lett. 49, 951 (1982), and R.G. Wheeler, K.K. Choi, A. Goel, R. Wisnieff, D.E. Prober, Phys. Rev. Lett. 49, 1674 (1982) and R.F. Kwasnick, M.A. Kastner, J. Melngailis, P.A. Lee, Phys. Rev. Lett. 52, 224 (1984).

4a. A.B. Fowler, A. Hartstein, R.A. Webb, Physica 117B-118B, 661 (1983), and A. Hartstein, R.A. Webb, A.B. Fowler, J.J. Wainer, Surf. Science, to be published.

5. M. Ya. Azbel', Sol. St. Comm. 45, 527 (1983).

5a. M. Ya Azbel', A. Hartstein, D.P. DiVincenzo, Phys. Rev. Lett. 52, 1641 (1984).

5b. A.D. Stone, M. Ya. Azbel', P.A. Lee, unpublished.

5c. M. Ya. Azbel', D.P.DiVincenzo, Sol. St. Comm. 49 949 (1984) and to be published.

5d. M. Ya. Azbel', M. Rubinstein, Phys. Rev. Lett. 51, 836 (1983).

6. D.P.DiVincenzo, M.Ya. Azbel', Phys. Rev. Lett. 50, 2102 (1983).

7. P.A. Lee, to be published.

8. E. Abrahams, P.W. Anderson, D.C. Licciardello, T.V. Ramakrishnan, Phys. Rev. Lett. 42, 673 (1979).

9. P.W.Anderson, D.J. Thouless, E. Abrahams,D.S. Fisher, Phys. Rev. B22, 3519 (1980), and P.W. Anderson, Phys. Rev. B23, 4828 (1981).

10. A.D. Stone, J.D. Joannopoulos, Phys. Rev. B25,1431 (1982).
10a. H.-L. Engquist, P.W. Anderson, Pnys Rev. B24, 1151 (1981), and M.Ya. Azbel', Phys. Lett. 78A, 410 (1980).
11. M.Ya. Azbel', Phys Rev. Lett. 46, 675 (1981).
12. M.Ya. Azbel', Phys. Rev. B28, 4106 (1983).
13. R. Landauer, Phil. Mag., 21, 863 (1970)
14. N.F. Mott, E.A. Davis, Electronic Processes in Non-crystalline Materials, Oxford Univ. Press, Oxford, 1971.
15. J. Kürkijarvi, Phys. Rev. B8, 922 (1973).
16. W. Brenig, G.H. Döhler, H. Hyszenau, Phil. Mag. 27, 1093 (1973).
17. M.Ya. Azbel', to be published.
18. R.F. Voss, J. Phys. C11, L923 (1978).
18a. M. Pollak, I. Riess, J. Phys. C9, 2339 (1976), and B.I. Shklovskii, Sov. Phys. Semicond. 10, 855 (1976).
19. U. Frisch, J.-L Gautero, to be published, and E. Akkermans, R. Maynard (preprint).
20. E. Guazzelli, E. Guyon, J. Physique-Lettres 44, L837 (1983).

Quasi One-Dimensional Transport in Narrow Silicon Accumulation Layers

C.C. Dean and M. Pepper*

Cavendish Laboratory, Cambridge, United Kingdom

*G.E.C. Research Laboratories, East Lane, Wembly, United Kingdom

A summary is presented of our work on conductivity and magneto-conductivity in narrow silicon accumulation layers. It is shown that when the inelastic length exceeds the sample width, one-dimensional localisation sets in. The temperature-dependence of the conductivity is in agreement with the theory of Thouless, and when the inelastic length exceeds the localisation length, conduction is by electron-electron scattering-assisted hopping. The strong localisation effect enhances the interaction correction, which is always two-dimensional.

1. Introduction

In this work we have used narrow accumulation layer devices on the (100) Si surface. The width of the channel could be varied by direct electrical means. The structure of the device is shown in Fig. 1, the substrate is 50Ωcm n-type and the n^+ and p^+ regions are formed by implantation. The SiO_2 is ≃ 1μm in thickness, this rather high value is due to the necessity of avoiding gate to $-p^+$ breakdown as the voltage on the p^+ regions is increased. The devices were 120μm length.

We attempted to use this structure with a p-type substrate and a doping gradient across the surface. The device would then initially turn on in a line in the centre, and the width could be altered with a substrate bias. However, it did not prove possible to obtain the necessary doping gradient and so the accumulation layer device was used. Care was taken to avoid movement of acceptors from the p^+ regions across the channel as this gave rise to strong conductance oscillations at temperatures up to ≃15K, like the oscillations in GaAs FET's [1-3] they did not alter position with the magnetic field. The devices used in this work only showed oscillations at ≃ 0.1K and these were slightly sensitive to a magnetic field. Experiments were performed on the conductivity and magneto-conductivity down to ≃ 0.1K in magnetic fields up to 12 Tesla.

The weak corrections to electron transport arising from quantum interference and the electron interaction are fairly well understood in three and two dimensions. [4-7]. In 3D the theory of the screening parameter F is not clear [8,9] but the temperature-dependence and magneto-resistance are in reasonable agreement with theory (excepting the possible electronic phase-transition in Si:Sb at very low temperatures) [10]. In 2D a magnetic field separates weak localisation and interaction corrections, and reasonable agreement is found between experiment and theory [11,12]. Decreasing $k_F l$ towards the strong localisation limit results in a large power law temperature-dependence of conductance [13]. It is not clear if the 2D wavefunctions fall away exponentially, as in scaling theory, or as a power law [14]. In this work the objective was to work in a regime where the relevant length scales exceed the sample width and a transition to 1D behaviour occurs [15,16]. As will be shown, this was ach-

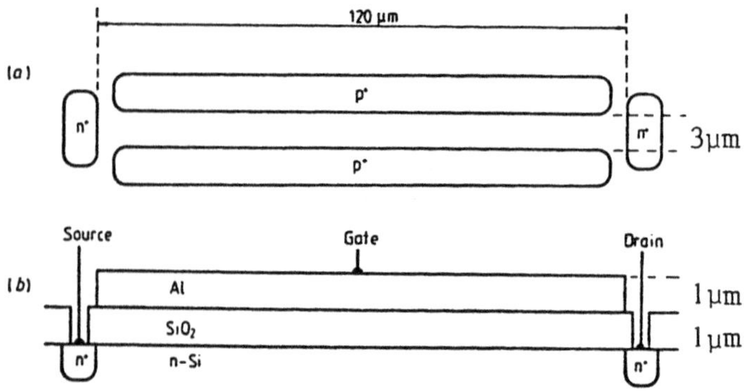

Fig. 1: Illustration of the device used in this work.

ieved with the inelastic diffusion length, but not the interaction length, thus a transition to 1D localisation was found with 2D interaction behaviour.

2. Theoretical Background

A system will show one-dimensional localisation if the inelastic diffusion length, L_i, exceeds the width; Thouless [17] showed that the localisation length L_o is the length of system for which the resistance is $\approx \hbar/e^2$. If the cross-sectional area is A, Thouless proposed that

$$L_o \simeq \frac{2A\,k_F^2\,l}{3\pi^2} \qquad (1)$$

where k_F and l are the Fermi wave vector and elastic mean free path respectively.

In this work we have investigated the transition to 1D behaviour in a conducting sheet. The length, L_o, of a sheet of Boltzmann conductor of width a and resistance $\hbar/e^2$ is given by:

$$\frac{e^2}{\hbar} = \frac{n_v e^2 k_F l}{2\pi\hbar}\,\frac{a}{L_o}\,, \qquad (2)$$

$$L_o = \frac{a n_v k_F l}{2\pi} \qquad (3)$$

where n_v is the valley degeneracy. If necessary, the factor n_v would also appear in the value L_o for the wire. The ratio of L_o (wire) to L_o (sheet) is now given, assuming the same values of k_F and l and writing $A = \pi a^2/4$, i.e. the diameter of the wire is the width of the sheet.

$$\frac{L_o\ (\text{wire})}{L_o\ (\text{sheet})} = \frac{k_F a}{3 n_v} \qquad (4)$$

If $n_v = 2$,

$$\frac{L_o\ (\text{wire})}{L_o\ (\text{sheet})} \simeq \frac{a}{\lambda_F} \qquad (5)$$

where λ_F is the wavelength at the Fermi energy. Thus, provided there is no strong one-dimensional quantisation of the density of states, i.e. one occupied sub-band, which is invariably the case, $\frac{a}{\lambda_F} >> 1$ and the one-dimensional effects are stronger in 2D than 3D systems because of the shorter localisation length. We note that in the system used here $\frac{a}{\lambda_F} \simeq 10 - 100$, typically $\simeq 50$ sub-bands are occupied although the electrons are in the ground state wave function in the direction normal to the $Si:SiO_2$ interface.

Thouless [18] has derived the following dependence of diffusion coefficient D on inelastic scattering time τ_i.

$$D = D_o \left[1 - 2 \frac{(D_o \tau_i)^{\frac{1}{2}}}{L_o} \tanh \left(L_o/2(D_o\tau_i)^{\frac{1}{2}}\right) \right] \tag{6}$$

D_o is the unperturbed (Boltzmann) diffusion coefficient.

For short inelastic times this gives the same first-order correction $\Delta\sigma_L$ to the Boltzmann conductance as other methods, i.e. Abrahams et al [4], Gorkov et al [5].

$$\Delta\sigma_L = \frac{-e^2 L_i}{\hbar} \tag{7}$$

At longer times $(D_o\tau_i >> L_o)$ $D \simeq L_o^2/12\tau_i$. This corresponds to a mode of conduction in which the electrons hop between localised states at a rate τ_i^{-1}, (the separation of states is much less than L_o). This mode of conduction has been predicted for both 2D and 1D transport [19,20] but the conditions for its observation in 2D are restrictive. In the system used here, the 1D regime can be entered with $k_F l >> 1$ and accessible temperatures, localisation lengths being up to a factor of 10^4 less than in 2D. The system has the advantage that measurements can be performed at higher temperatures where $L_i < a$ and the results can be compared to the well-established 2D theories for $k_F l > 1$.

We finally consider interaction effects, in this work the interaction length $(\hbar D/kT)^{\frac{1}{2}}$ was always much less than width. Consequently the appropriate correction $\Delta\sigma_I$ is given by

$$\Delta\sigma_I = \frac{\sigma}{k_F l} (1 - F) \ln (2\pi kT\tau h) \tag{8}$$

τ is the elastic scattering time and, for 2D transport in Si, $F \simeq 0.9$. The factor F becomes $\simeq 1/2$ in the presence of a magnetic field B such that $g\mu B > kT$. If $(\hbar D/kT)^{\frac{1}{2}} >$ width, the logarithmic correction is removed as the one-dimensional limit has been reached, a decrease in σ as $T^{\frac{1}{2}}$ is predicted, the correction $\Delta\sigma/\sigma_B$ is typically less than 10^{-3}.

3. Experimental Results

a. Sample Characterisation

In the absence of a squeezing voltage V_p we estimate the width of our channel to be $\simeq 1.5\mu m$ from device geometry. However, we wish to prove this and determine it in the presence of V_p; in all cases the width is a function of V_p and the gate voltage V_G. We also wish to know the carrier concentration accurately, and in order to determine this we plotted Shubnikov-de Haas oscillations by sweeping the magnetic field. The oscillations were observed at fields greater than $\simeq 1$ Tesla and the maxima and minima were periodic in B^{-1}. This enables us to extract the carrier concentration, n; for the values of V_{g_2} and V_p used in this work, n took values in the range $4 \cdot 10^{11} - 1.5 \cdot 10^{12} e^- cm^{-2}$, corresponding to values of Fermi energy in the range $\sim 2.5 - 10$meV.

The width of the accumulation layer can be estimated from the (perpendicular field) magneto-conductance in the temperature range 1.4K - 5K where the quantum interference is 2D. For values of $B > 1$ T the results can be fitted to the 2D expression of Hikami et al [21],

$$\Delta\sigma = \frac{n_v \alpha e^2}{2\pi^2 h} \left[\psi\left(\frac{1}{2} + \frac{\hbar}{4eBD\tau_i}\right) - \psi\left(\frac{1}{2} + \frac{\hbar}{4eBD\tau_i}\right) \right] \qquad (9)$$

ψ is the digamma function.

As we do not know the width of the system, a, we rewrite the above expression in terms of the conductance g and L_i, for $\tau < h/4eBD$ this becomes

$$\Delta g(B) - \Delta g(o) = \frac{a n_v \alpha e^2}{2\pi^2 \hbar L} \left[\psi\left(\frac{1}{2} + \frac{\hbar}{4eBL_i^2}\right) - \ln\left(\hbar / 4eBL_i^2\right) \right] \qquad (10)$$

requiring two independent fitting parameters $an_v\alpha$ and L_i.

Fig. 2 shows the agreement between theory and experiment with $V_g = 30$ volts, $T = 4.2$K, $an_v\alpha = 1.66\mu$m, and $L_i = 0.087\mu$m. We assume, by comparison with results on wider devices that $n_v\alpha = 1$, also suggested by Fukuyama's theory[22], hence $a = 1.66\mu$m considerably greater than L_i, expected on the basis of the fit to the 2D expression.

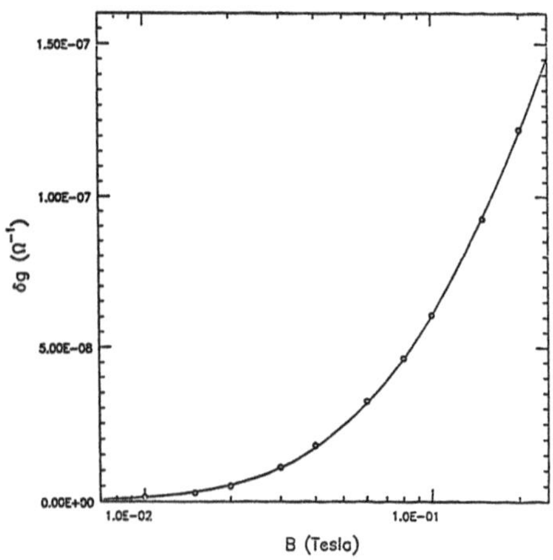

Fig. 2: Experimental magneto-conductance (dots) is plotted with the fitted theory (dashed line) for $V_g = 30$ volts, $V_p = 0$, $T = 4.25$K.

In order to calculate the 1D localisation length it is necessary to determine $k_F l$. We obtain this quantity from the flat value of the conductivity in the temperature region 1.4K - 5.0K, above this temperature it drops as a result of the temperature-dependence of elastic scattering. For $V_g = 30$ volts, quoted above, $k_F l \simeq 16$, a value considerably in excess of that used in two-dimensional localisation work.

b. ID Localisation

Fig. 3 shows the temperature-dependence of the conductance between 0.1K and 1.2K for the values of V_g stated. The measurements were at 70Hz using lock-in techniques and the values of electric field were sufficiently low to ensure Ohmic behaviour, being less than $10^{-2}Vm^{-1}$ at the lowest temperatures. The sheet conductivities were always greater than $10^{-5}\Omega^{-1}$ and if 2D behaviour occurred, the changes should be a few percent over this temperature range.

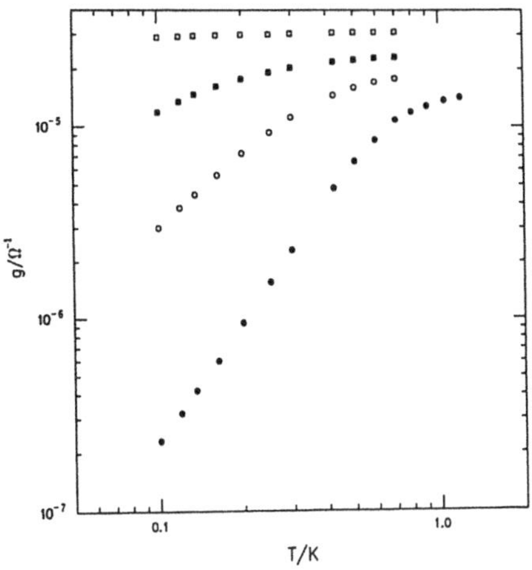

Fig. 3: Conductance versus temperature for V_p = 0, the gate voltage are • 30 volts, O 35 volts, ■ 40 volts and □ 50 volts.

In order to obtain further information on the transition from 2D to 1D behaviour we consider electron-electron scattering, which is the principal inelastic mechanism. For the values of $k_F l >> 1$ used here we can, for simplicity, neglect the disorder-induced T-term [23] and use the Landau-Baber expression; (We note that when L_i is of the order of the width the T correction is not present, the equivalent disorder-induced 1D correction is not applicable when the strong localisation is apparent),

$$\frac{1}{\tau_i} = \frac{A(kT)^2}{\hbar E_F} \quad (11)$$

where A is a constant of order unity. We would therefore expect a conductivity which varies as T^2 when conduction is by electron-electron scattering-induced "hopping", this is approximately found for V_g = 30 volts and T < 500mK. In this case A = 0.57 and L_i = 9.7μm at 0.2K considerably greater than the width, so pointing to 1D behaviour. Furthermore, we calculate L_o to be ≈8.0μm using the Boltzmann diffusivity, which is less than L_i so confirming that the 1D behaviour will be a strong and not a weak correction.

For V_g = 35 volts, a = 2.0 μm, $k_F l$ = 18 and L_o = 11.5 μm, if we use the Thouless expression with τ_i given by the Landau-Baber expression we find agreement for A = 3.7. This difference in A may result from the change in spatial overlap of electron states as L_o changes, the value of 3.7 is in good agreement with the value of 5 found in earlier 2D work.

We have not fitted higher values of V_g to theory as the measurements are not at sufficiently low temperatures. However, it is clear that results are in accord with theoretical expectation that the transition to scattering-induced hopping occurs at progressively lower temperatures as the width and $k_F l$ (hence L_o) increase.

Another interesting feature of the transition to one-dimensional behaviour was a large increase in the electron-phonon relaxation time, as determined from electron heating. This is discussed in detail elsewhere [16].

Finally, we point out that temperatures of the order of 0.5mK are required to observe variable range hopping in the conditions used here, this temperature corresponding to a typical energy-difference between spatially overlapping localised states.

c. Interaction Effects

As the interaction length $(hD/kT)^{\frac{1}{2}}$ is always less than the width, the interaction is always two-dimensional, and as $k_F l >> 1$, the correction is small. However, we find that the magneto-conductance is enhanced by the strong one-dimensional localisation. This will now be discussed.

The interaction correction $\Delta\sigma_I$ to the Boltzmann conductivity σ_B can be written

$$\frac{\Delta\sigma_I}{\sigma_B} = \frac{\hbar}{2\pi Dm} (1 - F) \ln (T/T_o) \tag{12}$$

T_o is a constant (for V_g = 30 volts, $T_o \sim 1.5K$), $F \simeq 0.9$ and the diffusion coefficient D is corrected for the presence of localisation. As F is near unity the interaction contribution to the temperature-dependence is negligible compared to that of localisation. If we consider the case of V_g = 30 volts, where $\sigma \alpha T^2$ at low temperatures, it is found that at 4.2K and 1.3K the magneto-conductance is due to the suppression of quantum interference and the values of $an_v\alpha$ are 1.77μm and 1.73μm respectively, i.e. in good agreement. However, at 0.4K the magneto-conductance changes sign above 0.1 Tesla with no observable effect below this field, the Shubnikov-de Haas periodicity is unchanged, indicating a constant carrier concentration.

We explain the results by suggesting that when $L_I > L_o$ the magnetic field acts as a minor perturbation on an exponentially decaying wavefunction, i.e. there is no quantum interpresence to be suppressed, and the principal effect of the magnetic field is due to interactions. In the weak perturbation limit Houghton et al [24] have shown that the incorporation of the localised D into the two-dimensional interaction expression is required to calculate the change in the Boltzmann conductivity. Here, the interaction correction is small, but the localisation is not weak. Nevertheless, using this procedure we find very good agreement between theory and experiment. The interaction magneto-conductance $\Delta g(B)$, when the localisation is taken into account, is given by (for $g\mu B > kT$).

$$\Delta g(B) = \frac{-F}{2} (g_o/g) \frac{an_v e^2}{2\pi^2\hbar L} \ln (B/B_o) \tag{13}$$

g_o is the Boltzmann conductance, g is the conductance, a and L are the sample width and length respectively, B_o is a constant. This expression is valid for B > 0.4 Tesla, this being the $g\mu B = kT$ condition, and predicts a gradient of $-6.7 \times 10^{-7}\Omega^{-1}$ which as shown in Fig. 4a is in very good agreement with the experiment, 4b shows the magneto-conductance at higher temperatures where the suppression of quantum interference is clear.

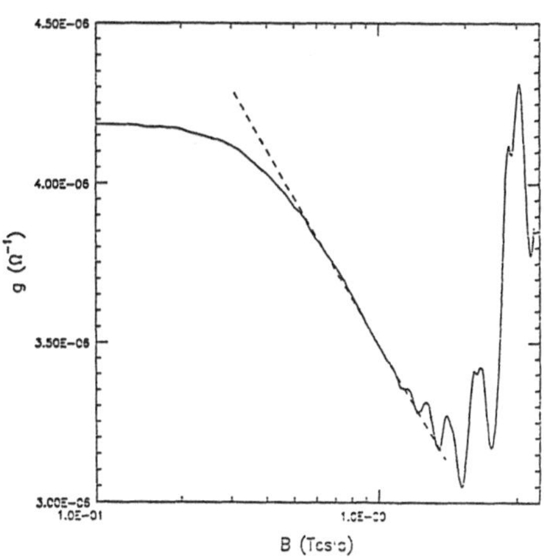

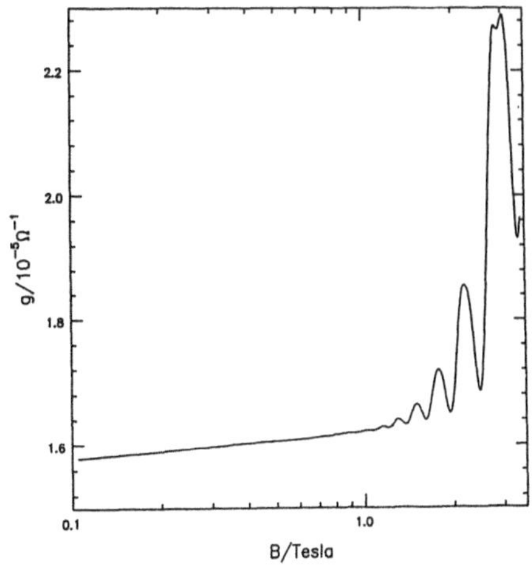

Fig. 4a: The perpendicular magneto-conductance for V_g = 30 volts and V_p = 0 is displayed for T = 0.4K, the dashed line is the gradient of the interaction effect discussed in the paper.

Fig. 4b: The perpendicular field magneto-conductance for V_g = 30 volts and V_p = 0 is displayed for T = 1.3K.

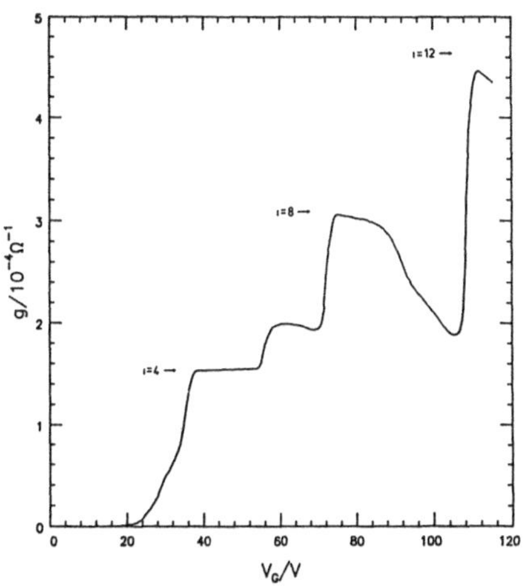

Fig. 5: The two terminal conductance is displayed as a function of gate voltage at T = 1.3K and V_p = 0 in a field of 10 Tesla.

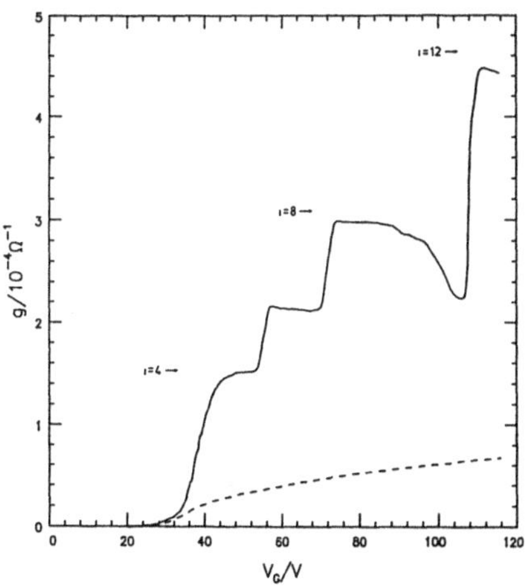

Fig. 6: The two terminal conductance is displayed versus gate voltage for T = 0.3K and V_p = 0 volts in a field of 10 Tesla. The conductance at B = 0 is shown as a dashed line.

d. Quantum Hall Effect

Previous work on two-dimensional systems showed that if all states below E_F were localised then the plateau of quantised Hall resistance disappears at low temperature [25].(This is in contrast to the normal case where the plateau becomes wider as the temperature drops). We have found the same behaviour due to the onset of one-dimensional localisation. The effect is shown in Figs. 5 and 6 where two terminal measurements are shown at 1.3K and 0.3K in a field of 10 Tesla. The i = 4 plateau is developed at 1.3K but has almost disappeared at 0.3K where, for these values of V_g, the conductivity is greatly reduced by the one-dimensional localisation. On the other hand, the i = 8 plateau develops further with decreasing temperature, the mobility is not sufficient for observation of the valley gap plateaux. This result indicates that the mechanism producing the localisation is not important for the removal of the quantisation, it also raises interesting questions concerning the existance of the QHE in a Hall geometry device as the temperature approaches zero.

Acknowledgements

This work was supported by S.E.R.C. and C.C.D. acknowledges an S.E.R.C. studentship. We would like to thank Professors Sir Nevill Mott, D.J. Thouless and M. Kaveh, Dr. W.J. Skocpol, Mr. M.C. Payne and Mr. R.P. Upstone for discussions.

References

1. Pepper, M. Microfabrication 4, 5. Published by Rutherford and Appleton Laboratories (1981)
2. Fowler, A.B., Hartstein, A. and Webb. R.A. Phys. Rev. Lett. 48, 196 (1982)
3. Pepper, M. J. Phys. C 12m K 617 (1979)
4. Abrahams, E., Anderson, P.W., Licciardello, D.C. and Ramakrishnan, T.V. Phys. Rev. Lett. 42, 673 (1979)
5. Gorkov, L.P., Larkin, A.I. and Khmelnitzkii, D.E. Jetp. Lett. 30, 228 (1979)
6. Kaveh, M. and Mott, N.F. J. Phys. c 14, L177 (1981)
7. Altshuller, B.L. Khmelnitzkii, D.E., Larkin, A.I. and Lee, P.A. Phys. Rev. B22, 5142 (1980)
8. Rosenbaum, T.F., Andres, K., Thomas A.S. and Bhatt, R.N. Phys. Rev. Lett. 45, 1723 (1980)
9. Long. A.P. and Pepper, M. J. Phys. C. 17, 3391 (1984)
10. Long, A.P. and Pepper, M. J. Phys. C. 17, L425 (1984)
11. Uren, M.J. Davies, R.A., Kaveh, M. and Pepper, M. J. Phys. C. 14. L395 (1981)
12. Bishop, D.J., Tsui, D.C. and Dynes, R.C. Phys. Rev. B26, 773 (1982)
13. Davies, R.A. and Pepper, M. J. Phys, C 15, L371 (1982)
14. Davies, R.A., Pepper, M. and Kaveh, M. J. Phys. C 16, L285 (1983)
15. Dean. C.C. and Pepper, M. J. Phys. C. 15, L187 (1982)
16. Dean, C.C. and Pepper, M. J Phys. C. In the press
17. Thouless D.J. Phys. Rev. Lett. 39, 1167 (1977)

18. Thouless, D.J. Solid State Comm. 34, 683 (1980)
19. Gogolin, A.A. Phys. Reports 86, 1 (1982)
20. MacKinnon, A. and Kramer, B. Z. Phys, B53, 1 (1983)
21. Hikami, S., Larkin, A.I. and Nagaoka, Y. Prog. Theor. Phys. 63, 707 (1980)
22. Fukuyama, H. Surface Sci. 113, 489 (1982)
23. Davies, R.A. and Pepper, M. J. Phys. C. 16, L353 (1983)
24. Houghton. A., Senna, J.R. and Ying, S.C. Phys. Rev. B25, 2169 (1982)
25. Pepper, M. and Wakabayashi, J. Phys. C 15, L 816 (1982)

Experiments on Localization and Interaction in a Strong Magnetic Field

G. Ebert

Physik-Department, Technische Universität München
D-8046 Garching, Fed. Rep. of Germany

1. Introduction

The observations of the quantum Hall effect (QHE) by v. KLITZING, DORDA, and PEPPER [1] and the fractional quantum Hall effect (FQHE) by TSUI, STÖRMER, and GOSSARD [2] have initiated much experimental and theoretical work in the area of two-dimensional systems in a strong magnetic field. Especially localization and interaction phenomena, which are considered to be essential for these two effects, are treated theoretically by many authors. Nevertheless, up to now, the QHE and in particular the FQHE are far from being understood.

In this paper, we review recent experimental work concerning localization and interaction phenomena in two-dimensional electron systems in a strong magnetic field, with emphasis on the integer QHE and the FQHE. In section 2, the fundamental physics of a two-dimensional electron gas in a strong magnetic field is briefly recalled. Section 3 reviews experiments related to localization phenomena carried out on silicon inversion layers as well as on GaAs-$Al_xGa_{1-x}As$ heterostructures. Finally, some recent experiments on the FQHE which is considered to originate from Coulomb interaction, are discussed in section 4.

2. Two-Dimensional Electrons

A two-dimensional electron gas (2DEG) exists e.g. at the Si-SiO_2 interface of a Si-MOSFET or at the interface of a GaAs-$Al_xGa_{1-x}As$ heterostructure, at present the two most important devices for the investigation of the electronic properties of two-dimensional systems.

In both cases, a large electric field confines the electrons within a narrow surface potential well, and the energy for their motion perpendicular to the interface becomes quantized in electric subbands E_0, E_1, E_2,... with typical energy separations of about 20 meV. Sometimes the number of electrons per unit area N_s and thereby the Fermi energy E_F can be varied by applying a gate voltage V_g across the sample. At low temperatures and not too high carrier densities N_s only the lowest subband E_0 is occupied, and the motion of the electrons perpendicular to the interface is completely suppressed. On the other hand, the motion parallel to the interface remains free (characterized by the effective mass m^*) and the carriers behave like a strictly two-dimensional electron gas.

In the presence of a strong magnetic field B ($\mu B > 1$, μ = electron mobility) perpendicular to the two-dimensional system, the electrons execute cyclotron motion with a cyclotron frequency $\omega_c = eB/m^*$, and the energy spectrum splits into discrete Landau levels

$$E_{n,s,v} = E_o + (n + \frac{1}{2})\,\hbar\omega_c + s\,\Delta s + v\,\Delta v \qquad (1)$$

where $n = 0,1,2,\ldots$, $s = \pm 1/2$, and $v = \pm 1/2$ denote the Landau, spin, and valley quantum numbers, respectively. In (1) also the spin-splitting $\Delta s = g\mu_B B$ (g = Landé factor, μ_B = Bohr magneton) and the valley-splitting Δv (only present in many-valley semiconductors) is taken into account. Each of these levels (n,s,v) has a degeneracy $N_L = eB/h$ (number of states per unit area). The filling factor i is determined by the equation

$$i = \frac{N_s}{N_L} = \frac{N_s h}{eB}\,. \qquad (2)$$

If the ratio N_s/B is chosen in such a manner (either by varying N_s via the gate voltage V_g of by varying the magnetic field B) that i is an integer, the scattering probability at $T = 0$ K should go to zero, since the empty states are separated by an energy gap from the occupied states. Qualitatively, a conductivity as shown in Fig. 1a is expected which can be explained as a diffusion of the center of the cyclotron orbits due to scattering processes. The experiments show that a vanishingly small conductivity σ_{xx} is observed, not only at discrete carrier densities but within a certain range of carrier densities around $N_s = iN_L$ (the valley-splitting for $n = 1$ is not completely resolved at the magnetic field of 19 T). This is explained by the existence of localized states in the tails of the Landau levels which do not contribute to the conductivity σ_{xx}.

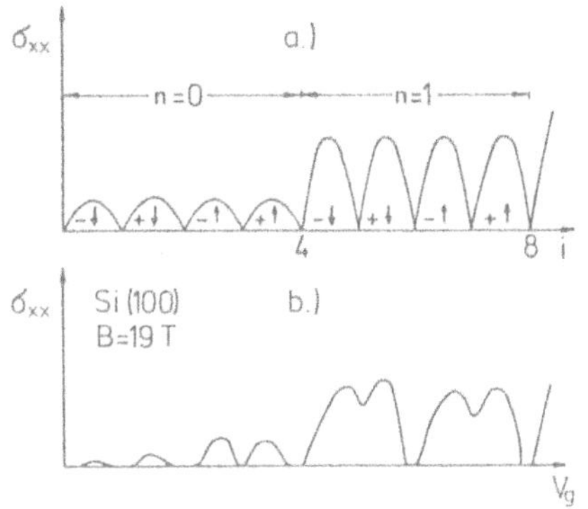

Fig. 1: Conductivity σ_{xx} in a strong magnetic field B as a function of the filling factor i or the gate voltage V_g respectively, for the first two Landau levels $n = 0$ and 1. a) σ_{xx} as it is expected theoretically for the case of short range scatterers. b) Corresponding experimental data for a (100)Si-MOSFET at T = 1.5 K. The signs +, -, and ↑,↓ characterize the valley- and spin levels.

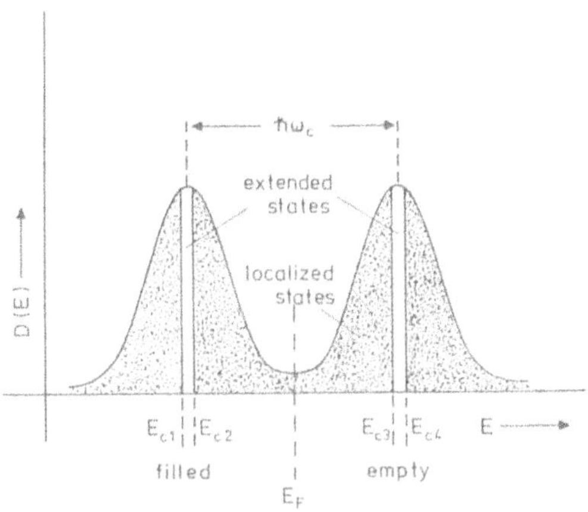

Fig. 2: Simple model for the density of states $D(E)$ of two Landau levels in the presence of disorder. Mobility edges E_c separate localized and extended states.

A qualitative explanation of the conductivity measurements is possible using a simple model for the density of states, as shown in Fig. 2. The localized tail states are separated from the extended states in the center by mobility edges E_c. A more detailed analysis of the energy distribution of extended and localized states is a subject of present theoretical work. ONO [3] used a self-consistent diagrammatic method to calculate the electron localization length for short-range scattering potentials and found that only the states in the vicinity of the center of the broadened Landau levels are practically extended. Qualitatively similar results were obtained by ANDO [4] applying a numerical Thouless number study, as well as by KRAMER and MACKINNON [5] and by AOKI and ANDO [6] using finite size scaling approaches, but quantitatively there are discrepancies. In the case of long range scatterers ($\ell \ll d$, $\ell = \sqrt{\hbar/eB}$ = magnetic length, d = range of scattering potentials) semi-classical percolation theory has been applied [7] and has led to an infinitely narrow energy-range where percolation is possible, but quantum effects may influence electron localization. In addition, it is not clear whether edge states [8,9], induced by the sample boundaries, play an important role as a source of extended states.

For an ideal system, with no scattering and without localized states, the classical behavior for the Hall resistance $R_H = B/N_s e$ is expected. If exactly i levels are completely filled with electrons, the Landau degeneracy eB/h leads to the well-known expression for the quantized Hall resistance

$$R_H = \frac{h}{ie^2} \tag{3}$$

It is shown theoretically [10-12] that the occupation of localized states does not change the Hall resistance, so that R_H shows plateaus with quantized values according to (3), if the magnetic field or the carrier density is varied. Simultaneously the conductivity σ_{xx} (no current in the electric field direction) and the resistivity ρ_{xx} (no voltage drop in the direction of the current) are zero. Experimentally, measurements on Hall devices (current direction defined by the geometry of the sample) are always related to the components ρ_{xx} and ρ_{xy} (in two dimensions R_H is identical with ρ_{xy}) of the resistivity tensor, whereas measurements on Corbino devices give information about the component σ_{xx} of the conductivity tensor. The different components are related by

$$\sigma_{xx} = \frac{\rho_{xx}}{\rho_{xx}^2 + \rho_{xy}^2} \qquad \sigma_{xy} = -\frac{\rho_{xy}}{\rho_{xx}^2 + \rho_{xy}^2} \tag{4}$$

3. Experiments Related to Localization

The number of localized states is closely related to the width of the plateaus in the Hall resistance. This idea is supported by the fact that the plateau width decreases with increasing sample mobility (due to a reduction of disorder), as shown in Fig. 3. The mobilities of the two devices are about 10^5 cm^2/Vs and 10^6 cm^2/Vs, respectively. At finite temperatures the plateau width is always reduced by thermal activation processes. Experiments carried out in the mK-temperature range [13-15] indeed show a dramatic increase of the plateau width. In Fig. 4, a nearly step-like behavior of ρ_{xy} and sharp peaks in ρ_{xx} are visible. The data which is taken from a GaAs-$Al_xGa_{1-x}As$ heterostructure (Hall device) at 8 mK indicates that the amount of localized states can be very large.

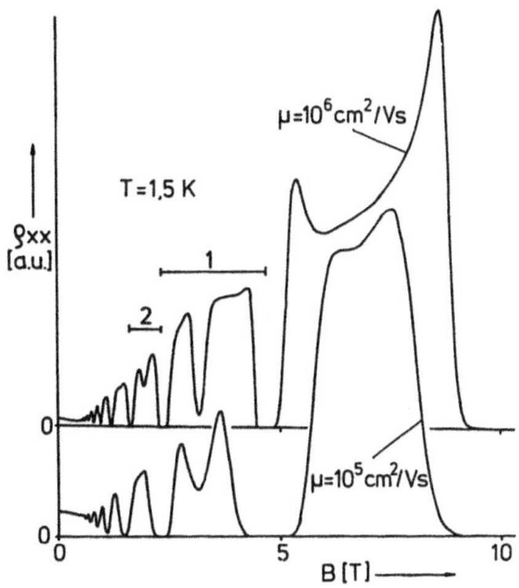

Fig. 3: Resistivity ρ_{xx} at T=1.5 K for two different GaAs-$Al_xGa_{1-x}As$ heterostructures as a function of the magnetic field. The two devices have nearly the same carrier densities but different carrier mobilities.

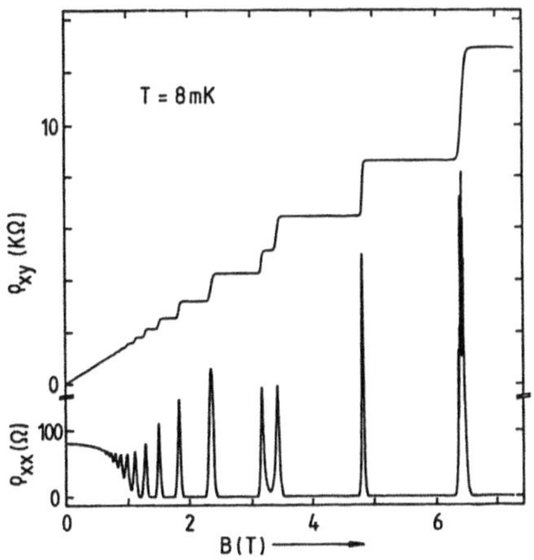

Fig. 4: ρ_{xx} and ρ_{xy} at T=8 mK for a GaAs-$Al_{0.29}Ga_{0.71}As$ heterostructure with a mobility of about 10^5 cm^2/Vs as a function of the magnetic field.

The result of a more detailed analysis [14] of the temperature-dependence of the Hall resistance is shown in Fig. 5. Here, complementary to the plateau width, the inverse slope of the Hall resistance $1/(d\rho_{xy}/dB)$ between two adjacent plateaus normalized to the corresponding classical value N_se/B for different samples with mobilities of about 40 000 cm^2/Vs and different quantum numbers n is plotted as a function of temperature. This quantity,which gives the number of extended states relative to the degeneracy of one Landau level,decreases with decreasing temperature. The important fact, however, is that an extrapolation to T = 0 leads to a finite value. This supports the theoretical suggestion that the range of extended states is very narrow - only a few percent of the states of one Landau level remain extended -, but that even at T = 0 not all the states are localized. However, experimental errors due to remaining pick-up signals which raise the electron temperature above the lattice temperature cannot be completely excluded.

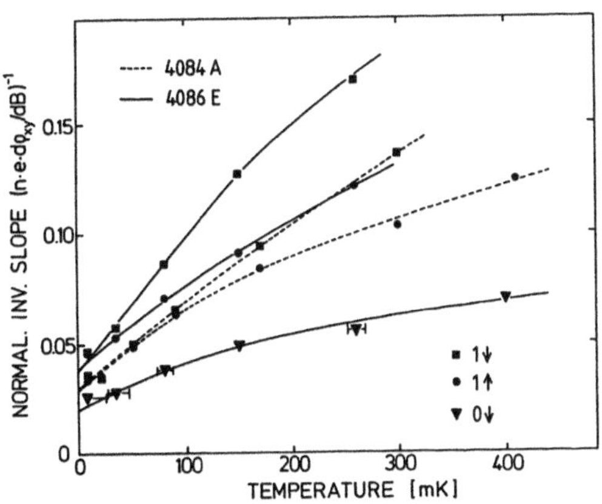

Fig. 5: Experimental data for the ratio of the classical slope $1/N_se$ to the slope $d\rho_{xy}/dB$ between adjacent plateaus of the Hall resistance for different Landau levels as a function of temperature. The two heterostructures have mobilities around 40 000 cm^2/Vs.

If the Fermi energy is located in the energy gap (or at least in a mobility gap) between two adjacent Landau levels (see Fig. 2), thermal activation processes cause a finite resistivity ρ_{xx} at non-zero temperature. Under symmetric conditions ($|E_F - E_{n+1}| = |E_F - E_n|$) a thermally-activated resistivity

$$\rho_{xx}(T) \sim \exp\left(-\frac{E_G}{2kT}\right) \qquad (5)$$

is expected with an activation energy E_G corresponding to the energy separation between the highest filled and the lowest empty Landau level. Such an activated behavior has been found for different two-dimensional systems, and Fig. 6 shows a typical result for a Si-MOSFET. The activation energy is within the experimental uncertainty proportional to the magnetic field (it should be noted that at much lower temperature deviations from the activated behavior occur which will be discussed separately). The measured E_G-values are close to the cyclotron energy $\hbar\omega_c$ as expected for a system where the spin- and valley-splitting Δs, Δv are negligibly small. This is not the case for Si-MOSFETs where already the spin-splitting is at least 20% of the cyclotron energy (see (1), with $g \gtrsim 2$). This means that the experimentally-determined activation energy is larger than expected from the energy spectrum calculated within the one-electron approximation. Similar results are obtained with GaAs-$Al_xGa_{1-x}As$ heterostructures too, and the activation energy in units of $\hbar\omega_c$ (with $m^* = 0.07\ m_e$) is plotted in Fig. 7 for different samples. The enhancement of the energy-splitting may originate from the exchange effect among electrons [16]. At very low temperatures, deviations from the activated conductivity are observed which are usually attributed to variable range-hopping (VRH). For a two-dimensional system with a wavefunction $\psi \sim \exp(-\alpha r)$ for the localized states, a temperature-dependence

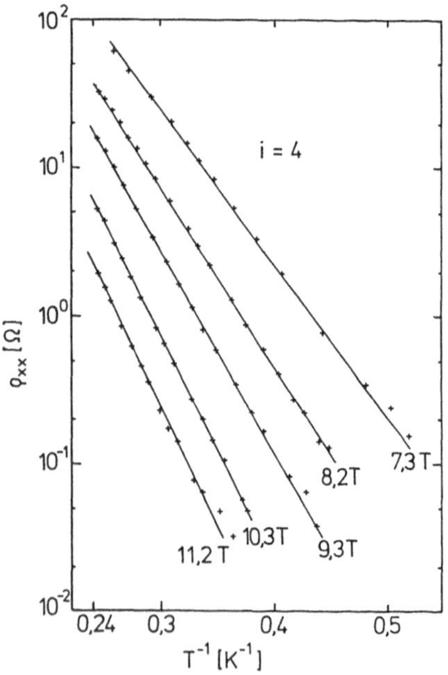

Fig. 6:
Temperature-dependence of ρ_{xx} at filling factor i = 4 for a (100)Si-MOSFET at different magnetic fields.

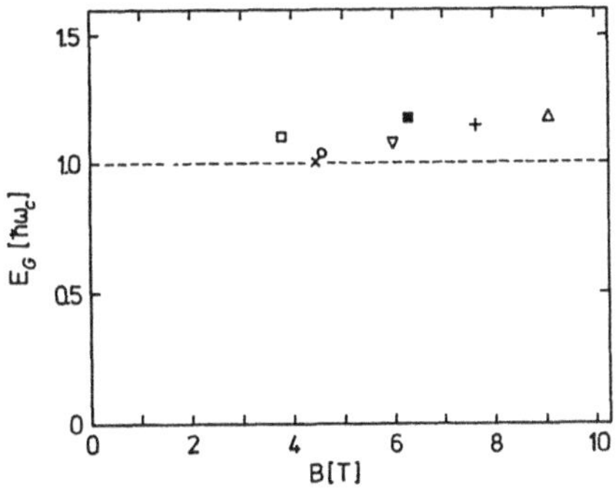

Fig. 7: Activation energies E_G deduced from the temperature-dependence of the resistivity ρ_{xx} according to (5) for different GaAs-$Al_xGa_{1-x}As$ heterostructures and filling factors i = 2 and 4 as a function of the magnetic field ($m^* = 0.07\ m_e$).

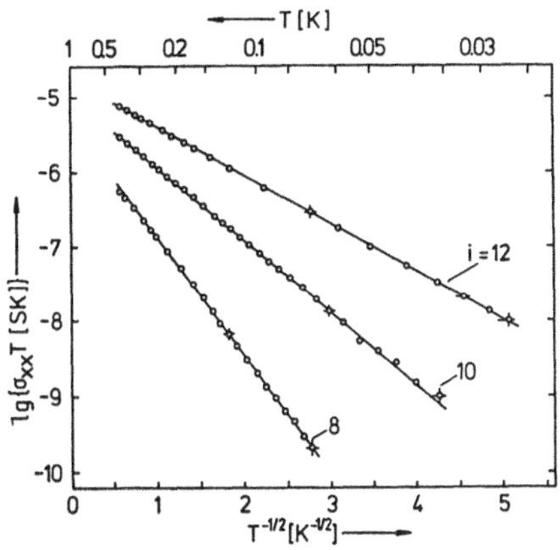

Fig. 8: Temperature-dependent conductivity σ_{xx} of a GaAs-$Al_xGa_{1-x}As$ heterostructure for different filling factors i and temperatures below 0.5 K.

$$\sigma \sim T^{-2/3} \exp - (T_0/T)^{1/3} \tag{6}$$

is expected [17]. In strong magnetic fields, however, the localized states wavefunction may decay according to $\psi \sim \exp(-r^2/2\ell^2)$ and the conductivity should change with temperature like

$$\sigma \sim T^{-1} \exp - (T_0/T)^{1/2} \tag{7}$$

as it is predicted by ONO [18]. Our experimental data are well-described by (7) as shown in Fig. 8 for different conductivity minima, whereas an interpretation on the basis of (6) was not possible [19]. The same temperature-dependence was found [15] for the 2DEG in $In_xGa_{1-x}As$-InP heterostructures too, but it should be noted that also a behavior according to (6) was observed by other authors [20]. EFROS and SHKLOVSKII [21] pointed out that a Gaussian-like localization cannot be used for the calculation of long-range hopping and that a Coulomb gap at the Fermi energy may be responsible for the observed temperature-dependence $\sigma \sim \exp - (T_0/T)^{1/2}$.

The contribution of variable range hopping processes to the Hall effect is negligibly small as deduced from an analysis of the deviation $\Delta\rho_{xy}$ of the ρ_{xy}-data from the quantized value as a function of temperature [22]. Figure 9 shows clearly that in the temperature range where ρ_{xx} is dominated by VRH, the deviation $\Delta\rho_{xy}$, which is proportional to the slope $d\rho_{xy}/dB$, shows activated behavior as expected, if the ρ_{xy}-component for hopping conductivity is negligibly small compared to the contribution of the extended states. Theoretical calculations [23] have shown that the contribution of VRH to ρ_{xy} can be described by a law like (7), but the prefactor and the value $1/T_0$ are much smaller than the corresponding values for ρ_{xx}, so that $\Delta\rho_{xy}$ is not strongly influenced by variable range hopping.

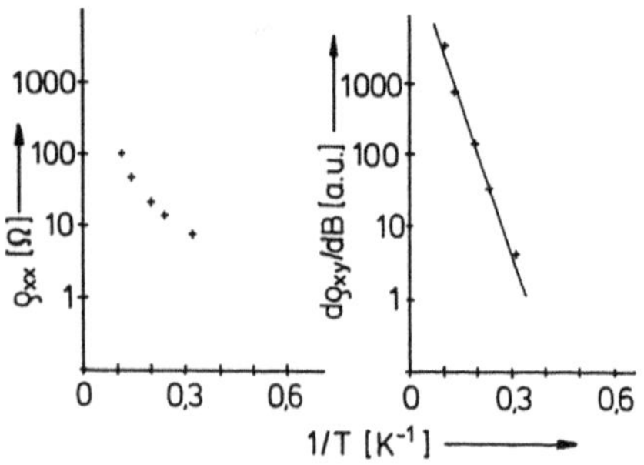

Fig. 9: Temperature-dependence of ρ_{xx} at a filling factor i = 4 and the corresponding slope of the Hall plateau $d\rho_{xy}/dB$ for a GaAs-$Al_xGa_{1-x}As$ heterostructure

4. Experiments Related to Interaction

Although many efforts were made in the last decade looking for electron-electron interaction-induced phenomena in inversion layers, the first experiment showing unambiguously the presence of such collective effects was the recent observation of the fractional quantum Hall effect (FQHE) by TSUI, STÖRMER, and GOSSARD [2]. In contrast to the integer QHE the Hall resistance of the 2DEG in high quality GaAs-$Al_xGa_{1-x}As$ heterostructures show plateau formation at some fractional filling factors $\nu < 1$, in the so-called magnetic quantum limit (mean electron spacing exceeds the cyclotron radius ℓ). Plateaus in ρ_{xy}, which are always in common with minima in ρ_{xx}, have been first observed at $\nu = 1/3$ and 2/3. Necessary conditions for the occurrence of the FQHE are high quality samples with low carrier densities and temperatures well below 1 K in combination with high magnetic fields. Up to now, ρ_{xx}-minima partly combined with more or less flat plateaus in ρ_{xy} have been found [24-26] at filling factors $\nu = 1/3$, 2/3, 1/5, 2/5, 3/5, 4/5, 2/7, 3/7, 4/7, 4/9, and 5/9, as shown in Fig. 10 /26/. Figure 10 also demonstrates that especially for the weak structures the effect is more pronounced in ρ_{xx} than in ρ_{xy}. Generalizing, structures related to the FQHE have been observed at filling factors $\nu = p/q$ (p and q are integers with q being odd) and are getting weaker with increasing q.

Except at very low filling factors $\nu \lesssim 0.1$ the original suggestion that the FQHE arises from a formation of an electronic crystal (Wigner crystal) is now considered unlikely. In the presently prevailing theoretical model [27] the electronic ground state is considered as an electron liquid at odd fractions of ν. Its fractionally-charged quasi-particle excitations are separated by energy gaps of the order of $0.01\ e^2/\varepsilon\ell$ which leads in connection with some localized states to the occurrence of plateaus in the Hall resistance. Within the experimental range of B, T, und μ, the plateaus for $\nu = 1/3$, 2/3, 2/5, 3/5, and 3/7 appear to be quite flat, and the relative deviation from the expected Hall resistance $h/(p/q)e^2$ is 10^{-3} or even smaller [26]. On the other hand, for weak structures, due to scattering contributions, the Hall resistance is no longer a good measure for the filling factor ν. In these cases, only the magnetic field position for the ρ_{xx}-minima can be used for the determination of ν which is sometimes not unambiguous. Especially, the 4/5-step is unexpectedly weak and broad compared with the 3/5 and 2/5-steps, and the experiments [24,28] show a sys-

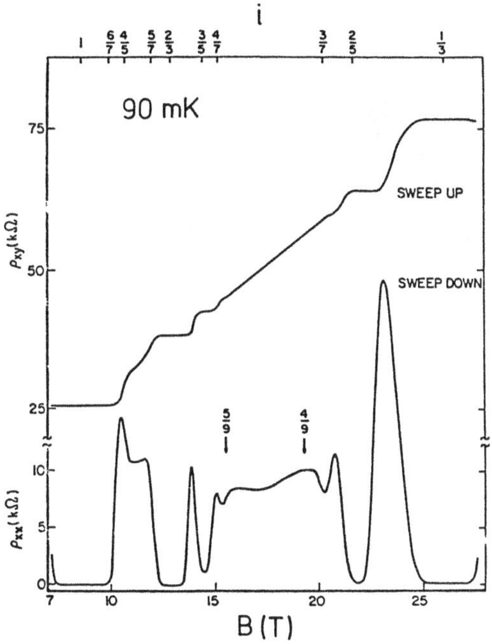

Fig. 10: ρ_{xx} and ρ_{xy} vs. B at T = 90 mK and filling factors i $\lesssim$ 1 for a high-quality GaAs-$Al_xGa_{1-x}As$ heterostructure [26].

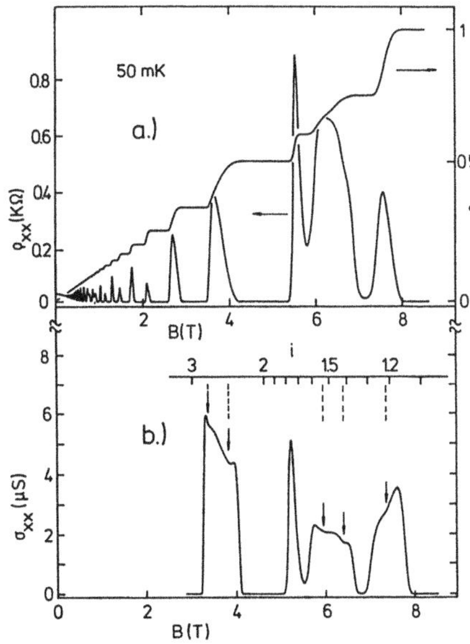

Fig. 11: Magnetic field dependence at T = 50 mK for the resistivity components ρ_{xx} and ρ_{xy} for a Hall device (a) and of the conductivity σ_{xx} for a Corbino device (b) from the same wafer. The high-quality GaAs-$Al_xGa_{1-x}As$ heterostructures have mobilities around 10^6 cm^2/Vs.

tematic shift towards lower ν-values between 3/4 and 4/5. In addition, a weak ρ_{xx}-minimum is observed around ν = 1/2. Therefore the question may arise whether there are, although much weaker, structures at filling factors p/q with even q.

Fractionally quantized Hall plateaus have also been found at filling factors $\nu > 1$ [24,28,29]. As shown in Fig. 11a, ρ_{xy} showed flat plateaus at $\nu = 4/3$ and 5/3 which are exactly quantized within 10^{-3} to their expected value [28]. Compared to the 1/3- and 2/3-steps, the temperature has to be as low as 50 mK, because the energy gaps at magnetic fields around 7 T are much smaller. In very high quality samples ($\mu \gtrsim 10^6$ cm^2/Vs) additional structures have been observed [28,29] at filling factors $1 < \nu < 3$ and, as shown in Fig. 11b, may be interpreted as $\nu = 6/5$, 7/5, 7/3, 8/3. In these experiments [28] Hall devices (Fig. 11a) as well as devices with Corbino geometry (Fig. 11b) have been used, and it turned out that the σ_{xx}-measurements on Corbino devices always have a higher resolution than the corresponding ρ_{xx}-data.

The occurrence of the QHE and even the FQHE is not restricted to the 2DEG in GaAs-$Al_xGa_{1-x}As$ heterostructures. STÖRMER et al. [30,24] found that also the two-dimensional hole gas in p-channel GaAs-$Al_{0.5}Ga_{0.5}As$ heterostructures shows plateau formation at integer as well as at fractional filling factors. Moreover, in very recent experiments [31,32] on high quality Si-MOSFETs with mobilities of about 40 000 cm^2/Vs structures at fractional filling factors $\nu = 2/3$, 4/3, 5/3, 7/3, 8/3, 6/5, and 7/5 were observed. In conclusion, these experiments show that the FQHE is a very general feature of two-dimensional charged carrier systems, and can be observed in higher Landau levels as well as in different materials, provided the sample quality is high enough. In general, the sample behavior is determined by the interplay between localization and interaction. With increasing sample quality (mobility) localization processes are more and more reduced, and interaction phenomena appear. Therefore, it is mainly a challenge for the sample manufacturer to get a deeper insight into the physics of the FQHE.

Acknowledgements

The valuable contributions of my colleagues K. v. Klitzing, N. Kleinmichel, H. Obloh, and B. Tausendfreund are gratefully acknowledged. In addition I would like to thank C. Probst, K. Andres, J.C. Maan, and G. Remenyi for making the low-temperature experiments possible. High quality devices were kindly provided by G. Weimann, K. Ploog, and G. Dorda. This work has been supported by the Deutsche Forschungsgemeinschaft and the Stiftung Volkswagenwerk.

References:

1 K. v. Klitzing, G. Dorda, and M. Pepper, Phys. Rev. Lett. 45, 494 (1980)
2 D.C. Tsui, H.L. Störmer, and A.C. Gossard, Phys. Rev. Lett. 48, 1559 (1982)
3 Y. Ono, J. Phys. Soc. Jap. 51, 2055 (1982)
Y. Ono, J. Phys. Soc. Jap. 51, 3544 (1982)
Y. Ono, J. Phys. Soc. Jap. 52, 2492 (1983)
4 T. Ando, J. Phys. Soc. Jap. 52, 1740 (1983)
T. Ando, submitted to J. Phys. Soc. Jap.
T. Ando, submitted to J. Phys. Soc. Jap.
5 B. Kramer and A. Mac Kinnon, Proc. Int. Seminar "Localization in Disordered Systems", Johnsbach b. Dresden (1983)
6 H. Aoki and T. Ando, preprint
7 S.V. Iordansky, Solid State Commun. 43, 1 (1982)
8 B.I. Halperin, Phys. Rev. B 25, 2185 (1982)

9 L. Schweitzer, B. Kramer, and A. MacKinnon , J. Phys. C17, 4111 (1984)
10 R.E. Prange, Phys. Rev. B 23, 4802 (1981)
11 H. Aoki and T. Ando, Solid State Commun. 38, 1079 (1981)
12 W. Brenig, Z. Physik B - Condensed Matter 50, 305 (1983)
13 M.A. Paalanen, D.C. Tsui, and A.C. Gossard, Phys. Rev. B 25, 5566 (1982)
14 G. Ebert, K. v. Klitzing, C. Probst, and K. Ploog, Solid State Commun. 44, 95 (1982)
15 A. Briggs, Y. Guldner, J.P. Vieren, M. Voos, J.P. Hirtz, and M. Razeghi, Phys. Rev. B 27, 6549 (1983)
16 G.A. Baraff and D.C. Tsui, Phys. Rev. B 24, 2274 (1981)
17 N.F. Mott and E.A. Davis, Electronic Processes in Noncrystalline Materials, 2nd ed. Clarendon Press, Oxford (1979)
18 Y. Ono, J. Phys. Soc. Jpn. 51, 237 (1982)
19 G. Ebert, K. v. Klitzing, C. Probst, E. Schuberth, K. Ploog, and G. Weimann, Solid State Commun. 45, 625 (1983)
20 D.C. Tsui, H.L. Störmer, and A.C. Gossard, Phys. Rev. B 25, 1405 (1982)
21 A.L. Efros and B.I. Shklovskii, J. Phys. C 8, L 49 (1975)
22 B. Tausendfreund and K. v. Klitzing, Surf. Sci. 142, 220 (1984)
23 K.I. Wysokinski and W. Brenig, Z. Physik B - Condensed Matter 54, 11 (1983)
24 H.L. Störmer, A. Chang, D.C. Tsui, J.C.M. Hwang, A.C. Gossard, and W. Wiegmann, Phys. Rev. Lett. 50, 1953 (1983)
25 E.E. Mendez, M. Heiblum, L.L. Chang, and L. Esaki, Phys. Rev. B 28, 4886 (1983)
26 A.M. Chang, P. Berglund, D.C. Tsui, H.L. Störmer, and J.C.M. Hwang, preprint
27 R.B. Laughlin, Phys. Rev. Lett. 50, 1395 (1983)
28 G. Ebert, K. v. Klitzing, J.C. Maan, G. Remenyi, C. Probst, G. Weimann, and W. Schlapp, submitted to J. Phys. C.
29 E.E. Mendez, Proc. IBM Europe Institute session on "Quantum Physics of Semiconductor Layers", Davos (1984)
30 H.L. Störmer, Z. Schlesinger, A. Chang, D.C. Tsui, A.C. Gossard, and W. Wiegmann, Phys.Rev. Lett. 51, 126 (1983)
31 V.M. Pudalov and S.G. Semenchinsky, Pisma ZhETP 39, 143 (1984)
32 M.G. Gavrilov, Z.D. Kvon, I.V. Kukushkin, and V.B. Timofeev, Pisma ZhETP 39, 420 (1984)

Localization and the Integral Quantum Hall Effect

Adrianus M.M. Pruisken

The Institute for Advanced Study, School of Natural Sciences
Princeton, NJ 08540, USA

I. Introduction

In the early theoretical investigations [1-4] of the Integral Quantum Hall Effect (IQHE) [5,6], it was recognized that the observed plateaus of quantized Hall conductance should be explainable from the two-dimensional, free electron gas subject to a constant magnetic background field and to a random impurity potential. The concept of localization due to the randomness turned out to be essential in the explanation of the width of the plateaus, as well as the precision of the quantization [7].

However, no microscopic theory on electronic disorder existed which supported the existence of extended electronic levels, which clearly were needed for macroscopic quantum transport, as well as aforementioned localized states. In view of the advances made in recent years on localization phenomena [8], it may have been Ando and Aoki [9] who were the first to realize that the one-parameter scaling ideas of Abrahams, Anderson Licciardello and Ramakrishnan [10] could not lead to an explanation of the effect. The theoretical justification for their ideas is commonly denoted by "weak localization" theory, in which one considers impurity scattering and its renormalization group extension in a certain classical limit, characterized by a small expansion parameter. The quantum fluctuations (maximally crossed diagrams) about the classical result for the conductance (Boltzmann Equation) are marginally relevant in two dimensions, and this is usually seen as the precursor effect to complete localization in the two and lower dimensional disordered electronic system.

It is well known that for problems in which time-reversal symmetry is broken, the logarithmic precursor effect shows up one order higher in the Feynman diagrammatic expansion [11]. The different versions [11,12] of such perturbative renormalization group analysis apparently all lead to essentially the same result as in the original Anderson model [8].

In a recent series of papers [13-18], this discrepancy between localization theory and the IQHE, as discovered experimentally by von Klitzing, Dorda and Pepper in 1980, has been resolved by field theoretic means. It has been shown that the usual weak localization results ought to be supplemented by non-perturbative, topological effects which are ultimately responsible for the breakdown of conventional localization and the appearance of extended electronic levels. These topological effects derive from a new, effective Lagrangian [13] which describes the long-distance scaling behaviour in the transport properties of the 2D system in the presence of a magnetic field.

The effective field theory contains two parameters, namely σ^0_{xx} and σ^0_{xy}, which are identified as the bare or unrenormalized magneto-, respectivily Hall, conductance, defined for a length equal to the magnetic or phase-coherence length μ. Hence, the disordered electronic system obeys two-parameter scaling in this case and the appropriate parameters are $\sigma_{xx}(\mu L), \sigma_{xy}(\mu L)$, dimensionless numbers measuring the conductance tensor in units e^2/h and defined for length scale L. The density of electronic states does not participate as a scaling parameter, and this aspect of the theory has been given a rigorous basis by Franz Wegner, who obtained an exact expression for the density of states in the lowest Landau level [19].

The parameters $\sigma^0_{\mu\nu}$ can further be given an interpretation in terms of the kinetic theory (Boltzmann Eq.). There are two different versions of such a semi-classical theory, known in the literature which are due to Ando [20] and Streda [21] and which conveniently describe the limits of weak and strong magnetic fields respectivily. These results derive from different versions of the linear response formulae for the conductance, where the effect of impurities is approximated via the insertion of a single and self-consistently determined relaxation time τ (Self-Consistent Born Approximation, SCBA). The different semi-classical approximations to the conductance tensor turn out to be equivalent, and form furthermore a good approximation to the bare parameters $\sigma^0_{\mu\nu}$.

Beyond the magnetic length μ, the semi-classical results cease to be valid, and the aforementioned effective field theory forms the microscopic theory which determines the length-scale dependence in the conductance via the renormalization group equations

$$\beta_{xx}(\sigma_{xx},\sigma_{xy})=\frac{\partial\sigma_{xx}}{\partial ln\mathrm{L}}\;;\qquad \beta_{xy}(\sigma_{xx},\sigma_{xy})=\frac{\partial\sigma_{xy}}{\partial ln\mathrm{L}}$$

It is the purpose of this paper to review some of the theoretical developments [13-18] which led to an analysis of the IQHE with the aid of renormalization group.

The identification of both σ_{xx} and σ_{xy} as renormalization group parameters, as well as the development of a renormalization group theory, derive to a large extend from the following two principles, each of which is beyond the semi-classical considerations:

First, the notion of a skipping orbit edge-current, the magnitude of which is determined by the interior of the system via a form of Stokes' theorem. This notion enables one to formulate the Hall conductance as a single particle problem which involves the scattering processes near the Fermi-level alone, rather than a many-body effect as it is usually understood.

Secondly, the insight that the long distance, scaling behaviour in the transport phenomena is governed by the Goldstone modes of a broken, continuous symmetry between the advanced and retarded wavefunctions, an issue discovered by Wegner in the context of conventional (Anderson) localization [22]. The notion of a continuous symmetry in the problem, however, goes far beyond its perturbative expansion and reaches out for conclusions which ordinary perturbation theory could never give.

In section II I will discuss the meaning of this symmetry in the context of a replicated generating functional for ensemble-averaged Green's functions, and show how the exact linear response formulae for the conductance can be obtained from unitary rotations between advanced and retarded waves. The Hall conductance is generated via realizations of the continuous symmetry which have non-trival topology. The notion of abovementioned edge-currents then results quite naturally in a topological classification of the continuous symmetry.

Topological excitations play an important role in the long distance scaling behaviour of the theory. The effective field theory, describing this behaviour, makes the IQHE a realization as well as an interpretation of the socalled ϑ-vacuum, a concept which originally arose in four-dimensional gauge theories. This will be discussed in section III. Renormalization group arguments will be reviewed, which relate localized bulk states at the Fermi-level to the quantization of Hall conductance in units of e^2/h. Furthermore, a renormalization group flow diagram will be presented for the conductance tensor, which shows in an explicit manner how the weak localization perturbative results are corrected for by the implementation of topological excitations (instantons). Finally, in section IV, I'll discuss a form of Stokes' theorem which relates the IQHE as a bulk-phenomenon to that of an edge-effect. This is a direct manifestation of the meaning of the continuous symmetry for the localization problem. Such a result is actually what is needed in order to establish the IQHE as a phenomenon which is independend of the sample geometry.

II. The Kubo formalism, boundary currents, generating functional and source terms

2.1 Introduction

For the microscopic evaluation of quantities like the density of states, conductance and participation ratio in the presence of an impurity potential, one usually makes use of the Kubo formalism, in which these quantities are expressed in terms of the Green's functions $[E-H\pm i0^+]^{-1}$ with the Hamiltonian given as

$$H = \frac{1}{2m}(\vec{p}+\frac{e}{c}\vec{A})^2+V(r) \ . \tag{1}$$

$V(r)$ is the random potential and $A_\mu=(0,Bx)$ represents the magnetic field B perpendicular to the plane. Without randomness, the energy-spectrum consists of a discrete set of highly degenerate Landau-levels with energy $\hbar\omega_c(n+\frac{1}{2})$; $\omega_c=eB/mc$ (cyclotron frequency). The number of states within a Landau-level equals eB/hc.

Instead of considering a single realization for the disorder, one thinks of an ensemble of impurity distributions over which the physical quantities are averaged. The starting point of the analysis is the construction of a generating function for above mentioned Green's functions via simple Gaussian integrals, whereby the averaged over quenched disorder is being performed employing the replica method, a proceedure which is fairly standard by now.

The functional integral is preferably represented by complex Grassmann (anticommuting) fields Ψ and $\overline{\Psi}=i\Psi^\dagger$ [13]

$$Z=\int D[V]P[V]\int D[\overline{\Psi}]\int D[\Psi]e^{L}$$

$$L=\int d^2r\sum_{p=\pm}\ \sum_{a=1}^{m}\overline{\Psi}_a^p(r)[E_p-H]\Psi_a^p(r)$$

$$E_p=E+is^p\eta;\quad \eta=0^+\ ,\quad s^p=sign(p)\ . \tag{2}$$

The integral over V together with the weight $P[V]$ represents the ensemble average over the impurity potential, contained in H. The sum over the index p accounts for the simultaneous average of advanced and retarded quantities, whereas the subscript a runs over m replicas.

The linear-response formulae for the conductance, to be discussed below, can be obtained from Z by adding the appropriate source terms to L. At this stage we only mention the averaged density of states $\rho(E)$ which can be obtained as

$$\rho(E)=\frac{i}{2\pi}\overline{\langle r\,|\,(E_+-H)^{-1}-(E_--H)^{-1}\,|\,r\rangle}$$

$$=\frac{i}{2\pi}\langle\overline{\Psi}_1^+(r)\Psi_1^+(r)-\overline{\Psi}_1^-(r)\Psi_1^-(r)\rangle \tag{3}$$

where the expectation <..> is with respect to L and where the bar stands for the average $\overline{O}\equiv\int D[V]\ P[V]\ O$.

As was discussed first by Wegner [22], the resulting formalism is a particularly suitable starting point for analyzing the long-distance scaling behaviour of the disordered system with renormalization group.

Before going into any detail, it might be worthwhile to emphasize three important aspects, upon which the discussion is heavily based. These aspects are:

(i) The one particle Green's functions are short-ranged (in the interior of the system). This aspect is a familiar one,as it is true for conventional problems of electronic disorder. It expresses the fact that the equilibrium quantities in the problem, like the density of electronic levels, are smooth analytic functions. As mentioned before, this aspect has recently be given a rigorous basis by Franz Wegner. For these quantities, it is sufficient to keep only one of the two indices p in the generating functional, by taking the number of replicas *a priori* zero in the other one.

(ii) The transport properties of the problem involve the simultanuous average of advanced and retarded Green's functions. In this aspect of the problem, there is a continuous symmetry which is spontanuously broken by the occurrence of a density of states. This symmetry refers to the invariance of L under global $U(2m)$ rotations on the Ψ, i.e. under those transformations, which leave the form $\sum_a(\overline{\Psi}_a^+\Psi_a^++\overline{\Psi}_a^-\Psi_a^-)$ invariant. The convergence term η in E_p breaks this symmetry except for the subgroup $U(m)\times U(m)$, ie the rotations in the $p=+,-$ sub-spaces $\sum\overline{\Psi}_a^+\Psi_a^+$ and $\sum\overline{\Psi}_a^-\Psi_a^-$ separately. The quantity conjugate to η is easily identified as the density of states ρ, where $\rho \propto \partial Z / \partial\eta$, so $\rho(E)$ spontanuously breaks the symmetry $\frac{U(2m)}{U(m)\times U(m)}$. It leads to the important insight that the macroscopic transport phenomena of the system are governed by the infrared singularities due to the massless (would be) Goldstone modes, i.e. the generators of the broken symmetry group. The power of this insight, as conjectured by Franz Wegner, is that one can proceed with renormalization group arguments, without *a priori* relying upon perturbation theory.

(iii) Finally, the aspect of another symmetry which is macroscopically broken and which is specific to the magnetic field problem, namely that of time-reversal invariance. This aspect is directly responsible for a new transport quantity in the problem, namely the Hall conductance or σ_{xy}.

Whereas the former two aspects are familiar ones in localization theory, it is less obvious to see how the Hall conductance fits into the formalism. Indeed, usually one thinks of the Hall conductance as a collective effect which involves contributions from all the electronic levels below the Fermi energy. The formalism introduced above only studies the propagators at a single energy E and upon this simplification, common to free-particle problems, the formulation of renormalization group ideas is heavily based. It is therefore important to realize that σ_{xy}, just like σ_{xx}, is explicitly determined by the properties of the eigenstates near the Fermi-energy alone. This will be the subject of the next section.

2.2 The Kubo-Streda formula

The dissipative d.c. conductance at T=0 can be written in the familiar form

$$\sigma_{xx} \times \frac{e^2}{h} = -\frac{\hbar e^2}{4\pi\Omega} \int_r \int_{r'} \overline{\tilde{\pi}_x (G^+(r,r') - G^-(r,r')) \tilde{\pi}_x (G^+(r',r) - G^-(r',r))} \tag{4}$$

where the velocity operator $\tilde{\pi}$ and propagator $G^{\pm}$ are given respectively by

$$\tilde{\pi}_\mu = \frac{1}{i\hbar}[r_\mu, H] = \frac{i}{m}(p_\mu - \frac{e}{c}A_\mu)$$
$$G^{\pm}(r,r') = \langle r \,|\, (E - H \pm i\eta)^{-1} \,|\, r' \rangle; \quad \eta \to 0^+ \tag{5}$$

and where Ω stands for the volume of the system.
On the other hand, the Hall conductance can be expressed as $\sigma_{xy} = \sigma_{xy}^I + \sigma_{xy}^{II}$ with

$$\sigma_{xy}^I \times \frac{e^2}{h} = -\frac{\hbar e^2}{4\pi\Omega} \int\int d^2r d^2r' \left[\overline{\tilde{\pi}_x G^+(r,r') \tilde{\pi}_y G^-(r',r) - \{x \leftrightarrow y\}} \right], \tag{6}$$

$$\sigma_{xy}^{II} \times \frac{e^2}{h} = \frac{ie^2}{4\pi\Omega} \int d^2r \, (r_x \tilde{\pi}_y - r_y \tilde{\pi}_x) \overline{(G^+(r,r) - G^-(r,r))} \; . \tag{7}$$

This form of the Hall conductance has been obtained first in the work by Smrcka and Streda [23] and has the desired property in that it contains a single energy E.

In a first attempt toward a better understanding of these expressions, one can approximate the effect of disorder in SCBA, i.e. insert a single relaxation time τ. In this approximation, Eqs (4) and (6) are related as

$$(\sigma_{xy}^I)^0 = -\omega_c \tau \sigma_{xx}^0 \; . \tag{8}$$

and reduce to the classical Drude-Zener result in the limit of weak fields ($\omega_c \tau \ll 1$).

The other quantity σ_{xy}^{II}, containing the magnetization operator $\varepsilon_{\mu\nu} r_\mu \tilde{\pi}_\nu$, vanishes in the same approximation. It turns out that this quantity, when written in the form of Eq.(7), gets its contribution only from the *extended edge-states*, which are not destroyed by (sufficiently weak) disorder (see section 2.3). In order to see how much this term contributes, one can go back to the representation for σ_{xy}^{II} which involves an integral over all energies below the Fermi-energy, as it originally came out of linear response theory:

$$\sigma_{xy}^{II} \times \frac{e^2}{h} = \frac{ie^2}{4\pi\Omega} \int_{-\infty}^{E} dE \int\int d^2r d^2r' \left[\overline{G^+(r',r)(r_x \tilde{\pi}_y - r_y \tilde{\pi}_x) G^+(r,r') - (G^+ \leftrightarrow G^-)} \right]$$

$$= -ec \int_{-\infty}^{E} dE \frac{\partial}{\partial B} \overline{\rho(E)} \tag{9}$$

with $\rho(E)$ the density of states. This result was first obtained by Streda [23]. Eqs (6) and (9) express the Hall conductance as a *bulk* quantity, which does not involve explicit contributions from the sample-edge. The degeneracy per Landau level equals eB/hc such that σ_{xy}^{II}, Eq.(9), is quantized in e^2/h for completely filled Landau levels. The quantity σ_{xy}^{I} vanishes in case the Fermi energy is located in a bulk density of states gap.

Notice that the previously mentioned aspect,which says that the equilibrium quantities are harmless, non-critical functions of energy (section 2.1,(i)), actually becomes a highly non-trivial statement which has severe consequences for the localization problem. Namely, Eq.(9) tells us that it is an *equilibrium* quantity, the density of states, which determines the fact that the Hall conductance passes through one unit as the Landau-level is gradually filled up with electrons (we imagine here the strong magnetic field limit, such that $\hbar\omega_c$ is much larger that the Landau level broadening due to impurities). This result is incompatible with the conventional localization lore, which says that all states in 2D are localized. Under these latter circumstances, macroscopic electron transport is impossible,and the full conductance tensor vanishes everywhere [9].

Consequently, Eq.(9) implies that conventional localization *has* to break down somewhere in the Landau-level, namely there where the full Hall conductance passes through one unit. A satisfactory answer to this problem can only be found in a (scaling) theory ,which involves both the dissipative conductance and the Hall conductance simultaneously and explicitly. This, in turn, requires a better understanding of the expression for σ_{xy}^{II}, Eq.(7), as will be discussed next.

2.3 Boundary currents and Stokes theorem I

Eq.(7) can be understood from the ensemble averaged current density of the states near the Fermi level

$$j_\mu(E,r) \equiv i\hbar \overline{\tilde{\pi}_\mu[G^+(r,r) - G^-(r,r)]}$$

$$\equiv \hbar \sum_j \overline{\Psi_j(r) \tilde{\pi}_\mu \Psi_j^\dagger \delta_\eta(E - E_j)} \tag{10}$$

where E_j and Ψ_j denote the eigen energies and the corresponding wavefunctions for a single realization of the impurity ensemble. In ref.[13] it has been shown that this quantity has some unusual properties,in that it acts like an operator in our continuum model. Formally one can write

$$j_\mu(E,r) = \sigma_{xy}^{II} \varepsilon_{\mu\nu} \nabla_\nu + O(\nabla^2) \tag{11}$$

where σ_{xy}^{II} is given by Eq.(9). One can easily check that Eqs (9) and (7,11) are consistent with one another in the sense that in the combination $j_\mu r_\nu$ the operator j_μ acts on r_μ. The derivation of Eq.(11) follows from operating Eq.(10) with $\int_{-\infty}^{E} dE \frac{\partial}{\partial E'}$, as is done in Eq.(9), and subsequently studying the long-distance behaviour of the resulting two-particle propagators via a systematic gradient expansion. All this is done for an infinite system and under the notion that averaged combinations of only advanced (retarded) propagators are short-ranged. Eq.(11) can now be given the following interpretation. We can think of this result as being due to an *orbital*, *conserved* and *fictitious* current flow surrounding each point in space. The size of each orbit is approximately equal to the phase-coherence length, the length beyond which the gradient expansion of Eq. (11) holds. Such an interpretation is sensible, because a current-density distribution of the type

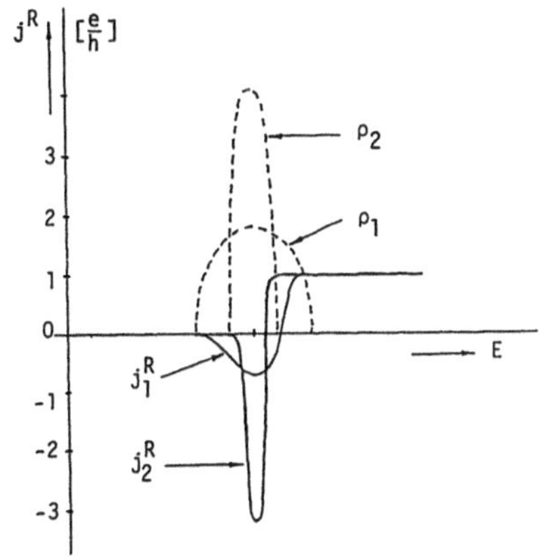

Fig.1 The density of states and the static edge-current in the lowest Landau level in SCBA.

$$j_{\mu}(r)=\varepsilon_{\mu\nu}\nabla_{\nu}\varphi(r-r_0) \tag{12}$$

with φ an arbitary, radially symmetric and short-ranged function, fullfills this description and gives rise to an expansion exactly of the type as in Eq.(11) as far as its effect at large length-scales is concerned.

Furthermore, the total current through each point is of course zero,as one has to add the contribution from each orbit to which a certain point belongs. Although the current is fictitious, the resulting local dia-magnetic moment $\Delta(r)=\varepsilon_{\mu\nu}j_{\mu}(E,r)r_{\nu}$ is not,and equals a constant, namely σ_{xy}^{II}. From this we can deduce the effect of making the system subject to boundaries. Namely, because $\Delta(r)$ has to vanish outside the sample, a real edge-current will flow according to

$$j_{\mu}^{R}(E,r)=\varepsilon_{\mu\nu}\nabla_{\nu}\Delta(r) \tag{13}$$

which says that the total amount of current flowing along the the perimeter, equals σ_{xy}^{II} (Eq.9). The latter is a bulk quantity,and hence the amount of boundary current is actually determined by the bulk. Formally,we can write this in terms of the Stokes' theorem

$$\int d^2r\, j_{\mu}\Lambda_{\mu}=\sigma_{xy}^{II}\int_C dr_{\mu}\Lambda_{\mu} \ . \tag{14}$$

Notice that the sequence of arguments above actually are those of paramagnetism but in reverse: in paramagnetism one can replace the sum of the atomic magnetic moments by an effective and fictitious edge-current; in our diamagnetic system we have replaced the fictitious orbital currents in the bulk by a real edge-current.

In Fig. 1 the edge-current, thus obtained, is plotted as a function of energy and as calculated in SCBA for the lowest Landau level and for two values of the impurity parameter g (section III). The result has the same behaviour as the exact result obtained from Wegner's expression for the density of states for the white noise potential [19]. We observe that for a filled Landau level the edge-current is quantized in units e/h and the consequences for the Hall conductance have been discussed first by Halperin [4]. Nevertheless, for the states within the Landau level, the edge-current varies as the negative of the density of states and quantization is far from being obvious. Indeed, the edge-current as considered here is an equilibrium quantity and has nothing to do with the IQHE. If the system is taken out of thermodynamic equilibruim by a perturbing field, then according to linear response theory, one has to account for the other quantity in the Hall conductance, namely σ_{xy}^{I}. If the states near the Fermi-level are localized in the interior, then a non-zero Hall conductance can only be a result of the extended edge states at the Fermi-level. Clearly, the equilibrium current at the edge and near the Fermi-level needs a large correction in order to get back to quantization. This merely reflects the rather strong difference between the equilibrium and non-equilibrium properties of the system.

What has been discussed above is a form of Stokes theorem, which relates the equilibrium properties of the bulk (Eq. 9) to the occurrence of a boundary current. A similar relation holds for lattice systems, and recent numerical work has shown this to be valid rather precisely even for systems with a size on the order of 10 cyclotron radii. This, as well as a discussion on the diamagnetic properties, will be given elsewhere [24].

2.4 Source Terms

As mentioned in the introduction, the non-equilibrium aspects of the problem involve a continuous symmetry which is spontaneously broken by a positive density of states. In this section we establish a direct relation between this symmetry and the conductance formulae discussed above. This is done via evaluating the generating function Z (Eq.2) in the presence of a slowly-varying back-ground field $\hat{U}$ which is an element of the group $U(2m)$. For convenience, we will replace the kinetic term in the Hamiltonian by the Hermitian combination $\pi_\mu \pi^\mu$ with

$$\pi_\mu = \frac{1}{\sqrt{2m}}\left(-\frac{\overleftarrow{\nabla}_\mu}{i} - \frac{e}{c}A_\mu\right) \qquad \pi^\mu = \frac{1}{\sqrt{2m}}\left(\frac{\overrightarrow{\nabla}_\mu}{i} - \frac{e}{c}A_\mu\right) \ . \tag{15}$$

Z in the presence of the background field $\hat{U}$ is defined as

$$Z \Rightarrow Z[\hat{U}] = \int D[V]P[V]\int D[\bar{\Psi}]\int D[\Psi]e^{L[\hat{U}]}$$
$$L[\hat{U}] = \int d^2r \sum_{p,a} \bar{\Psi}_a^p(r)\left[(E-V(r))\hat{1} + i\eta\hat{s} + \hat{U}^{-1}\pi_\mu\pi^\mu\hat{U}\right]_{ab}^{pp'}\Psi_b^{p'}(r) \tag{16}$$

where we have introduced the matrix notation

$$\hat{1} = \delta_{pp'}\delta_{ab}; \qquad \hat{s} = s^p\,\delta_{pp'}\delta_{ab}; \qquad \hat{U}_{ab}^{pp'}(r) \in U(2m) \ . \tag{17}$$

According to the rules of Grassmann integration, we can simplify the result somewhat more by writing

$$Z[U] = \int D[V]P[V] \exp Tr \ln[(E-V(r))\hat{1} + i\eta\hat{s} + \hat{U}^{-1}\pi_\mu\pi^\mu\hat{U}] \tag{18}$$

where the trace (Tr) is over spatial as well as the p,a indices. Notice that $Z[\hat{U}]$ will be different from Z by virtue of the spatial dependence in $\hat{U}$. The difference $[Z[\hat{U}]-Z]_{\eta\to 0}$ teaches us something about the dynamics of the (would be) Goldstone modes in the problem. Indeed, we can show that the difference involves the exact expressions for the conductance. We introduce a specific unitary matrix

$$\hat{U}(r) = [e^{ij_x\hat{\tau}_x x + ij_y\hat{\tau}_y y}]_{ab}^{pp'} \ ; \qquad \hat{\tau}_\mu = \delta_{ab}\delta_{a1}[\tau_\mu]^{pp'} \tag{19}$$

After staightforward expanding in the "sources" j_μ in Eq.(19) it can be shown [16]

$$[\ln Z[\hat{U}] - \ln Z]_{\eta\to 0} = \sigma_{\mu\mu} j_\mu^2 \mathrm{L}^2 + i\,\sigma_{xy} j_x j_y \mathrm{L}^2 + O(j^3) \tag{20}$$

where L^2 is the sample size and $\sigma_{\mu\nu}$ are given by Eqs (4,6,7).
Hence, the effect of these background fields $\hat{U}$ is similar to that of a true electric field. Field configurations of the type of Eq.(19) form an example of "topologically" non-trivial realizations of the Wegner-symmetry and the Hall conductance in Eq.(20) is only generated via such configurations. The meaning of topology can easiest be understood from the existence of boundary currents in the problem. Notice hereto that the equation above implies that the system (i.e. Z) is still sensitive to the perturbing field, even if the Fermi level is located in a density of states gap ($\sigma_{xx}\to 0, \sigma_{xy}\to\sigma_{xy}^{II}$). This phenomenon is clearly a result of the previously discussed extended edge-states, which are sensitive to the value of $\hat{U}$ near the boundary of the system. Apparently, the "broken" symmetry remains significant, despite the fact that the order parameter in the

problem, the bulk density if states, vanishes. This is an important difference with conventional problems of electronic disorder, and we will next discuss several consequences of this observation.

2.5 Boundary currents and topology

First, the arguments above suggest that Z in this case will be invariant under those background fields, which leave the boundary current unperturbed. For such field configurations it is required that $\hat{U}$ reduce to an arbitrary $U(m)\times U(m)$ gauge near the boundary of the system. It is a rather simple matter to see that this is true indeed and this will lead to a topological classification of the fields $\hat{U}$, which will play a crucial role in the statistical processes of diffusion.

By virtue of the fact that the Fermi energy E is located in a density of states gap, we can alternatively evaluate $Z[\hat{U}]$ via a gradient expansion, which is appropriate for the assumption of slowly varying fields $\hat{U}$.

$$\hat{U}^{-1}\pi_\mu\pi^\mu\hat{U}=\pi_\mu\pi^\mu+\pi_\mu D_\mu+D_\mu\pi^\mu+D_\mu D_\mu \quad ; \quad D_\mu=\frac{i}{\sqrt{2m}}\hat{U}^{-1}\nabla_\mu\hat{U} \tag{21}$$

The gradient expansion amounts to expanding the $Trln$ in Eq.(18) in the quantity D_μ. It turns out that only the term linear in D_μ is important

$$Z_{GAP}[\hat{U}]=Z_{GAP}[\hat{1}]e^{\int d^2r\sum_{p,a}<(\pi_\mu+\pi^\mu)g^p(r,r)>[D_\mu(r)]^{pp}_{aa}+\dots} \tag{22}$$

where

$$g^p(r,r')=<r\,|(E+i\eta s^p-\pi_\mu\pi^\mu-V(r))^{-1}|\,r> \tag{23}$$

and where the expectation $<\dots>$ is with respect to $Z[\hat{U}=\hat{1}]$. In the limit of zero number of replica's, the expectation in Eq.(22) precisely correspond to the expression for the current density in Eq.(10). We can now fruitfully make use of of the previous discussion on edge- currents, section 2.3, and write Eq.(22) as

$$Z_{GAP}[\hat{U}]=Z_{GAP}[\hat{1}]e^{R(\hat{U})+\cdots} \quad ; \quad R(\hat{U})=\tfrac{1}{2}\sigma^{II}_{xy}\int_C dr_\mu\sum_{p,a}\left[\hat{U}^{-1}\nabla_\mu\hat{U}\right]^{pp}_{aa}s^p \tag{24}$$

with σ^{II}_{xy} given in Eq.(9). According to section 2.2, this quantity is integer-valued. The resulting integral is along the perimeter of the system. With the aforementioned boundary condition on $\hat{U}$, only the U(1) part of the $U(m)\times U(m)$ gauge contributes to the integral, i.e. the relative phase between the advanced and retarded blocks. Consequently

$$R(\hat{U})=2\pi i\,\sigma^{II}_{xy}\times[integer]$$

Substituting this result back in Eq.(24), we conclude that Z is indeed invariant under these special background fields.

In order to see the meaning of this result, notice that the states near the Fermi-level actually constitute a current-loop along the perimeter of the system. The contributing parts of $trD_\mu\hat{s}$ in Eqs (21,22) are nothing but a gauge-transformation performed on these physical boundary states; the surface integral in Eq.(24) therefore can be identified with an (adiabatic) change in magnetic flux through the area enclosed by the current-loop. The invariance derived above therefore correctly describes the invariance of the free energy of these states under a change of a flux- quantum through the loop. On the other hand, for the system as a whole, the change in flux is related to a change in total number of *bulk* electrons below the Fermi-level. For this reason, the invariance transformation has a special meaning within many-body theory (i.e. Fermi-statistics).

In order to maintain the relation between the bulk and the edge of the system, it is important that the $\hat{U}$ has the necessary smoothness such Stokes' theorem can be applied. For compact unitary rotations, as considered here, this is generally true [17] and Eq.(24) and be written as

$$R(\hat{U})=\frac{1}{2}\sigma_{xy}^{II}\int d^2r\,\varepsilon_{\mu\nu}\sum_{p,a}\nabla_\nu\left[\hat{U}^{-1}\nabla_\mu\hat{U}\right]_{aa}^{pp}s^p \tag{24a}$$

Eq.(24a) can be evaluated one step further and written as an expression for the bulk

$$R(\hat{U})=-\frac{1}{8}\sigma_{xy}^{II}\int d^2r\,\varepsilon_{\mu\nu}Tr\,\hat{Q}\nabla_\mu\hat{Q}\nabla_\nu\hat{Q}; \qquad \hat{Q}=\hat{U}\hat{s}\hat{U}^{-1} \tag{24b}$$

The quantity $\hat{Q}$ is an element of the coset space $\frac{U(2m)}{U(m)\times U(m)}$. It reduces to a constant ($\hat{s}$) at the boundary, so for its group manifold, the two-dimensional space can be thought of as being compactified to the sphere S^2 with the edge contracted to a point. The results above form a verification of the formal homotopy result $\pi_2\left[\frac{U(2m)}{U(m)\times U(m)}\right]=Z$ and the winding number (topological charge)

$$q(\hat{Q})=\frac{i}{16\pi}\int d^2r\,tr\,\varepsilon_{\mu\nu}\hat{Q}\nabla_\mu\hat{Q}\nabla_\nu\hat{Q} \tag{25}$$

is the integer Z associated with each homotopy sector [25,14,17].

Just as in the discussion of the static current in section 2.3, the meaning of Stokes' theorem can be formulated by saying that it translates a many-body aspect of the problem (Fermi-statistics) in terms of a single-particle problem. Phase factors of the type of Eqs (24) with $\sigma_{xy}^{II}\to\sigma_{xy}^{0}$ become important as statistical factors in the process of diffusion (localization). The above discussed invariance for gap-states is a perfect symmetry in the problem, which should be retained as a valid limit in the formulation of the localization problem.

This aspect forms the justification for the use of Grassmann variables, and consequently compact symmetries in the original definition of the generating function, Eq.(2). A well known and alternative formulation involves ordinary Boson fields [13], forwhich the continuous symmetry is non-compact. In this case, the contact with the gap-physics as discussed above cannot so easily be made because of two reasons; namely, there doesnot exist a formulation which is free of divergence problems in case the order parameter $\rho(E)$ vanishes [26,27] and secondly, the validity of Stokes' theorem which connects Eqs (24), is in general violated. It may, however, not be a surprise that boson fields are not capable of dealing with this problem because of the special role of Fermi statistics in this case, which is absent in the conventional localization theory.

III. The localization problem

3.1 Introduction

In ref.[13] an effective Lagrangian has been derived describing the non-equilibrium properties of the disordered system, and which serves as the basis for further analysis with renormalization group. The topological aspects require a rather subtle treatment and the essential ingredients are unavoidably overlooked by the standard discussions, which for the major part rely upon ordinary perturbation theory.

The method of ref.[13] is based upon earlier work by Lothar Schaefer, and the present author [26] in the context of the original Anderson model for electronic disorder. For computational convenience, one considers a white noise distribution

$$P[V]=N^{-1}e^{-\frac{1}{2g}\int d^2r V^2(r)} \tag{26}$$

in which case the integration over randomness in Eq.(2) can be performed explicitly, leading to a four-point interaction

$$L'=\int d^2r\left[\sum_{p,a}\bar{\Psi}_a^p(r)[E_p-\pi_\mu\pi^\mu]\Psi_a^p(r)+\frac{1}{2}g\left[\sum_{p,a}\bar{\Psi}_a^p(r)\Psi_a^p(r)\right]^2\right] \tag{27}$$

The starting point of the discussion is the introduction of hermitian, composite fields $Q_{ab}^{pp'}(r)$ which replace the combination $\bar{\Psi}_a^p\Psi_b^{p'}$ in the four-point term in L via a Gaussian transformation. This socalled Q-field method had been introduced first by Wegner [22] and leads to a formalism in which the symmetries in the problem are naturally incorporated. The formalism allows one furthermore to make contact with the various diagrammatic expansion techniques which have been introduced by various schools of thought. The equivalence between the conventional diagrammatic and field theoretic descriptions has been analyzed in detail by Schaefer and Wegner [27], McKane and Stone [28] and Hikami [11].

The IQHE however, cannot be analyzed diagrammatically and not even by resummation techniques involving large orders in perturbation theory as applicable for e.g. the density of states. Although this insight was implicit in the derivation of ref [13], it might be worthwhile to provide some more detail on this point. The introduction of aforementioned fields Q leads to a partition function, equivalent to Eq (2)

$$Z=\int D[Q]\int D[\bar{\Psi}]\int D[\Psi]e^{L(\bar{\Psi},\Psi,Q)}$$

$$L(\bar{\Psi},\Psi,Q)=\frac{1}{2g}Tr\,Q^2+Tr\,\bar{\Psi}_a^p[(E-\pi_\mu\pi^\mu)\hat{1}+s+iQ]_{ab}^{pp'}\Psi_b^{p'} \tag{28}$$

The resulting integral for the Ψ is Gaussian and can be performed explicitly, leading to a description in terms of Q alone

$$Z=\int D[Q]e^{L[Q]}$$

$$L[Q]=-\frac{1}{2g}Tr\,Q^2+Tr\ln[(E-\pi_\mu\pi^\mu)\hat{1}+\hat{s}+iQ] \tag{29}$$

The naive way to proceed is to look for the saddle-point of the Lagrangian and treat the fluctuations order by order in perturbation theory. The saddle-point equation can be obtained by replacing $Q_{ab}^{pp'}(r)\rightarrow\delta_{ab}\,\delta_{pp'}\,q^p$

$$-\frac{1}{g}q^p+\frac{i}{2\pi\alpha}\sum_n\frac{1}{E-\hbar\omega_c(n+\frac{1}{2})+s^p+iq^p}=0 \tag{30}$$

where the imaginary part of iq^p is proportional to the density of states in saddle point approximation, $\rho^0(E)$. In Eq.(30), use has been made of the eigenfunctions $\xi_{n,k}(r)=e^{iky}\varphi_n(x+\alpha k)$ of the operator $\pi_\mu\pi^\mu$ with φ_n the harmonic oscillator functions and with $1/(2\pi\alpha)$ identified as the degeneracy per Landau level.

One may next proceed and formulate a type of loop-expansion about the saddlepoint by replacing $Q\rightarrow q^p+\delta Q$, a proceedure which is equivalent to the 1/N expansion of Oppermann and Wegner [29]. A sign of concern should be mentioned at this stage, which is that the boundaries of the system and hence the effect of edge-currents are not naturally taken into account by the saddle-point equation. In order to hammer this point down, let us summarize the results of the saddle-point analysis. Taking into account the fluctuations to quadratic order, one obtains

$$L(Q)=L(q^p)-\int d^2r\int d^2r'\sum_{pp'}\sum_{ab}\delta Q_{ab}^{pp'}(r)\,M^{pp'}(r,r')\,\delta Q_{ba}^{p'p}(r')+O(\delta Q^3)$$

$$M^{pp'}(r,r')=+\frac{1}{2g}\delta(r-r')-\frac{1}{2}G_0^p(r,r')G_0^{p'}(r',r) \tag{31}$$

where G_0^p denote the mean-field propagators

$$\begin{aligned}G_0^p(r,r')&=\langle r\,|(E-\pi_\mu\pi^\mu+s^p+iq^p)^{-1}|\,r'\rangle\\&=\sum_{k,n}\frac{\varphi_n(x+k\alpha)\varphi_n^*(x'+k\alpha)}{E-\hbar\omega_c(n+\frac{1}{2})+s^p+iq^p}e^{ik(y-y')}\end{aligned} \tag{32}$$

The matrix elements $M^{pp'}$ can easily be shown to be translationally and rotationally invariant if one assumes an infinite system. After straightforeward algebra, one can show that the matrix elements M^{+-} have the diffusive, long wavelength behaviour

$$M^{+-}(k)=\frac{1}{2}\int d^2r M^{+-}(r)e^{ik\cdot r}\sim\frac{1}{2}\sigma_{xx}^0\,k^2+\eta\rho^0(E) \tag{33}$$

with σ^0_{xx} and ρ^0 the mean field conductance and density of states respectivily. The inverse of this matrix element is usually referred to as the diffusion propagator.
On the other hand, the matrix elements diagonal in p,p' are "massive"

$$\mathrm{Re}[M^{pp}(k)]>0 \quad \text{for all k.} \tag{34}$$

Going beyond the Gaussian approximation, the above-mentioned loop-expansion will lead to "dressed" diffusion propagator in which the bare conductance acquires a logarithmic correction [11,12]

$$\sigma^0_{xx} \to \sigma^0_{xx} - \frac{1}{8\pi^2 \sigma^0_{xx}} \ln \mu L \ , \tag{35}$$

in agréement with the non-linear σ-model calculations (section 3.2).
However, there is a loop-hole in the apparantly innocent series of arguments as sketched above. Namely, the long-distance behavior of the matrix elements M^{+-} turns out to be described by

$$M^{+-}(r,r') = \frac{1}{2}\sigma^0_{xx}\delta(r-r')\nabla^2_\mu + \frac{1}{4}\bar{\sigma}^0_{xy}\varepsilon_{\mu\nu}\overleftarrow{\nabla}_\mu \delta(r-r')\overrightarrow{\nabla}_\nu \tag{36}$$

The additional term in Eq.(36) is a surface term and can only be seen from Eq.(31) by explicitly taking into account the edges in the saddlepoint equation. This result can be arrived at in a fashion, similar to the discussion on edge-currents, section 2.3. Hence, the diffusion propagator doesnot exhibit the usual type of behaviour,and the full effect of the surface term can only be discussed in a way which goes beyond the loop-expansion as discussed here. More general, the assumption of the simple behaviour of Eq.(33) for the diffusion propagator in *any* approximation scheme is bound to give the wrong answer [30].

As a final comment, the parameter $\bar{\sigma}^0_{xy}$ in Eq.(36) turns out be a very bad mean field approximation to the Hall-conductance. Apparantly,the relation between bulk and edge effects is badly approximated in mean field theory.

3.2 Effective Fieldtheory

A formulation which is free of these difficulties is inspired by the notion that for $\eta \to 0$ the mean field Lagrangian and hence the saddle-point is invariant under the insertion $q \equiv \delta_{ab}\delta_{pp'}q^p \Rightarrow [T^{-1}\hat{q}T]^{pp'}_{ab}$ with $T \in U(2m)$. Hence,there is a saddle-point manifold and the transverse fluctuations δQ^{+-} are identified as infinitesimal fluctuations along that manifold. This allows one to formally split the Q-field into massive (P) and critical (T) components

$$\begin{aligned} Q(r) &\to T^{-1}(r)P(r)T(r) \\ P(r) &= \delta_{pp'}P^p_{ab}(r) = P^\dagger(r) \\ T(r) &\in U(2m) \end{aligned} \tag{37}$$

such that these components can be studied individually and at their own right. The aim is to formulate an effective Lagrangian in terms of the T by formally integrating over P

$$e^{L[T]} = \int D[P]I[P]e^{-\frac{1}{2g}TrQ^2 + Tr\ln[E\hat{1} - T\pi_\mu\pi^\mu T^{-1} + i\eta T s T^{-1} + iP]} \tag{38}$$

which can be achieved by cumulant expansion about the theory $L[T\equiv\hat{1}]$. The latter only contains the fluctuations P, which are harmless and short-ranged,as can be explicitly checked via the saddle-point method, described above. It is nevertheless important to keep the complete P in the problem because of the edge-effects, which can only be related to those of the bulk in an exact theory. This will avoid problems of the type described under Eq.(32).
Following the analysis of ref.[13], $L[T]$ can be obtained via a gradient expansion in the field T similar to the analysis of section 2.5.

$$L[T] = -\frac{1}{4}\sigma^0_{xx}\int_r tr\nabla_\mu\tilde{Q}\nabla_\mu\tilde{Q} + \frac{1}{8}\sigma^0_{xy}\int_r tr\,\varepsilon_{\mu\nu}\tilde{Q}\nabla_\mu\tilde{Q}\nabla_\nu\tilde{Q} - \eta\rho(E)\int_r tr\,\tilde{Q}\hat{s} \tag{39}$$

where $\tilde{Q}=T^{-1}\hat{s}T$ belongs to the group $\frac{U(2m)}{U(m)\times U(m)}$. The quantities $\sigma^0_{\mu\nu}$ are expressed in terms of complicated Kubo-type formulae [13] and are to a good approximation given in the SCBA [20]. The quantity $\rho(E)$ in the symmetry-breaking term is the exact density of states in the problem.

The resulting Lagrangian is the most general, renormalizable field theory that one can write down, which respects the symmetries of the original replicated generating functional. The presence of the magnetic field in the problem is contained in the second term in Eq.(39) which breaks the "pseudo" parity symmetry $(x,y)\to(-x,y)$. In the following sections the implications will be discussed for the electronic system.

3.2.1 The conductance in SCBA

The mean field results for $\sigma_{\mu\nu}$ follow from evaluating the expressions, derived in [13], in the saddlepoint approximation (Eq.30). The saddle-point equation is recognized as the self-consistent equation for the self-energy, intensively studied by Ando [20], under the identification

$$iq^p = E_0 + \frac{i}{\tau} s^p$$

where τ stands for the relaxation time and E_0 the shift in Fermi energy. Adapting the more conventional notation, the mean field result for the Hall conductance becomes

$$\sigma^0_{xy} \times \frac{e^2}{h} = -ec\frac{\partial N_0}{\partial B} - \omega_c \tau \sigma^0_{xx} \times \frac{e^2}{h} \approx -\omega_c \tau \sigma^0_{xx} \quad \text{for} \quad \omega_c \tau \ll 1 \tag{40}$$

where N_0 is the density of electrons, i.e. the density of states in saddle point approximation, integrated over all energies below the Fermi energy. Eq.(40) reduces to the semi-classical Drude-Zener result in the limit of weak magnetic fields ($\omega_c \tau \ll 1$). It turns out that Eq.(40) can also be casted in the form

$$\sigma^0_{xy} \times \frac{e^2}{h} = -ec\frac{N_0}{B} + \frac{1}{\omega_c \tau}\sigma^0_{xx} \times \frac{e^2}{h} \approx -N_0 \frac{ec}{B} \quad \text{for} \quad \omega_c \tau \gg 1 \tag{41}$$

which conveniently describes the strong field limit ($\omega_c \tau \gg 1$). In this latter case, σ^0_{xy} varies as the filling-fraction of the partly occupied Landau level plus the number of fully occupied levels. The dissipative conductance can be written as

$$\sigma^0_{xx} \times \frac{e^2}{h} = \frac{(E+E_0)\rho_0 e^2 \tau}{m} \frac{1}{1+(\omega_c \tau)^2} \tag{42}$$

which reduces for weak magnetic fields to the familiar semi-classical result

$$\sigma^0_{xx} \times \frac{e^2}{h} = \frac{N_0 e^2 \tau}{m} \frac{1}{1+(\omega_c \tau)^2} \ ; \quad \omega_c \tau \ll 1 \ . \tag{43}$$

For strong fields, the maximum in the dissipative conductance occurs at the center of the Landau level and is given by

$$[\sigma^0_{xx}]^{\max} \times \frac{e^2}{h} = \frac{2e^2}{\pi h}(n+\tfrac{1}{2}) \ ; \quad \omega_c \tau \gg 1 \tag{44}$$

independent of magnetic field and strength of impurity potential. A simple interpretation of the strong field results in terms of the kinetic theory have been given by Ando and Uemura [20]. More details on the saddle-point approximation will be given elsewhere.

3.2.2 Conventional Localization

The first term in L is a generalization of the more familiar CP^{N-1} model,which can be obtained by putting $\tilde{Q}\in\frac{U(N)}{U(N-1)\times U(1)}$. This field theory is well-known to apply for problems inwhich time-reversal symmetry is broken, i.e. spin-flip scattering and ran-

dom complex matrix models, and the results are similar to the one with orthogonal symmetries, describing potential scattering (Anderson model).
The renormalizability of the theory to all orders in perturbation theory is contained in ref.[31]; the basic result is contained in the renormalization group equation ($\varepsilon=d-2$) [32,33]

$$\frac{\partial t}{\partial \ln L}=-\varepsilon t+2mt^2+2(m^2+1)t^3+(3m^2+7)m^2t^4+\ldots \tag{45}$$

with $t=4\pi/\sigma_{xx}$. At this stage, the analytic continuation to $m\to 0$ can be performed. The theory is asymptotically free in d=2 for all $m\geq 0$ and this gives rise to the believe in "weak localization" at short length scales and "strong localization" at large length scales. In $2+\varepsilon$ dimensions there is a fixed point t_c of order $\sqrt{\varepsilon}$ as $m\to 0$. The initial value t^0 for scaling is via σ_{xx}^0 a smooth function of energy E such that this fixed point can be associated with a "mobility edge", E_c, separating metallic ($\sigma_{xx}\neq 0$) from insulating ($\sigma_{xx}=0$) behaviour. Near E_c one has the scaling behaviour of the conductance

$$\sigma_{xx}\sim L^{d-2}|E-E_c|^s f_0\left[\frac{\eta}{|E-E_c|^{d\nu}}\right] \tag{46}$$

with correlation length exponent $\nu=-1/\beta'(t_c)$ and with conductivity exponent $s=(d-2)\nu$ which in $2+\varepsilon$ dimensions are given by

$$\frac{1}{\nu}=2\varepsilon+O(\varepsilon^2); \qquad s=\tfrac{1}{2}+O(\varepsilon).$$

These arguments emphasize a strong analogy with the Heisenberg ferromagnet [22]. Characteristic, however, for the localization problem are the rather unconventional, critical fluctuations in the "order parameter", $\rho(E)$. The density of states itself is non-critical; this is correctly reproduced by the field theory, Eq.(39), because the conjugate, critical operator $\tilde{Q}\hat{s}$ has a vanishing critical exponent as $m\to 0$ [32].
Local fluctuations in $\rho(E;r)$ on the other hand, equal the inverse of the participation ratio of the states at energy E and these should diverge as one approaches E_c from the metallic side. These fluctuations are described by a critical operator bilinear in the $\tilde{Q}$ with a negative critical exponent [34]. This peculiar aspect only shows up in the limit of vanishing number of field components, and this has recently been checked for the unitary model in a $2+\varepsilon$-expansion [35]. Specifically, the scaling behaviour

$$\overline{(\rho(E;r)-\bar{\rho})^2}\sim|E-E_c|^{-\mu_2}f_1\left[\frac{\eta}{|E-E_c|^{d\nu}}\right] \tag{47}$$

is characterized by a critical index μ_2 which in $2+\varepsilon$ dimensions is given by $\mu_2=\frac{1}{\sqrt{2\varepsilon}}+O(\sqrt{\varepsilon})$ which is to be compared with that for Anderson localization [34] $\mu_2=2+O(\varepsilon)$. This completes the picture for conventional localization.

3.2.3 Localization in a magnetic field

The second term in L is the topological invariant, discussed in section 2.5 and already noted in several related σ-models [36] and introduced for the unitary case in [25]. This term is a 2-d realization of the ϑ-term, an issue which originally arose in 4-d gauge-theories [37]. The way in which the resulting Lagrangian resolves the localization problem in the IQHE is a rather unique field theoretic issue and in refs [14-18] a more or less complete discussion has been presented. The central point here is to see how the additional topological term can change the results of the perturbative renormalization group analyses as described above.

The first thing to remark is that the functional integral is restricted to only those field configurations $\tilde{Q}$ which have integer-topological charge $q(\tilde{Q})$, Eq.(25). This is a result of the existence of boundary currents in the underlying non-critical theory, described by the fields P in Eq.(38). Under these circumstances, the theory makes furthermore a smooth connection with the previously discussed situation inwhich the Fermi level lies in a density of states gap.

i) In order to detect the dependence of the theory on σ^0_{xy}, one clearly has to take into account all the topological sectors q in the theory. According to the Polyakov inequality,

$$tr(\nabla_x \tilde{Q} + i\tilde{Q}\nabla_y \tilde{Q})^2 \geq 0 \tag{48}$$

there is an upperbound to the statistical weight, given to each topological sector

$$\int d^2r tr \nabla_\mu \tilde{Q} \nabla_\mu \tilde{Q} \geq 16\pi |q(\tilde{Q})| \quad . \tag{49}$$

This means that the topologically non-trivial sectors in the theory involve an Action of at least $4\pi\sigma^0_{xx}$; such contributions to the Greens' functions typically involve exponentials $\exp(-4\pi\sigma^0_{xx} \pm 2\pi i \sigma^0_{xy})$ which are unobservable in the weak localization, $1/\sigma^0_{xx}$ asymptotic series.

Field configurations forwhich the Polyakov inequality is satisfied as an equality are called "instantons". These are minima for each topological sector, and form a subset of solutions to the Euler-Lagrange equations of motion. In the weak localization limit, we may expect that the functional integral can be treated by saddle-point methods and go beyond the usual weak localization methodology by including instantons as non-perturbative vacuum states. Much is known about the topological properties of the excitations in the closely related and popular O(3) non-linear σ-model [38], which can be obtained by putting m=1. The topological content of the effective Lagrangian, Eq. (39), is independent of m and the semi-classical methods parallel those employed for the more simple m=1 theory. The specific instanton solutions for the general, unitary case and the way their contribution to the free energy survives in the replica limit $m \to 0$ have been discussed in ref.[16].

ii) There is a general expectation that a phase-transition occurs for σ^0_{xy} = half-integer. This has been anticipated in the end of section 3.2 on the basis of the Kubo-Streda formula, and this is what is needed in order to make the localization picture consistent. It is, however, encouraging that this expectation holds for general number of field components and clearly is not specific for the replica limit $m \to 0$. An example forms the exactly soluble large-N limit of the CP^{N-1} model [39], for which the ϑ- or σ^0_{xy}-dependence is similar to that of the massive Schwinger model [40]. A first order phase transition occurs with a gusp in the free energy, and this result persists for finite N in a 1/N expansion [41]. The same results have been recently obtained in a strong coupling coupling expansion for finite N [42]. This might be suggestive for what happens in the electronic system, but the evidence is not strong enough as of yet.

In ref.[16] a duality approach has been given which argues for a phase-transition for σ^0_{xy} near half integer values for the general unitary case by adapting 't hooft's arguments, originally given for SU(N) gauge theories, for the 2-d system. In this approach, the boundary conditions on the fields Q are relaxed such as to allow for field configurations with half-integer topological charge. Finite Action saddle-point solutions with this topology are also expected in a suitable, compactified space such as the projective plane in which opposite points on the edge are identified. These twisted boundary conditions can be employed in order to probe for localization in the interior of the system and the spirit is similar to the work by Edwards and Thouless [43] on conventional localization. The duality argument becomes meaningfull within a Hamiltonian or transfer matrix approach to the functional integral, in the context of which the issue of the ϑ-vacuum originally arose. These arguments will not be repeated here and we proceed in a fashion which is somewhat closer to the physics of the IQHE.

iii) A less familiar point is the "renormalization" of the conductance-parameters. This aspect is particularly meaningfull for the disordered electronic system and doesnot appear in field theoretic context. Indeed, the large N results of the CP^{N-1} model shows that only one renormalization constant is needed in in order to absorb the ultra-violet divergencies in the theory. From this point of view there seems nothing indicative of the IQHE.

In order to resolve this question, let us go back to the source-term formalism introduced in section 3.4. The conductance follows from the response of the free energy to a slowly varying rotation between advanced and retarded waves. In the formulation of the Lagrangian we have retained those fluctuations in the theory which directly couple to those long-wavelength backgound fields, whereas the short-wavelength components P have been integrated out. Following up the method of section 2.4, the conductance can be obtained from the effective Lagrangian $L[T]$, Eq.(39), by replacing Q by $\hat{U}^{-1}Q\hat{U}$, leaving the symmetry breaking term unchanged. In this way an expression can be obtained for the "renormalized" conductance in terms of correlations of the Q field

$$\sigma_{xx} = \sigma^0_{xx} - \tfrac{1}{2}\sigma^0_{xx} tr \langle \hat{\tau}_x \tilde{Q}(r) \hat{\tau}_x \tilde{Q}(r) \rangle \\ - \tfrac{1}{2}(\sigma^0_{xx})^2 \Omega^{-1} \int_r \int_{r'} \langle \left[tr \hat{\tau}_x \tilde{Q}(r) \nabla_x \tilde{Q}(r) \right] \left[tr \hat{\tau} \tilde{Q}(r') \nabla_x \tilde{Q}(r') \right] \rangle \tag{50}$$

$$\sigma_{xy} = \sigma^0_{xy} - \sigma^0_{xx} tr \langle \varepsilon_{\mu\nu} r_\nu \tilde{Q}(r) \nabla_\mu \tilde{Q}(r) \hat{\tau}_z \rangle \\ - \tfrac{1}{2}(\sigma^0_{xx})^2 \Omega^{-1} \int_r \int_{r'} \varepsilon_{\mu\nu} \langle \left[tr \hat{\tau} \tilde{Q}(r) \nabla_\mu \tilde{Q}(r) \right] \left[tr \hat{\tau}_y \tilde{Q}(r') \nabla_\nu \tilde{Q}(r') \right] \rangle \tag{51}$$

where Ω equals the volume and $\hat{\tau}_\mu$ is defined in Eq.(19). The expectation $\langle ... \rangle$ is with respect to $L[T]$, Eq.(39).

The importance of this result is contained in the following three aspects. First, the contact with the situation of a gap in the bulk density of states is explicitly present, i.e. $\sigma_{xx} \to \sigma^0_{xx} \to 0$, $\sigma_{xy} \to \sigma^0_{xy} \to integer$. This condition is what we have to impose by the arguments given in section 2.3.

Furthermore, by considering the effect of the transformation $\tilde{Q}(x,y) \to \tilde{Q}(y,x)$ one concludes that the bare and renormalized conductances are related as

$$\sigma_{xx} = \sigma^0_{xx} + c_0 + \sum_{n=1}^{\infty} c_n \cos(2\pi n \sigma^0_{xy}) \tag{52}$$

$$\sigma_{xy} = \sigma^0_{xy} \qquad + \sum_{n=1}^{\infty} \vartheta_n \sin(2\pi n \sigma^0_{xy}) \tag{53}$$

where c_n, ϑ_n are functions of σ^0_{xx} only and originate from the topologial sectors with charge n. These coefficients typically contribute with factors $exp(-4\pi n \sigma^0_{xx})$. On the other hand, conventional perturbation theory tells something about c_0 alone.

Finally, Eqs (50,51) make explicit what we mean by renormalization of the parameters $\sigma_{\mu\nu}$. One can argue that if the theory has a mass gap, then $\sigma_{xx}=0$ and $\sigma_{xy}=integer$. Let us first consider the situation $\sigma^0_{xy}=0$. In this case, the topological invariant in L is immaterial and this situation is correctly reproduced in Eqs(52,53). Furthermore, asymptotic freedom implies that free energy is not dependend upon the very long wavelength components in Q, namely those components which vary slowly over a length given by the correlation length or localization length. Alternatively, we can say that the free energy is insensitive to the insertion of slowly varying background fields. Such a statement can be explicitly verified in the large N limit [39,32] or within the renormalization group analysis as given by Polyakov [31].

If it is reasonable to assume that no phase transition occurs for small, non-zero σ^0_{xy}, then the free energy must remain insensitive to slowly varying background fields with arbitrary integer topological charge. In particular, because the theory describes finite-range correlations, we can perform a gradient expansion in the background field and obtain non-trivial results. The equivalent of Eq.(16) becomes

$$Z[\hat{U}] = \int D[\tilde{Q}] e^{L[T,\hat{U}]}$$

$$L[T,\hat{U}] \doteq -\frac{1}{4}\int_r tr \nabla_\mu [\hat{U}^{-1}\tilde{Q}\hat{U}] \nabla_\mu [\hat{U}^{-1}\tilde{Q}\hat{U}] + 2\pi i \sigma^0_{xy} q[\hat{U}^{-1}\tilde{Q}\hat{U}] - \eta\rho(E)\int_r tr \tilde{Q}\hat{s} \tag{54}$$

The only form which is compatible with the symmetries is the form of L, Eq.(39), itself

$$Z[\hat{U}]_{\eta\to 0} = Z[\hat{U}=\hat{1}]_{\eta\to 0} e^{-\frac{1}{4}\sigma_{xx}\int d^2r tr \nabla_\mu V \nabla_\mu V + 2\pi i \sigma_{xy} q[V] + O(\nabla^3 V)} \tag{55}$$

with $V=\hat{U}^{-1}\hat{s}\,\hat{U}$. The parameters $\sigma_{\mu\nu}$ are identified as the renormalized parameters, Eqs.(50,51), as can be explicitly checked by inserting the special form of $\hat{U}$, namely Eq.(19). Abovementioned invariance now implies that the dissipative part, σ_{xx}, has to vanish but σ_{xy} must be equal to an integer. This, then, is the IQHE.

Notice that the arguement only breaks down if a phase transition occurs, i.e. the system describes long range correlations such that the gradient expansion is no longer justified. It is clear, however, that a phase-transition *has* to occur if σ_{xy}^0 passes through ½. Namely, by the arguments presented above, $\sigma_{xy}^0=\delta\gtrsim 0$ scales toward $\sigma_{xy}=0$, whereas $\sigma_{xy}^0=1-\delta$ scales toward $\sigma_{xy}=1$. Then there must exist a critical value $|\delta|<1$ for which the argument breaks down and hence *de* localization occurs.

The arguements presented above are quite general and not specific tc the limit of zero replica's and to the Gaussian white-noise potential. In [16] a formulation has been presented which is free of these restrictions.

3.2.4 Dilute instanton gas

The scaling toward quantization of the Hall conductance is a property of taking the thermodynamic limit but it need not show up as a divergence in the ultraviolet behaviour of the theory. Similarly, there is no continuously diverging length-scale associated with the first-order phase-transition found in the large-N CP^{N-1} model and in the finite N, strong-coupling expansion in [42]. From the point of view of the IQHE, the interesting properties are in the infrared and the physically relevant renormalization group scheme is the block-spin type of approach in which the fastly varying components in Q are integrated over giving rise to an effective Lagrangian in which the slowly varying ones are retained and with renormalized coupling constants $\sigma_{\mu\nu}^0$. In such a scheme, there would be an infrared stable fixed point associated with the points $(\sigma_{xx},\sigma_{xy})$ = (0, integer) per the above constructed quantization argument. Furthermore, the first order singularity in the free energy along the line $\sigma_{xy}^0=½$, as suggested by the models above, would be described in such a renormalization group scheme by a fixed point on that line with a relevant critical exponent in the σ_{xy}-direction equal to the dimensionality d=2.
Hence, the parts in the scaling diagram, relevant for the IQHE are already present far away from the replica limit $m\to 0$.

As mentioned before, in the weak localization limit one can expect saddlepoint methods to work which then explicitly show how non-perturbative contributions change the scaling behaviour. Recently, an approximate renormalization group analysis of the type described above, has been performed [18] in the socalled dilute instanton gas approximation [44]. In this approximation, one saturates the functional integral with independent and widely separated instantons of unit charge $q=\pm 1$. The single instanton is characterized by its position, size and furthermore by its orientation within the coset space $\frac{U(2m)}{U(m)\times U(m)}$. A dilute gas of such objects gives rise to a contribution to the partition function

$$Z_{INST} = \sum_{n_+,n_-} \frac{e^{-4\pi\sigma_{xx}^0(n_++n_-)+2\pi i\sigma_{xy}^0(n_+-n_-)}}{n_+!\;n_-!}(\Omega D_0)^{n_++n_-} \tag{56}$$

where the factor ΩD_0 orignates from the one-loop fluctuations about the instanton solutions, which contains furthermure the integration over the single instanton degrees of freedom. The factor Ω, the size of the system, is singled out and is a result of the arbitrary position. The factor D_0 still contains the integration over the arbitary scale size ρ and a volume factor as a result of the orientation of the instanton within the coset space.

The well-known draw-back of the semi-classical method for scale-invariant theories is the infrared divergence in D_0 due to the arbitrary scale-size ρ of the instanton [37,38]. This gives rise to a singular contribution to the free energy F

$$F_{INST}\propto \Omega\int \frac{d\rho}{\rho^3}e^{-4\pi\sigma_{xx}(\rho)}\cos(2\pi\sigma_{xy}^0) \tag{57}$$

with $\sigma_{xx}(\rho)$ identified as the perturbative scale-size dependent conductance, the one-loop result obtained from Eq.(45)

$$\sigma_{xx}(\rho)=\sigma_{xx}^{0}-\frac{m}{2\pi}\ln\rho. \tag{58}$$

This divergence for large scale-sizes obscurifies the exact meaning of individual instantons somewhat and indicates that the gas of instantons, rather than being dilute, is strongly interacting and dense. This indicates furthermore that the terms in the series of Eq.(52,53) do not have a meaning individually and have to be summed over to infinite order.

However, in the replica limit $m\to 0$ the integration over scale-sizes gives rise to an ultra-violet divergent contribution to the free energy, a result which is due to the fact that the one-loop correction to the perturbative β-function vanishes in this case. This means that the dominant contribution comes from small size instantons and the diluteness-ansatz of the instanton gas is justified in a rather large region from weak to intermediately strong localization. This is a lucky coincidence which can be exploited in order to proceed with a sensible renormalization group scheme, in which the small-size instantons and length scales are eliminated and the longer ones are kept. For this purpose, one can evaluate the partition function in the back-ground of a very large instanton, which schematically takes on the following form

$$Z'_{INST}=e^{(-4\pi\sigma_{xx}^{0}+2\pi i\sigma_{xy}^{0})}\times\sum_{n_+,n_-}\frac{e^{\left[-4\pi\sigma_{xx}^{0}(n_++n_-)+2\pi i\sigma_{xy}^{0}(n_+-n_-)+\sigma_{xx}^{0}(I_{++}n_++I_{+-}n_-)\right]}}{n_+!\;n_-!}. \tag{59}$$

The Action of the single, back ground field instanton has been singled out and the interaction of the dilute gas with the back ground field is denoted by I_{++} for instantons and I_{+-} for anti-instantons. The latter is positive definite, saying that instantons of opposite charge attract each other, whereas I_{++} can be neglected in the same order of approximation [38,45]. Evaluating the sum, one concludes that the interaction I_{+-} gives rise to the renormalization

$$\sigma_{xx}^{0}\to\sigma_{xx}^{0}-\sigma_{xx}^{0}\,\frac{I_{+-}}{8\pi}e^{-4\pi\sigma_{xx}^{0}}\cos(2\pi\sigma_{xy}^{0}) \tag{60}$$

$$\sigma_{xy}^{0}\to\sigma_{xy}^{0}-\sigma_{xx}^{0}\,\frac{I_{+-}}{4\pi}e^{-4\pi\sigma_{xx}^{0}}\sin(2\pi\sigma_{xy}^{0}) \tag{61}$$

This leads to a heuristic explanation of the first terms in the series of Eqs(52,53) with the important addition that the coefficients I_{+-} are positive definite.

When fluctuations about the instantons are taken into account, the elimination of contributions from small length-scales can be completed, leading to the non-perturbative renormalization group equations [18]

$$\beta_{xx}=\frac{\partial\sigma_{xx}}{\partial\ln L}=\frac{-1}{8\pi^{2}\sigma_{xx}}-(\sigma_{xx})^{3}\overline{D}e^{-4\pi\sigma_{xx}}\cos(2\pi\sigma_{xy}) \tag{62}$$

$$\beta_{xy}=\frac{\partial\sigma_{xy}}{\partial\ln L}=-(\sigma_{xx})^{3}\overline{D}e^{-4\pi\sigma_{xx}}\sin(2\pi\sigma_{xy}) \tag{63}$$

where $\overline{D}$ is a positive constant. The diluteness of the instanton gas in this case means that the above results can be trusted qualitatively over a rather large region for σ_{xx} [18]. This furthermore makes it meaningfull to analytically continue the renormalization group flow to smaller values of σ_{xx} where the instanton gas becomes increasingly dense. The results are plotted in Fig. 2 under the assumption that the fixed-point at finite σ_{xx} exists [46,47]. The dashed parts of the flow-lines at very small values of σ_{xx} are inserted by hand. The lines indicated as (—.—) are the location of the bare parameters $\sigma_{\mu\nu}^{0}$ (section 3.2.1) for the strong magnetic field limit.

An important question which remains is the nature of the phase transition near half-integer values for σ_{xy}^{0}. The first order phase transition, mentioned above implies that the localization length stays finite as one approaches the singular point but there is as of yet no reason to believe that such a result extrapolates all the way down to the replica limit $m\to 0$.

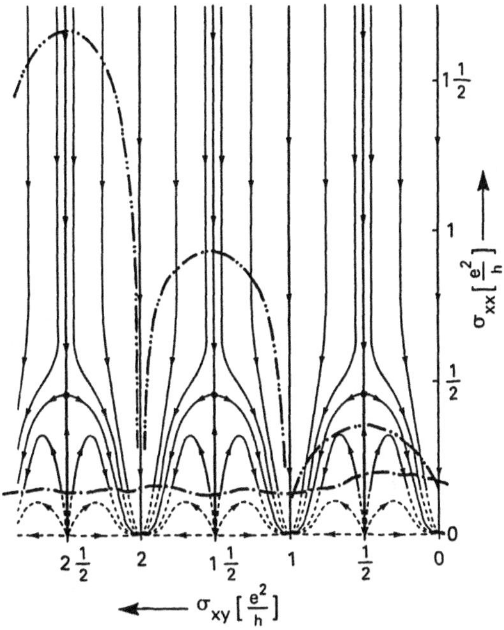

Fig.2 The renormalization group flow-diagram in the dilute instanton gas approximation.

The results discussed above are amenable to experimental verification. Recent experimental work on InGaAs-InP heterostructures at finite temperatures [48] has established the relevance of the two parameter scaling theory for the real physical system. The data are consistent with the existence of a single singular energy in the Landau level and with the existence of the fixed points at non-zero σ_{xx} in Fig.2. The results are very encouraging for further experimental as well as theoretical research. A discussion on experimental relevance, as well as on the recent attempts to incorporate the Fractional Quantum Hall Effect in in the flow-diagram of Fig.2 [49], will not be pursued here.

IV. Boundary currents and Stokes' theorem II

Going back to the expression for the local current density, Eq.(16), we can ask what the above discussed relation between localization and continuous symmetry actually implies for the electronic current flow in the system in response to external perturbations. Notice hereto that Eq.(16) can be written as

$$j_\mu(E,r)=\frac{i\hbar}{\sqrt{2m}}\,tr\,\overline{(\pi_\mu+\pi^\mu)\hat{G}(r,r)\tau_z} \tag{64}$$

where

$$\hat{G}(r,r')=\langle r|\left[(E-H)\hat{1}+i\eta\tau_z\right]^{-1}|r'\rangle \tag{65}$$

The concept of localization can now be understood by saying that *no* macroscopic current response can result from replacing

$$i\eta\tau_z \rightarrow i\eta\, T^{-1}(r)\tau_z\, T(r) \tag{66}$$

for $T\in U(2)$, slowly varying with respect to the localization length and characterized by integer topological charge. The consequences of this statement can be worked out via performing a gradient expansion in ∇T of the type discussed before. This results in [16]

$$j_\mu(E,r;T)=\sigma_{xx}\,tr\,V(r)\nabla_\mu V(r)\tau_z+\tfrac{1}{2}\sigma_{xy}\,tr\,\varepsilon_{\mu\nu}\nabla_\nu V(r)\tau_z+O(\nabla^2 V(r)) \tag{67}$$

where $V(r)=T^{-1}(r)\tau_z T(r)$ and the $\sigma_{\mu\nu}$ precisely given by the exact expressions, Eqs(4,9).

If the total current in, say, the y-direction has to vanish then clearly $\sigma_{xx}=0$ is a necessary condition which is consitent with the *ansatz* of localization. For the term involving the Hall-conductance, we have

$$J_y \equiv \int_0^L dx j_y(E,r;T) = \sigma_{xy}\left[\tilde{V}(L,y) - \tilde{V}(0,y)\right] \tag{68}$$

with $\tilde{V}$ equal to $\frac{1}{2} tr V\tau_z$. For integer topological charge for the field T, V is constant along the boundary and the expression vanishes.

Clearly, a macroscopic current J is only flowing by the imposition of non-trivial boundary-conditions on the field T. But because the state in the interior is localized, the response comes entirely from extended edge-states in this case. The equation above is precisely the response formula for the Hall current as a consequence of a *potential difference* between the edges of the system. The expression for the Hall conductance, however, is purely bulk quantity. In the context of linear response theory, this quantity relates the current response in the interior of the system to the switch-on of an *internal electric field*. In this case, the Hall current is carried by the extended bulk electronic level below the Fermi-surface. The equivalence between the different processes is therefore contained in the last equation (68).

ACKNOWLEDGEMENTS

The author is grateful for numerous conversations with many collegues. In particular, mention must be made of discussions with E. Abrahams, P.W. Anderson, H. Aoki, R. Dashen, M. Kosterlitz, B. Kramer, A.I. Larkin, H. Levine, S.B. Libby, A. MacKinnon, R. Oppermann, N. Seiberg, L. Schaefer, L. Schweitzer, D.C. Tsui, F.Wegner and P.B. Wiegmann. The generosity of the Physikalisch Technische Bundesanstalt (PTB Braunschweig, F.R.G.) for offering visiting appointments is appreciated. This work has been supported in part by the U.S. Department of Energy under grant No. DE-AC02-76ER02220 and by the Office of the Naval Research under grant No. N00014-80-C-0657.

1 R.B. Laughlin, Phys. Rev. B **23** 5632 (1981).

2 D.J. Thouless, J. Phys. C **14** 3475 (1981).

3 R.E. Prange, Phys. Rev. B **23** 4802 (1981); R.E. Prange and R. Joynt, Phys. Rev. B **25** 2943 (1982)

4 B.I. Halperin, Phys. Rev. B **25** 2185 (1982).

5 K. von Klitzing, G. Dorda and M. Pepper, Phys. Rev. Lett. **45** 494 (1980).

6 D.C. Tsui and A.C. Gossard, Appl. Phys. Lett. **38** 550 (1981); See also: "Fifth Int. Conf. on Electronic Properties of Two Dimensional Systems", Univ. Oxford, UK (1983).

7 B.I. Halperin, Helv. Phys. Acta **56** 75 (1983).

8 **Anderson Localization** , Springer Series in Solid State Sciences **39** , edited by Y. Nagaoka and H. Fukuyama (1982).

9 H. Aoki and T. Ando, Solid State Communications **38** 1079 (1981).

10 E. Abrahams, P.W. Anderson, D.C. Licciardello and T.V. Ramakrishnan, Phys. Rev. Lett. **42** 673 (1979).

11 S. Hikami, Phys. Rev. B **24** 2671 (1981); See also Ref.[8].

12 K.B. Efetov, Sov. Phys. JETP **55** 521 (1982); K.B. Efetov, A.I Larkin and D.E. Khmelnitskii, Sov. Phys. JETP **52** 568 (1980); K. Juengling and R. Oppermann, Z. Phys. B **38** 93 (1980); S. Hikami, "Anderson Localization of the Two Dimensional Electron in a Random Potential under a Strong Magnetic Field", preprint (1984); T. Streit, J. Physique Lett. **45** L713 (1984).

13 A.M.M. Pruisken, Nucl. Phys. B235FS[11] 277 (1984).

14 H. Levine, S.B. Libby and A.M.M. Pruisken, Phys. Rev. Lett. **51** 1915 (1983).

15 H. Levine, S.B. Libby and A.M.M. Pruisken, Nucl. Phys. B240[FS12] 30 (1984).

16 H. Levine, S.B. Libby and A.M.M. Pruisken, Nucl. Phys. B240[FS12] 49 (1984).

17 H. Levine, S.B. Libby and A.M.M. Pruisken, Nucl. Phys. B240[FS12] 71 (1984).

18 A.M.M. Pruisken, "The Dilute Istanton Gas as the Precursor to the Integral Quantum Hall Effect", SDR and IAS preprint (July 1984).

19 F. Wegner, Z. Phys. B **51** 279 (1983).

20 For references, see: T. Ando, A. Fowler and F. Stern, Rev. Mod. Phys. **45** 437 (1982)

21 P. Streda, J. Phys. C **15** 1717 (1982)

22 F. Wegner, Z. Phys. B **38** 207 (1979) and this conference.

23 L. Smrcka and P. Streda, J. Phys. C **10** 2153 (1977).

24 B. Kramer, A.M.M. Pruisken and L. Schweitzer, in preparation

25 A.J. MacFarlane, Phys. Lett. B **82** 239 (1979); H. Eichenherr and M. Forger, Nucl. Phys. B **155** 381 (1979); R. Pisarski, Phys. Rev. D **15** 3385 (1979).

26 A.M.M. Pruisken and L. Schaefer, Nucl. Phys. B *200* 20 (1982); Phys. Rev. Lett. *46* 490 (1981).

27 L. Schaefer and F. Wegner, Z. Phys. B **38** 113 (1980).

28 A.J. McKane and M. Stone, Ann. of Phys. **131** 36 (1981).

29 R. Oppermann and F. Wegner, Z. Phys. B **34** 327 (1979).

30 S. Hikami, Phys. Rev. B **29** (1984).

31 A.M. Polyakov, Phys. Lett. B **59** 79 (1975); E. Brezin, J. Zinn-Justin and S.C. LeGuillou, Phys. Rev. B **14** 2614 (1976).

32 E. Brezin, S. Hikami and J. Zinn-Justin, Nucl. Phys. B **165** 528 (1980).

33 S. Hikami, Phys. Lett. B **98** 208 (1981).

34 F. Wegner, Z. Phys. B **36** 209 (1979).

35 A.M.M. Pruisken, Phys. Rev. B, to be published (Januari 1985).

36 A.A. Belavin and A.M. Polyakov, JETP Lett. **22** 245 (1975).

37 A.M. Polyakov, Phys. Lett. B **59** 82 (1975); A.A. Belavin, A.M. Polyakov, A.S. Schwartz and Y.S. Tyupkin, Phys. Lett. B **59** 85 (1975); G. 't Hooft, Phys. Rev. Lett. **37** 8 (1976); for a review on the subject, the reader is referred to R. Rajaraman, **Solitons and Instantons**, North Holland (1982).

38 V.A. Fateev, I.V. Frolov and A.S. Schwarz, Nucl. Phys. B **154**; B. Berg and M. Luescher, Nucl. Phys. B **190** [FS3] 412 (1981); B. Berg and M. Luescher, G. 't Hooft et al (Cargese 1979), (Plenum Press, New York, 1980); M. Luescher, Nucl. Phys B **200** [FS4] 61 (1982); B. Berg and M. Luescher, Comm. Math. Phys. **69** 57 (1979).

39 E. Witten, Nucl. Phys. B **149** 285 (1979).

40 S. Coleman, Ann. Phys. **101** 239 (1983).

41 I. Affleck, privat communication.

42 N. Seiberg, Phys. Rev. Lett. **53** 637 (1984).

43 J.T. Edwards and D.J. Thouless, J. Phys. c *5* 807 (1972)

44 C. Callan, R. Dashen and D. Gross, Phys. Rev. D **20** 3279 (1979); R. Jackiw, C. Rebbi, Phys. Rev. Lett. **37** 172 (1976).

45 D.J. Gross, Nucl. Phys B *132* 439 (1978).

46 The same diagram as fig.2 has been conjectured in: D.E. Khmel'nitzkii, JETP Lett. *38* 552 (1983) as a pedagogical guide to the theory of refs [13-17].

47 The renormalization group equations where also proposed in: V.G. Khizhnik and A.Yu. Morozov, JETP Lett. *39* 240 (1984); the author is endebted to Dr D.E. Khmel'nitzkii for bringing this reference to his attention.

48 H.P. Wei, A.M. Chang, D.C. Tsui, A.M.M. Pruisken and M. Razeghi, "Localization in the Quantized Hall Regime", preprint.

49 R.B. Laughlin, M.L. Cohen, M.J. Kosterlitz, H. Levine, S.B. Libby and A.M.M. Pruisken, "Scaling of Conductivities in the Fractional Quantum Hall Effect", preprint.

Quantum Hall Effect and Additional Oscillations of Conductivity in Weak Magnetic Fields

D.E. Khmel'nitskii

L.D. Landau Institute for Theoretical Physics, Chernogolovka
Moscow District, 142432, USSR

In the framework of scaling hypothesis for localization of 2D-electrons in a magnetic field B it is shown that at T=0 conductivity σ_{xx} has maxima at

$$B_n^{(1,2)} = \frac{mc}{e\hbar}\left\{\frac{E_F}{2n+1} \pm \left[\left(\frac{E_F}{2n+1}\right)^2 - \left(\frac{\hbar}{\tau}\right)^2\right]^{\frac{1}{2}}\right\} \quad (1)$$

i.e. besides oscillations of the Shubnikov type with maxima of σ_{xx} at

$$B_n^{(1)} \simeq \frac{mc}{e\hbar}\frac{E_F}{n+\frac{1}{2}} \quad (2)$$

there is the same number of additional oscillations at

$$B_n^{(2)} = \frac{mc}{e}\frac{\hbar}{E_F\tau^2}\left(n+\tfrac{1}{2}\right) . \quad (3)$$

The Hall conductivity

$$\sigma_{xy} = \frac{e^2}{2\pi\hbar}\, n \quad (4)$$

at $B_n^{(1)} < B < B_{n-1}^{(1)}$ and at $B_{n-1}^{(2)} < B < B_n^{(2)}$

gives the number of delocalized states at $E < E_F$ with energies E_n given by the interpolating relation

$$E_n = \hbar\Omega\,(n+\tfrac{1}{2})\,[1+(\Omega\tau)^{-2}] . \quad (5)$$

At $B < B_0$, when $r_H > l$ (r_H is the magnetic length, l is the mean free path), all states with $E < E_F$ are localized.

1. The quantization of the Hall conductivity discovered experimentally in 1980 [1] was almost immediately interpreted as a result of the Anderson localization of electronic states in a random impurity potential [2,3,4]. It was necessary to suggest [3] that, provided disorder is weak, i.e. $\Omega\tau \gg 1$ ($\Omega = eB/mc$ is the cyclotronic frequency, τ is the mean free time), at least one state at each Landau level should be localized. The evident contradiction between this and generally accepted suggestion that all states in a 2D-system are localized [5] has recently been settled by LEVIN, LIBBY, and PRUISKEN [6]. They have shown that the effective statistical sum Z of this problem contains the Hall conductivity in the form

$$\exp\left\{i\sigma_{xy}\frac{4\pi^2\hbar}{e^2}N\right\} , \quad (6)$$

where N is integer due to topological reasons, and therefore, σ_{xy} changing by an integer multiple of $e^2/2\pi\hbar$, does not change Z at all. They have also shown that at

$$\sigma_{xy} = \frac{e^2}{2\pi\hbar}(n+\tfrac{1}{2}) \tag{7}$$

where n is integer, the dissipative conductivity $\sigma_{xx} \neq 0$, and, under these conditions, the state $E = E_F$ is delocalized.

Thus, the necessary agreement has been achieved, but it has turned out that under the Fermi level there are $n \; E_F/\hbar\Omega$ delocalized states, n increasing with decreasing magnetic field. It is evident that at B=0 all states with $E < E_F$ should be localized. This paper aims at solving this paradox.

2. In Ref. [7] a scaling theory of electron localization in magnetic field has been developed. The σ_{xx} and σ_{xy} dependencies on the scale L are given by the renormalization group equations

$$\frac{d\sigma_{xx}}{d\xi} = \beta_{xx}(\sigma_{xx}, \sigma_{xy}); \; \frac{d\sigma_{xy}}{d\xi} = \beta_{xy}(\sigma_{xx}, \sigma_{xy}) \tag{8}$$

where β_{xx} and β_{xy} are periodic functions of σ_{xy}, their period equal to $e^2/2\pi\hbar$. The phase diagram of the system (8) is shown in Fig. 1. Initial conditions, $\sigma_{xx}^{(0)}$ and $\sigma_{xy}^{(0)}$, at ln L=0 should be obtained by means of the kinetic equation. In particular, it is convenient to use the following interpolating formula for $\sigma_{xy}^{(0)}$:

$$\sigma_{xy}^{(0)} = \frac{ne^2\tau}{m} \bullet \frac{\Omega\tau}{1+(\Omega\tau)^2} \tag{9}$$

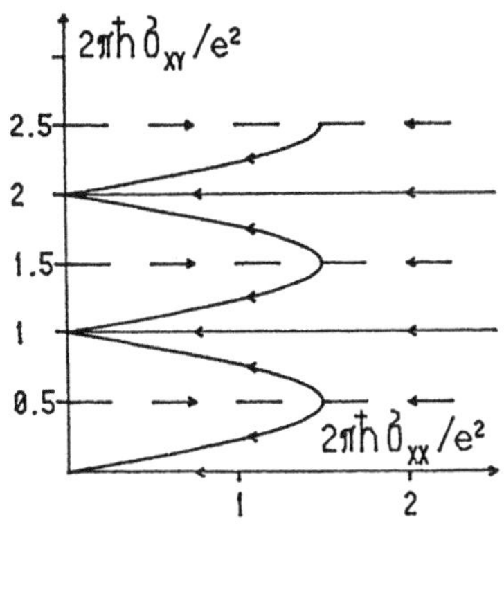

Fig. 1: The phase diagram for the renormalization group equations (8)

The result of the renormalization of σ_{xy} and σ_{xx}, i.e. the field dependence of observed values of σ_{xy} and σ_{xx} at T=0, is shown in Fig. 2a, b. As seen from Fig. 2b, besides the Shubnikov type oscillations with maxima at fields

$$B_N^{(1)} \simeq \frac{mc}{\hbar e}\,\frac{E_F}{n+\frac{1}{2}} \tag{10}$$

there is the same number of additional oscillations with maxima of σ_{xx} at

$$B_n^{(2)} \simeq \frac{mc\hbar}{e}\,\frac{1}{E_F\tau^2}\,(n+\tfrac{1}{2}) \tag{11}$$

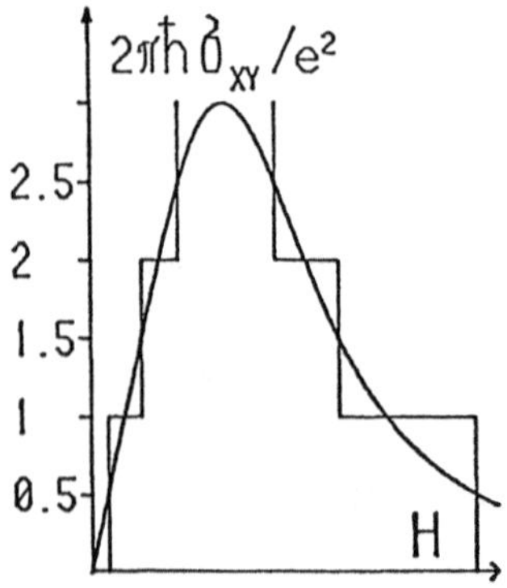

Fig. 2a: Steps of σ_{xy}

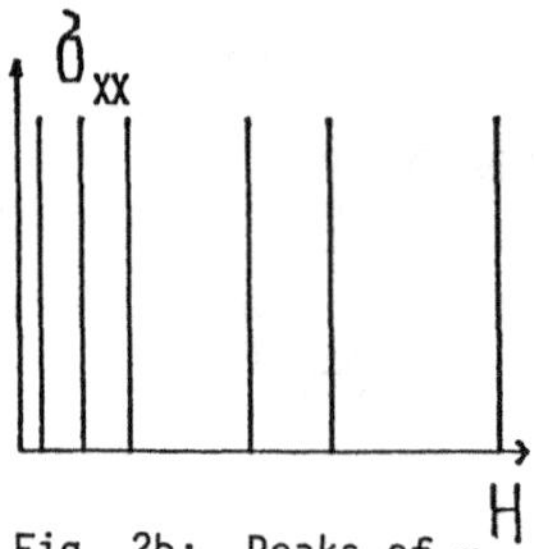

Fig. 2b: Peaks of σ_{xx}

This result may be interpreted as follows: at $\Omega\tau \geq 1$ a delocalized level is associated with each Landau subband. When the energy of the delocalized level coincides with the Fermi level, σ_{xx} has a peak. At $\Omega\tau < 1$ the broadening of Landau levels is larger than the distance between the levels. Under these conditions, the energies of delocalized states increase with decreasing field and once more coincide with E_F, resulting in additional oscillations. The number of delocalized states with $E < E_F$ is governed by the value of the Hall conductivity measured in the units of $e^2/2\pi\hbar$. Using (9) and expressing electronic concentration through E_F, we obtain the following interpolating formula for the energy of the n-th delocalized state[1]:

$$E_n \simeq \hbar\Omega(n+\tfrac{1}{2})\ [1 + \frac{1}{(\Omega\tau)^2}] \tag{12}$$

The lowest delocalized level crosses E_F at such fields, when

$$r_H = (c\hbar/eB)^{\frac{1}{2}} \sim 1 = v_F\tau\ . \tag{13}$$

At $1 < r_H$ all states with $E < E_F$ are localized.

3. The additional oscillations are not easily observed, since it is necessary that, due to weak localization, the conductivity

$$\sigma_{xx} \sim \frac{E_F\tau}{\hbar}\ \frac{e^2}{\hbar} \tag{14}$$

should decrease down to the value of order $e^2/\hbar$. However, if σ_{xy} and σ_{xx} are measured in a 2D-system, for which $E_F\tau/\hbar \sim 1$ and σ_{xy} has a maximum of order $e^2/\hbar$ at $T \sim 10$ - 20 K, the function $\sigma_{xy}(B)$ is expected to exhibit a plateau. The beginning and end of this plateau should coincide with the σ_{xy} maxima (of the Shubnikov type at large B and additional at small B).

1) This relation has also been suggested by LAUGHLIN (private communication).

References

1 K. von Klitzing, G. Dorda and M. Pepper: Phys. Rev. Lett. **45**, 494 (1980)
2 R.B. Laughlin: Phys. Rev. B **23**, 5632 (1981)
3 B.I. Halperin: Phys. Rev. B **25**, 2185 (1982)
4 H. Aoki and T. Ando: Solid St. Comm. **38**, 1079 (1981)
5 E. Abrahams, P.W. Anderson, D.C. Licciardello, and T.V. Ramakrishnan: Phys. Rev. Lett. **42**, 673 (1979)
6 H. Levine, S.B. Libby and A.M.M. Pruisken: Phys. Rev. Lett. **53**,1915 (1983)
7 D.E. Khmel'nitskii: Pis'ma ZhETF **38**, 454 (1983)

Electron-Phonon Scattering in Dirty Metals

Albert Schmid

Institut für Theorie der Kondensierten Materie, Universität Karlsruhe
Postfach 6380, D-7500 Karlsruhe, Fed. Rep. of Germany

In a clean metal, experiment and theory agree that inelastic electron-phonon collisions occur at a rate $(kT)^3/\hbar^3\omega_D^2$. In a dirty metal the theory predicts a reduction of the collision-rate by a factor of $k_F\ell(kT/\hbar\omega_D)$, provided that this factor is small; but in other cases there may also be an increase of this rate due to a participation of transverse phonons. The results obtained recently in weak localization experiments do not agree satisfactorily with this theory.

1. Introduction

Consider a system of degenerate fermions and allow for a coupling to a set of harmonic oscillators. Then, the fermions undergo transitions between their single particle eigenstates. Of course, in thermal equilibrium, the occupation numbers do not change; but if an extra fermion is inserted, it will be scattered out of its state by a rate which depends on its energy E (measured from the Fermi level) and on the temperature T. This inelastic collision-rate can be found from a golden rule calculation;one obtains

$$\frac{1}{\tau(E,T)} = 4\pi \int_0^\infty d\omega\, W(\omega)[N(\hbar\omega) + \tfrac{1}{2}f(\hbar\omega-E) + \tfrac{1}{2}f(\hbar\omega+E)], \qquad (1)$$

where N and f are the Bose and Fermi functions, respectively. The quantity $W(\omega)$ measures the bare probability of a collision with energy-transfer $\hbar\omega$; it is proportional to the square of the coupling and the density of states of the oscillators.

Note that the inelastic collision-rate counts each event irrespective of the size of energy transfer $\hbar\omega$. However, the effective rate appropriate to the process in consideration ('transport collision-rate') differs substantially only in particular cases.

A.) In case of a metal, lattice vibrations or phonons may be put in correspondence to harmonic oscillators. In a simple Debye model, we have $W_{e-p}(\omega) = \lambda\omega^2/\omega_D^2$, where ω_D is the Debye frequency and λ a dimensionless constant of order unity. Then, one finds from (1) the rates $1/\tau_{e-p} \sim (\lambda/\hbar^3\omega_D^2)\,\mathrm{Max}\{k^3T^3, |E|^3\}$, provided that $\mathrm{Max}\{\} \ll \hbar\omega_D$.

Physically, this rate corresponds to the emission or absorption of a real phonon in an electronic transition. In terms of diagrams, it may be represented as shown in Fig. 1.

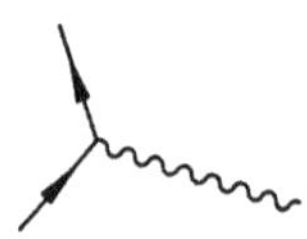

Figure 1

B.) We may equally well apply (1) to other collective oscillations, for instance, to plasmons. Though their resonant frequency is very high, there is also some damping which allows for a finite spectral density even at low frequencies. A calculation (jellium model) leads to $W_{e-e}(\omega) = \hbar\omega/8\varepsilon_F$, where ε_F is the Fermi energy; and

$1/\tau_{e-e} \sim (1/\hbar^2\varepsilon_F)\ \mathrm{Max}\{k^2T^2,\ E^2\}$.

Physically, the damping of a plasmon is accounted for by its decay in a particle-hole pair. Thus, we may represent these processes by the diagram of Fig. 2.

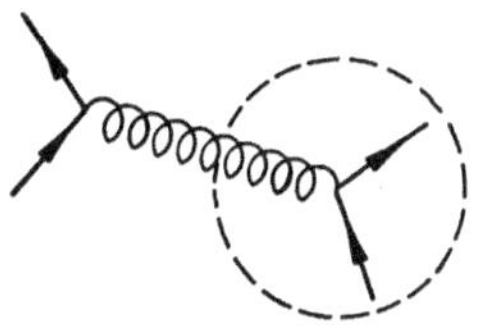

Figure 2

The part of the diagram in the bubble to the right represents plasmon decay; as a whole, the diagram depicts electron-electron scattering by a screened Coulomb interaction. This will be the subject of the lecture by ABRAHAMS [2].

C.) There is an analogous process involving phonons, which may also decay in particle-hole pairs. Thus, high frequency (=large wave vector) phonons contribute to the spectral density at low frequencies (scattering) by virtual phonons, Fig. 3). Incidentally, their contribution to the scattering rate [1] is of the same order as the plasmon contribution.

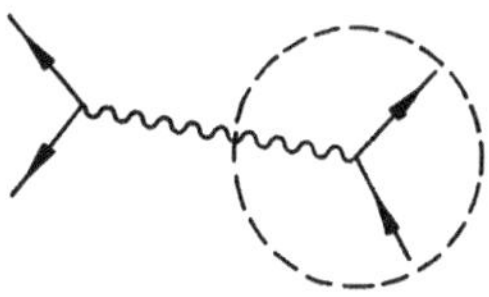

Figure 3

2. Experimental Observations

There are observable transport processes and relaxation phenomena where either (A) elastic and inelastic collisions, or (B) only inelastic collisions contribute.

A.) The electrical resistivity [3] is a well-known example of the former type. At low temperatures - which we are interested in - the phonon contribution can only be seen in sufficiently clean metals. Essentially, the resistivity depends on the momentum relaxation rate; and it is for this reason,and since the sound velocity c is so much smaller than the Fermi velocity v_F that the transport collision-rate of the phonons is proportional to T^5 (Bloch-Grüneisen) at low temperatures.

Electron-electron collisions conserve momentum and hence, they do not affect the resistivity. However, this argument becomes invalid when umklapp processes play a role.

More conclusive evidence is provided by the electronic thermal resistance, which depends on momentum as well as on energy relaxation. Thus, this resistivity includes contributions which depend on $1/\tau_{e-p}$ and $1/\tau_{e-e}$ as discussed in Sec. 1.

Of course, impurity scattering also contributes in proportion to its rate $1/\tau_{imp}$. In dirty metals and at low temperatures, this contribution is overwhelmingly large. Careful observations reveal deviations [4] from the rule (named after Matthiessen) that these resistivities simply add.

There are beautiful experiments on the resonance spectrum of surface Landau levels which allow one to extract the collision rate for selected positions on the Fermi surface [5]. The cubic temperature-dependence characteristic for phonon scattering has been observed, but there is also an anisotropy (by a factor of 10 in aluminium) which demonstrates the importance of band structure in clean samples.

B.) The relaxation process of the order parameter in a superconductor depends in important cases almost exclusively on inelastic scattering. Since the order parameter is a complex quantity, we have a transverse and a longitudinal relaxation, also referred to as charge and gap relaxation. Near the transition temperature, both rates are of the same order $1/\tau_Q$, $1/\tau_\Delta \sim \Delta/kT_c\ \tau(0,T)$. Unfortunately, as experiments done away from T_c are not very conclusive, the inelastic scattering rate can only be measured at $T=T_c$. These relaxation processes and other phenomena of non-equilibrium superconductivity are presented in the lecture by MOOIJ and KLAPWIJK [6].

In the theory of superconductivity, $W_{e-p}(\omega)$ is known as the effective phonon density of states $\alpha^2(\omega)F(\omega)$ which can be obtained from tunneling spectroscopy. It seems that at lower frequencies at least, $\alpha^2(\omega)F(\omega)$ increases with increasing impurity scattering [7]. This effect is very pronounced in amorphous metals; it should be borne in mind, however, that tunneling spectroscopy is not very reliable at low frequencies.

Weak localization means an interference of pairs of electronic partial waves returning to the starting point and differing in the sense of circulation. If phase-coherence, that means time-reversal symmetry, is broken at a rate $1/\tau_\varphi$, the change in sheet conductance is given by

$$\delta g_\square = -(e^2/2\pi^2\hbar)\ln\tau_\varphi/\tau_{imp}.$$

Inelastic scattering is one of the processes that destroy phase-coherence. They contribute to $1/\tau_\varphi$ with an effective rate which is not much different [9] from $1/\tau(0,T)$. It is expected that electron-phonon collisions dominate in experiments done at higher temperatures, whereas electron-electron collisions are important at lower temperatures.

3. Ultrasonic Attenuation

Since it is difficult to calculate the inelastic collision-rate of electrons in presence of large elastic scattering, we will first study the decay of phonons. At low temperatures and low frequencies, this decay is dominated by interaction with electrons. The result which we will obtain will eventually allow us to determine $1/\tau_{e-p}$.

The physics of phonon decay, or equivalently, of ultrasonic attenuation has been explained quantitatively by PIPPARD [10]. His calculation can be interpreted most conveniently in terms of a Boltzmann equation for the electrons. In the presence of a lattice distortion, $\vec{u}(\vec{r},t) \propto \exp i(\vec{q}\vec{r}-\omega t)$, this equation is marked by two important features.

A.) There is an electric field $-\nabla\varphi$ acting on the electrons, and the potential φ is such that charge neutrality is maintained. In quasistatic approximation, one finds

$$e\varphi = \frac{n}{2N(0)} \operatorname{div}\vec{u} \ , \tag{2a}$$

where $n/2\,N(0) = (2/3)\varepsilon_F$.

B.) Since the lattice defects move in phase with the distorted lattice, the impurity scattering is elastic only in the frame of reference which moves locally with the lattice. Therefore, and in case of isotropic scattering, one has a collision integral

$$K = \frac{1}{2N(0)\tau_{imp}} \int \frac{d^3p'}{(2\pi)^3}\, \delta(\varepsilon_{\vec{p}} - \varepsilon_{\vec{p}'})\, [f_{\vec{p}} - f_{\vec{p}'}] \quad ; \qquad \varepsilon_{\vec{p}} = \frac{1}{2m}(\vec{p} - m\dot{\vec{u}})^2 \quad . \tag{2b}$$

Eventually, the phonon decay-rate $1/\tau_b(q)$ - the branch index is b - is found as the ratio of energy dissipated per unit time in the electronic system to the energy stored in the lattice wave.

Due to the fact that $c_b \ll v_F$, the result depends on $q\ell = q\, v_F \tau_{imp}$ rather than on $\omega\tau_{imp}$. Explicitly, one obtains

$$\frac{1}{\tau_b(q)} = \frac{\pi}{6} \frac{Zp_F}{Mc_b}\, \omega(q)\phi_b(q\ell) \quad , \tag{3}$$

where Z and M are the valency and the mass of the ions. $\phi_b(q\ell)$ measures the strength of the coupling of the phonons to the electrons; it is given by

$$\phi_L(x) = \frac{2}{\pi} \left[\frac{x \operatorname{at} x}{x - \operatorname{at} x} - \frac{3}{x}\right] \rightarrow \begin{cases} 8x/5\pi \\ 1 \end{cases} \tag{4}$$

$$\phi_T(x) = \frac{3}{\pi} \frac{2x^3 + 3x - 3(x^2 + 1)\operatorname{at} x}{x^4} \rightarrow \begin{cases} 6x/5\pi \\ 6/\pi x \end{cases}$$

where the upper and low limits apply to $x \ll 1$ and $x \gg 1$, respectively.

Transverse phonons do not couple in a clean metal since the momentum transferred to electrons is perpendicular to the direction of polarization. The finite coupling for $q\ell \sim 1$ can be explained by loss of momentum conservation in the presence of impurity scattering.

C.) There is a compensation effect in the case of longitudinal phonons. Observe that there is a decrease of coupling with increasing impurity concentration, which is about a factor $q\ell$ for $q\ell \ll 1$. This behaviour should be put in contrast to results one would obtain if only either mechanism A) or B) would have been taken into consideration. In both cases, the coupling would appear enhanced, roughly by a factor

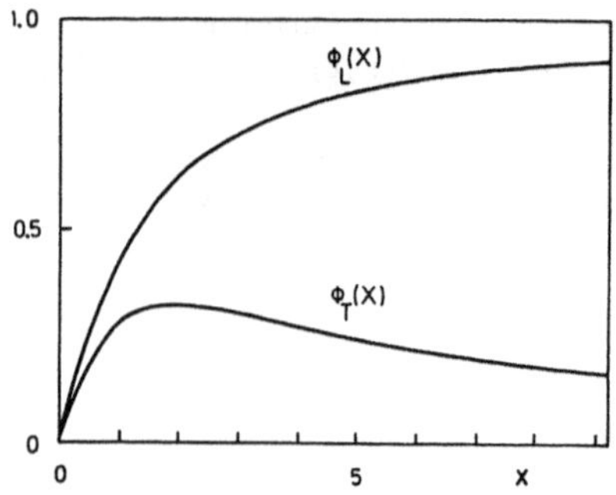

<u>Fig. 4:</u>
Electron-phonon coupling as a function of $x=q\ell$

$(q\ell)^{-1}$. Thus, a substantial cancellation of contribution takes place if the physics is done correctly.

Such a cancellation occurs also in calculations based on microscopic theories. There is a rather long list of papers where the authors have failed to collect carefully all important contributions - and I have to confess to having allowed myself once or twice such a negligence.

Pippard's theory has been cast in a microscopic form by TSUNETO [11]. A prominent feature of his approach is a canonical transformation,which physically corresponds to a transition in a frame of reference where the lattice is at rest. In this frame, impurity scattering is elastic and the standard impurity technique [1] can be applied. Further procedure can be understood as follows [12] . The phonon decay-rate can be calculated as the imaginary part of the phonon self-energy.

Its most important contribution is contained in a particle-hole bubble shown in Fig. 5. There, the cross-hatched electron-phonon vertex to the right, is dressed with impurity scattering and Coulomb interaction.

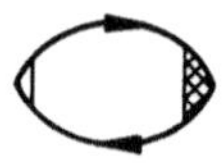

<u>Figure 5</u>

4. Electron-Phonon Collisions

In calculating the inelastic electronic decay-rate, one has to make sure that both mechanisms 3 A.) and 3 B.) are included. Again, the remarks in 3 C.) on compensation are of importance. But there is an additional point to be observed.

In microscopic theory, the electronic decay-rate is twice the imaginary part of the electronic self-energy. In standard impurity technique, Σ consists of a phonon and impurity contribution as shown in Fig. 6.

<u>Figure 6</u>

In presence of strong impurity scattering,however, it is not possible to separate the inelastic contribution [12].

Therefore, one has to ask a more specific question: where do the electrons go if they are scattered out of a single particle state? Only the events where the energy changes have to be counted.

Such a detailed information is contained in the two-particle Green's function,which in the present case is in the form of a ladder diagram. There, the rungs are formed either by blocks of impurity lines or by units [12] which contain phonon lines and which are dressed suitably by impurity lines as shown in Fig. 7.

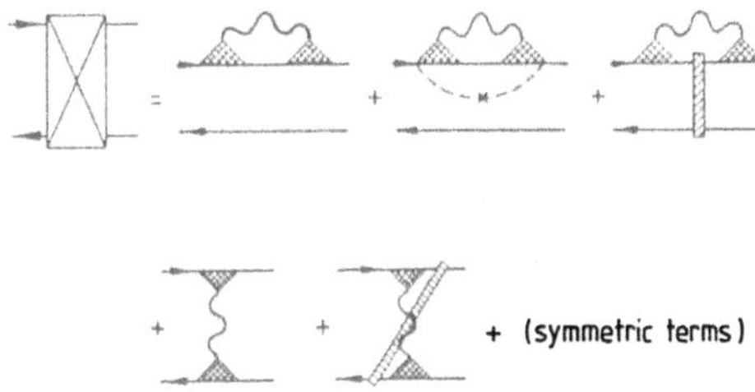

Figure 7

The calculations according to this scheme are rather involved. Furtunately, there is a considerably simpler approach that leads to the same result [8] . If one uses a representation on the basis of single-particle eigenstates in presence of impurities, the elastic scattering does not show up in transitions and all the scattering observed is inelastic.

The final result is of the form

$$W_{e-p}(\omega) = \frac{1}{8}\Lambda \frac{\omega^2}{\omega_1^2}\left[\phi_L\left(\frac{\omega\ell}{c_L}\right) + 2\eta^4\phi_T\left(\frac{\omega\ell}{c_T}\right)\right], \quad \text{where} \tag{5}$$

$$\omega_1 = c_L k_F \; ; \quad \Lambda = \frac{Z\hbar\omega_1^2}{3mMc_L^4} \quad ; \quad \eta = \frac{c_L}{c_T} \quad . \tag{6}$$

One finds $\eta > 0$ in all cases; furthermore, ω_1 and Λ are of the order ω_D and 1, respectively. Specifically, $\Lambda = 1$ for a jellium model. Obviously, (5) is valid only in a region where frequencies and wave vectors of the phonons are related linearly; hence we require $\omega \ll \omega_1/\eta$.

The collision-rate $1/\tau_{e-p} = 1/\tau(0,T)$ can be calculated according to (1,4,5). In limiting cases, it is equal to

$$\frac{1}{\tau_{e-p}} = \begin{cases} \dfrac{7\pi}{4}\zeta(3)\Lambda\dfrac{(kT)^3}{\hbar^3\omega_1^2} \; ; & k_F\ell \gg \dfrac{\hbar\omega_1}{kT} \\[2ex] \dfrac{\pi^4}{10}\Lambda k_F\ell\dfrac{(kT)^4}{\hbar^4\omega_1^3}\left[1 + \dfrac{3}{2}\eta^5\right] \; ; & k_F\ell \ll \dfrac{\hbar\omega_1}{\eta kT} \end{cases} \tag{7}$$

In the clean limit, we recover essentially the relations quoted in 1 A.)

The quantity B(T) defined by

$$\frac{1}{\tau_{e-p}} = \Lambda\frac{(kT)^3}{\hbar^3\omega_1^2} B(T) \tag{8}$$

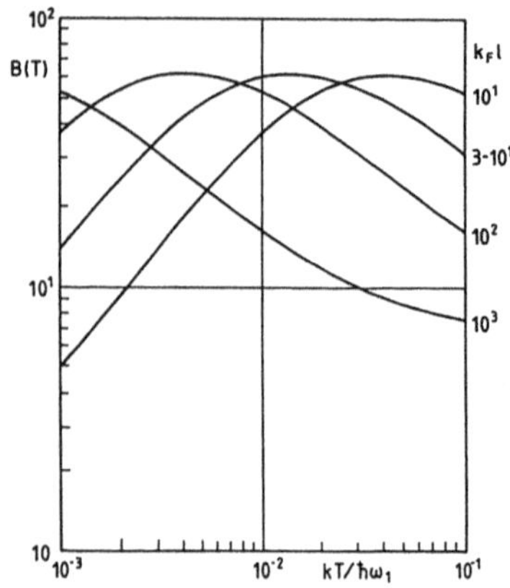

Figure 8: Electron-phonon collision-rate as a function of temperature. B(T) defined by (8) and $kT/\hbar\omega_1$ are shown on the vertical and horizontal axis, respectively. Parameter is the mean free path expressed by $k_F\ell$. Note that the different branches can be mapped by horizontal shifts. The limiting value far to the right is $B=7\pi\zeta(3)/4$

has been calculated in the case $\eta=2$. It is plotted in Fig. 8 as a function of $kT/\hbar\omega_1$ for various parameters $k_F\ell$. For illustration, consider Mg and Al at T=10K.

Mg). Here, Z=2; $c_L = 6\cdot10^5\text{cm sec}^{-1}$; $\omega_1 = 0{,}82 \cdot 10^{14}\text{sec}^{-1}$ ($T_1 = \hbar\omega_1/k = 630\text{K}$); $\Lambda= 1$. For $k_F\ell = 10$, we find B(T=10k)=50, and $1/\tau_{e-p} = 1{,}6 \cdot 10^{10}\ \text{sec}^{-1}$.

Aℓ). Here, Z=3; $c_L = 6{,}4 \cdot 10^5\ \text{cm sec}^{-1}$; $\omega_1 = 1{,}1 \cdot 10^{14}\ \text{sec}^{-1}$ ($T_1 = 850\text{K}$); $\Lambda= 2$. For $k_F\ell = 100$, we find B(T=10k) = 50, and $1/\tau_{e-p} = 1{,}8 \cdot 10^{10}\ \text{sec}^{-1}$.

We compare the Mg-value with Bergmann's result in his experiments on weak localization. He finds [13] for $k_F\ell \sim 10$ the rate $1/\tau_{e-p}(10k)=1\cdot10^{11}\ \text{sec}^{-1}$ and furthermore, it is found that $1/\tau_{e-p} \propto T^\alpha$ with $\alpha \sim 2$ in the range 5k ... 25k; and that there is only an insignificant dependence [14] on $k_F\ell$. In the present theory, the rate of 10k is about one order of magnitude smaller. In addition $\alpha \sim 3{,}5$; and one may observe that the rate has a tendency to increase with $k_F\ell$.

Concerning Aℓ, there are various experimental results available which agree or disagree more or less. For sake of definiteness, we consider the results of PROBER et al [15] . They find for $k_F\ell \sim 100$ the rate $1/\tau_{e-p}(10k) \sim 0.9\cdot10^{10}\text{sec}^{-1}$ and $\alpha \sim 3$. In the present theory, the rate is about a factor of two larger, and $\alpha \sim 2{,}5 \ldots 2{,}8$.

5. Discussion

Fortunately, the conductivity changes due to weak localization depend strongly on an applied magnetic field. Therefore, it is possible to separate this effect experimentally and to determine the inelastic scattering rates rather accurately. There is now a considerable amount of data on $1/\tau_{e-p}$ for various materials. Still, there are differences between various experiments which may perhaps be explained by differences in sample preparation. In the preceding section, two "typical" experiments have been compared with the present theory without satisfactory results: discrepancies by one order of magnitude can be found.

The theory may easily be blamed for neglecting various important pieces of physics. Perhaps the most serious defects are: (i) neglecting the real structure of the metals and (ii) neglecting changes in this structure by film preparation.

(i) It is known that there is a substantial coupling of transverse phonons to the electrons by umklapp processes in clean metals. However, nothing seems to be known about changes in this effect when impurities are added to the metal.

(ii) It is known that many quenched condensed metal films are of a structure similar to an amorphous state. Consider, for instance Ga, where the ratio of sound velocities $\eta = c_L/c_T$ increases from 1.7 in the crystalline phase to 4.75 in the amorphous phase [16]. This change, which occurs at approximately constant c_L, might lead to drastic increase (perhaps proportional to η^3) in $1/\tau_{ep}$. However, only experiments will lead to information on such structural changes.

There are other simplifying assumptions in the theory which can be removed without changing its basic structure. They are as follows.

In a disordered solid, the low frequency phonon modes have admixtures of fast spatial fluctuations, that is, of high wave vectors [17]. In leading order of impurity concentration, one finds an insignificant contribution to $1/\tau_{e-p}$ which is of order $k_F\ell(kT)^6/\hbar^6\omega_1^5$ at low temperatures.

As it is known that plasmon exchange (1 B.) is strongly enhanced in dirty metals [18], one may wonder about corrections in (longitudinal) phonon exchange (1 C.). However, the impurity effects cancel for the following reasons. There is the known increase by a factor $(q\ell)^{-1}$ at the vertex; but at the same time, the spectral density, that is the damping, is decreased by a factor $(q\ell)$.

The form, in which the theory has been presented here, applies strictly to three-dimensional samples. Concerning the phonon modes in weak localization experiments, it should be borne in mind that the metal films are supported by a massive substrate. On the other hand, the screening of space charges by electrons is essentially two-dimensional. Though drastic changes are not expected, it is desirable to analyse this situation more carefully.

Eventually, I wish to summarize the essential features of the present theory. It is a quasiclassical theory which includes band structure effects only insofar as they can be expressed in terms of an effective mass. The quasiclassical nature follows immediately by comparing it with Pippard's theory. More indirectly, we may observe that the microscopic calculations are based on a random phase type-approximation of the Coulomb interaction and on the standard impurity technique for Green's function. Both methods are more or less equivalent to approximations of the quasiclassical type. Therefore, the relations quoted previously are restricted to cases where

$$\ell^{-1} \ll k_F \; ; \; q \ll k_F \; ,$$

which is $k_F\ell \gg 1$; $\hbar\omega_D/kT \gg 1$.

The same restrictions apply to the theory of weak localization. Indeed, all known relations can be derived quantitatively from (single particle) wave functions of the type

$$\psi \;=\; \Sigma_{\vec{x}_t} \exp \frac{i}{\hbar} \int dt \; L(\vec{x}_t) \; ,$$

where L is the Lagrangian and $\vec{x}_t$ the possible classical paths of the electrons. It is also implied that the set of classical paths for randomly distributed impurities can be found from the solutions of the Boltzmann equation. Furthermore, it is also possible [19] to incorporate consistently phase-breaking due to inelastic collisions in this ansatz, which reproduces all the results discussed here.

I wish to thank W. Eiler and J. Rammer for useful discussions.

References and Footnotes

1. A.A. Abrikosov, L.P. Gorkov, and I. Ye. Dzyaloshinski: Quantum field theoretical methods in statistical physics. (Pergamon Press, Oxford 1965).
2. E. Abrahams, Inelastic scattering due to electron-electron interaction. This volume.
3. J.M. Ziman, Principles of the theory of solids (Cambridge, 1965); Ch. 7.
4. J. Bass, Adv. Phys. 21, 431 (1972).
5. T. Wegehaupt and R.E. Doezema, Phys. Rev. B 18, 742 (1978).
6. J.E. Mooij and T.M. Klapwijk, Inelastic lifetime of conduction electrons as determined by nonlocalization methods. This volume.
7. G. Bergmann, Phys. Rep. 27C, 161 (1976), See also Ref. 8.
8. B. Keck and A. Schmid, J. Low Temp Phys. 24, 611 (1976).
9. First of all, only collisions with energy-transfer $\omega \geq 1/\tau_\varphi$ contribute. This modification is irrelevant for phonons. Secondly, an energy average $<\ell n\tau(E,T)>$ in the range $|E| \lesssim kT$ should be computed. Only changes of order 10^{-1} are expected.
10. A.B. Pippard, Phil. Mag. 46, 1106 (1955).
11. T. Tsuneto, Phys. Rev. 21, 402 (1960).
12. A. Schmid, Z. Phys. 259, 421 (1973).
13. G. Bergmann, Phys. Rev. B 28, 515 (1983) (see also: Phys. Rep. 107, 1 (1984)).
14. R.P. Peters and G. Bergmann, Dependence of the phase coherence time on the resistivity in disordered films. This conference.
15. D.E. Prober, S. Wind and P. Santhanan, Fluctuation and localization effects in quasi-one-dimensional metallic structures. This volume.
16. W. Dietsche, H. Kinder, J. Mattes and H. Wühl, Phys. Rev. Lett. 45, 1332 (1980).
17. R.J. Elliott, J.A. Krumhansl, and P.L. Leath, Rev. Mod. Phys. 46 465 (1974).
18. A. Schmid, Z. Phys. 271, 251 (1974). E. Abrahams, P.W. Anderson, P.A. Lee and T.V. Ramakrishnan, Phys. Rev. B 24, 6783 (1981).
19. J. Rammer and A. Schmid, Destruction of phase-coherence by electron-phonon interaction in weakly-disordered conductors. This conference and to be published.

Electron Scattering Times in Thin Metal Films Experimental Results

C. Van Haesendonck, M. Gijs, and Y. Bruynseraede

Laboratorium voor Vaste Stof-Fysika en Magnetisme, Katholieke Universiteit Leuven, B-3030 Leuven, Belgium

1. Introduction

The discovery of weak-localization effects in two-dimensional metallic systems [1] has given a new impetus to the study of transport properties of disordered conductors [2]. These effects are a direct consequence of the quantum interference between diffusively-scattered electron waves. Due to this interference, the back-scattering probability for the electron is modified during the time interval $0 < t < \tau_\phi$ with τ_ϕ the characteristic time, after which the phase of one of the interfering electron waves is changed. This phase-breaking time is a combination of the inelastic scattering time τ_{in} (due to electron-phonon and electron-electron scattering) and the magnetic scattering time τ_s (due to interaction with magnetic moments):

$$\tau_\phi^{-1} = \tau_{in}^{-1} + 2\tau_s^{-1} \qquad (1)$$

The time τ_ϕ during which the back-scattering is built up is temperature-dependent, and the weak-localization is observable as an extra-resistance which increases logarithmically with decreasing temperature. ALTSHULER et al. [3] showed, however, that a weak logarithmic divergence of the resistance in thin metal films can also be explained by electron-electron interaction which is enhanced by elastic impurity scattering. Hence, it is not possible to extract information about τ_ϕ (T) from the logarithmic temperature-dependence of the resistance, which has now been observed in numerous thin films of metals and alloys [2].

HIKAMI et al. [4] showed that the back-scattering is only enhanced when $\tau_{in}(T) \ll \tau_{so}$, where τ_{so} is the typical time after which the conduction electron spin is reversed due to spin-orbit scattering. For $\tau_{in}(T) \gg \tau_{so}$ (strong spin-orbit scattering or at very low temperatures) localization theory [4] predicts a decrease in the back-scattering intensity which results experimentally in a logarithmic decrease of the resistance with decreasing temperature. Probably due to the important influence of the electron-electron interaction, this anti-localization has not been observed in resistance versus temperature measurements.

The influence of the different scattering mechanisms upon weak-localization can, however, be probed unambiguously [4] by measuring the magneto resistance (MR) in a small magnetic field (B < 1 T) perpendicular to the film plane.

For a normal (= non-superconducting) metal film with a resistance per square $R_\square$ and an elastic life-time τ_{el}, the anomalous normalized MR due to weak localization is given by [4]:

$$\frac{\delta R_{\square}(B,T)}{R_{\square}^2(B=0,T)} = (R_{\square}(B,T) - R_{\square}(B=0,T))/R_{\square}^2(B=0,T)$$

$$= \frac{e^2}{2\pi^2\hbar}\left[\psi\left(\frac{1}{2}+\frac{B_o}{B}\right) + \ln\left(\frac{B}{B_o}\right) - \frac{3}{2}\psi\left(\frac{1}{2}+\frac{B_1(T)}{B}\right) - \frac{3}{2}\ln\left(\frac{B}{B_1(T)}\right)\right.$$

$$\left. + \frac{1}{2}\psi\left(\frac{1}{2}+\frac{B_\phi(T)}{B}\right) + \frac{1}{2}\ln\left(\frac{B}{B_\phi(T)}\right)\right] \qquad (2)$$

In eq. 2 the characteristic fields B_o, $B_1(T)$ and $B_\phi(T)$ are defined as:

$$B_o = B_{el} + B_{so} + B_s$$
$$B_1(T) = B_{in}(T) + \frac{4}{3}B_{so} + \frac{2}{3}B_s \qquad (3)$$
$$B_\phi(T) = B_{in}(T) + 2\,B_s$$

Fitting of the experimental MR data with eq. 2 and eq. 3 produces values for the fields B_o, $B_1(T)$, $B_\phi(T)$. Reliable values for the physically relevant fields $B_{in}(T)$, B_{so} and B_s, can, however, only be obtained if assumptions are made concerning the temperature-dependence of $B_{in}(T)$. Each characteristic field B_K is directly related to a characteristic electronic scattering time τ_K via the diffusion constant D:

$$B_K = \hbar/4eD\tau_K \qquad (4)$$

It is clear that eq. 4 only allows to calculate the scattering times if the diffusivity in the thin metal film is known (see Sect. 2). On the other hand, the characteristic diffusion lengths $L_K = (D\tau_K)^{1/2}$ can always be obtained unambiguously from the MR curves through the relation:

$$L_K = (\hbar/4eB_K)^{1/2} \qquad (5)$$

Since theoretical calculations always provide a value for the scattering times [5], a quantitative comparison between experiment and theory is only possible if the elastic diffusion process is well-defined.

In this paper we discuss the determination of D in thin metal films and the influence of the different scattering times on localization. An overview of the published τ_ϕ - and τ_{so}-values measured in two-dimensional films is presented.

2. Calculation of the Diffusion Constant

For the derivation of eq. 2, HIKAMI et al. [4] assumed a homogeneous metal structure for which the elastic scattering is entirely due to lattice defects and/or impurity atoms. While for quench-condensed films measured in situ at liquid helium temperature [5, 6] the resistance is mainly caused by defects in the lattice structure, most of the elastic scattering in films fabricated at room temperature comes from the diffuse scattering at the (oxidized) surface or at (oxidized) grain boundaries. Moreover, eq. 2 shows that a reliable determination of the characteristic scattering times requires metal films with $R_{\square} \geq 100\ \Omega/\square$ if a sensitive but classical resistance measuring technique is used. For quench-condensed films there is no difficulty in obtaining homogeneous films with a large $R_{\square}$-value and with randomly distributed scattering centers. Most of the weak-localization experiments are, however, performed on thin metal films condensed at room temperature and subsequently exposed to air. Under these circumstances most metal films with $R_\square \gtrsim 100\ \Omega/\square$ are nearly discontinuous and can be inhomogeneous on a scale comparable to the characteristic diffusion lengths $L_\phi(T)$ and L_{so}. This inhomogeneity of the film structure hinders considerably the determination of the scattering times via eq. 4. Indeed, in order to calculate the diffusion constant $D = 1/3\ v_F l_{el}$ (we use the three-dimensional expression for D since $k_F d \gg 1$) one has to estimate the elastic mean free path l_{el}. The value of l_{el} can be either calculated from the film thickness and $R_\square(4.2\ K)$ $(= l_{el}(d))$ or from the resistance ratio $RR(= l_{el}(RR))$. For homogeneous metal films with $1\ \Omega/\square \lesssim R_\square \lesssim 100\ \Omega/\square$ corresponding to $50\ Å \lesssim d \lesssim 200\ Å$ one expects $l_{el}(d) \simeq l_{el}(RR)$.

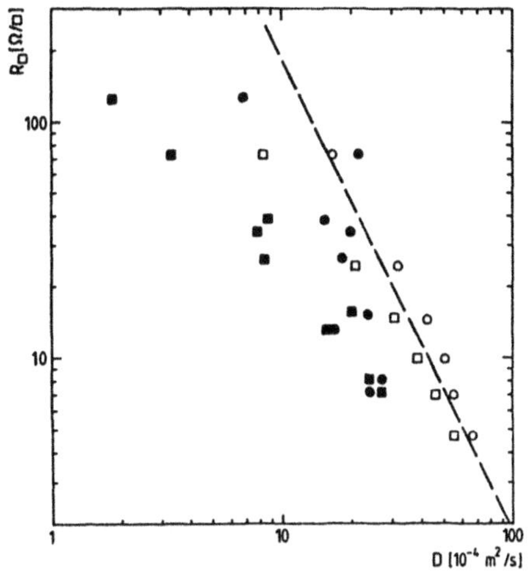

Fig. 1: The sheet resistance $R_\square(4.2\ K)$ as a function of the diffusion constant D for Cu-films (open symbols) and Mg-films (closed symbols) evaporated on a glass substrate held at room temperature. The diffusion constant was calculated using $l_{el}(RR)$ (circles) and $l_{el}(d)$ (squares)

In Fig. 1 we have plotted $R_\square$ versus D on a double logarithmic scale for Cu [7] and Mg [8] films condensed on a glass substrate held at room temperature. For each film the values of D calculated from $l_{el}(d)$ and $l_{el}(RR)$ are shown. Analyzing first the data obtained for the Cu-films, it is clear that:

(i) for films with $R_\square \lesssim 20\ \Omega/\square$ the sheet resistance is proportional to D^{-2} and l_{el}^{-2}, indicating that the $R_\square$-value is determined by surface scattering and $l_{el} \propto d$ as predicted by FUCH's theory [9]. This is also confirmed by the fact that $R_\square$ is a linear function of d^{-2}. Therefore, in this low-resistance regime, the values of D obtained from $l_{el}(d)$ and $l_{el}(RR)$ are comparable and the diffusion constant is a well-defined quantity independent of the length scale. It follows that the information about the scattering times obtained from eqs. 2, 3 and 4 is meaningful;

(ii) for $R_\square > 50\ \Omega/\square$ the Cu films approach the discontinuity limit and $R_\square$ diverges exponentially with decreasing thickness. This behaviour is caused by the inhomogeneity of the film structure (microscopic cracks, cluster formation etc...) producing an $l_{el}(d)$-value substantially smaller than $l_{el}(RR)$. This difference is due to the fact that $l_{el}(d)$ is a macroscopic quantity (influence of the cracks is included), while $l_{el}(RR)$ reflects the microscopic elastic scattering in the conducting parts of the film. This explains also why $l_{el}^{-2}(d)$ is no longer proportional to $R_\square$ for $R_\square > 50\ \Omega/\square$ and why $l_{el}(RR)$ deviates from the expected behaviour when the $R_\square$-value is mainly determined by the inhomogeneities in the film structure.

It should be noted that the difference between $l_{el}(RR)$ and $l_{el}(d)$ can also be explained by an enhancement of the electron-phonon scattering in the thinnest films resulting in an increase of $l_{el}(RR)$ [10]. Since it is not clear whether the macroscopic D-value (calculated from $l_{el}(d)$) or the microscopic value (calculated from $l_{el}(RR)$) has to be used in eq. 4, experiments on Cu films with $R_\square > 50\ \Omega/\square$ can not probe the $R_\square$-dependence of the characteristic scattering times.

Analyzing now the $R_\square$ versus D data for the Mg films, it is clear from Fig. 1 that even for low-resistance films the $R_\square$-value is determined by the inhomogeneous structure. Experimentally, it is very difficult to obtain homogeneous Mg films by condensing on a substrate held at room temperature. (All the Mg films presented in Fig. 1 have approximately the same film thickness $d \simeq 100$ Å). The films with the highest $R_\square$-value have many inhomogeneities (cracks, local thickness fluctuations, oxidation etc.) so that $l_{el}(d) \ll l_{el}(RR)$. The fact that $l_{el}(RR)$ is nearly constant supports the idea that the change in $R_\square$ is not caused by a variation of the surface scattering. It should be noted that since Mg oxidizes easily, the difference between $l_{el}(d)$ and $l_{el}(RR)$ may be due to the presence of a granular-like film structure. In that case $l_{el}(d)$ describes the macroscopic electronic transport between the Mg grains, while $l_{el}(RR)$ strongly depends upon the electronic conduction within the grains. Our experiments therefore clearly indicate that it is difficult to study the $R_\square$-dependence of the scattering times in Mg films condensed at room temperature. Moreover, the inhomogeneity scale (size of the cracks or the grains) can become larger than the scale L_B which is imposed by the magnetic field B:

$$L_B = (\hbar/4eB)^{1/2} \tag{6}$$

This implies that eq. 2 can not be used to analyze the experimental MR data when the magnetic field is so high that L_B is smaller than the scale of the inhomogeneities.

We conclude this section by indicating that eq. 2 may also be used to describe the anomalous MR in the fluctuation regime ($T \gg T_c$) of a superconducting film. As one approaches T_c (without entering the critical region where $(T-T_c)/T_c \ll 1$), an extra positive magnetoresistance appears due to the destruction of the Maki-Thompson fluctuations [11]:

$$\frac{\delta R_\square(B,T)}{R_\square^2(B=0,T)} = \frac{e^2}{2\pi^2\hbar}\,\beta(T)\,\left[\ln\left(\frac{B}{B_\phi(T)}\right) - \psi\left(\frac{1}{2} + \frac{B_\phi(T)}{B}\right)\right] \tag{7}$$

where the temperature-dependent parameter β is a universal function of T/T_c. Using the sum of eq. 2 and eq. 7 one can describe the observed anomalous MR in thin Al films above T_c [12]. Moreover, the values derived for the inelastic scattering time τ_{in} agree very well with the values obtained from non-localization measurements [13]. In the case of superconducting films, a value for the elastic mean free path can be obtained from measurements of the upper critical field H_{c2}. This $l_{el}(H_{c2})$-value is an average of the elastic scattering on the scale of the superconducting coherence length (typically 1000 Å for a dirty Al film). Using the $l_{el}(H_{c2})$-value to calculate the diffusion constant, it may be possible to study the $R_\square$-dependence of the characteristic scattering times even in inhomogeneous (granular) superconducting films. It must be noted that only those MR-data can be used for which L_B is larger than the size of the inhomogeneities.

In the next three Sections we discuss the influence of inelastic, magnetic and spin-orbit scattering on the weak-localization (perpendicular MR) in thin metal films. For this analysis we will mainly rely on the results obtained at Leuven in thin Cu and Mg films.

3. Influence of the Inelastic Scattering Time

Figure 2 shows some typical magnetoresistance data for a pure Cu film (starting material MRC 99.999 % pure Cu) with $R_\square(4.2\ K) = 234\ \Omega/\square$ and $d = 50$ Å. Using eq. 2 and $B_o \simeq B_{el}$, B_1 and B_ϕ as fitting parameters an excellent agreement between theory and experiment can be obtained up to $B = 2$ T. The predicted saturation of the MR for $B \geqslant B_{el}$ can be neglected since $B_{el} = 3\hbar/4el_{el}^2 \simeq 500$ T for this Cu film. Hence, the theoretical analysis only allows to determine the fields $B_1(T)$ and $B_\phi(T)$. The magnetic and spin-orbit scattering are expected to be temperature-independent, while the inelastic scattering depends strongly on temperature following a power law:

$$\tau_{in}(T) \propto T^{-P} \qquad (8)$$

where the exponent P depends on the dominating inelastic scattering mechanism.

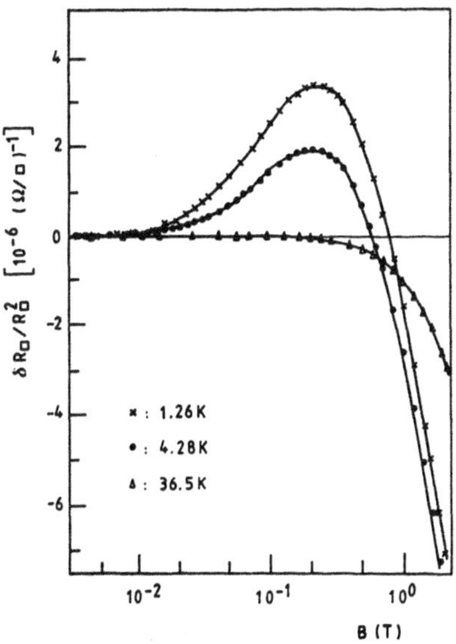

Fig. 2: Temperature-dependence of the normalized magnetoresistance for a Cu film with $R_\square(4.2\ K) = 234\ \Omega/\square$ and $d = 50$ Å. The full curves are calculated with eq. 2. The scattering times at 4.2 K are:

$\tau_{in} = 2.5 \times 10^{-11}$ s

$\tau_{so} = 0.45 \times 10^{-11}$ s

$\tau_s = 5 \times 10^{-11}$ s

Information about the temperature-dependence of τ_{in} is obtained from a plot of $B_\phi = B_{in} + 2B_s$ as a function of temperature (see Fig. 3). The full curve was calculated using the relation:

$$B_\phi(T) = 27 \times 10^{-4}\ T + 0.14 \times 10^{-4}\ T^3 + 101 \times 10^{-4} \qquad (9)$$

The temperature-independent part of $B_\phi(T)$ may be identified as the magnetic scattering term $2B_s$ in eq. 3. The presence of about 1 ppm residual magnetic impurities in the Cu film (most likely Cr and Fe) may explain this

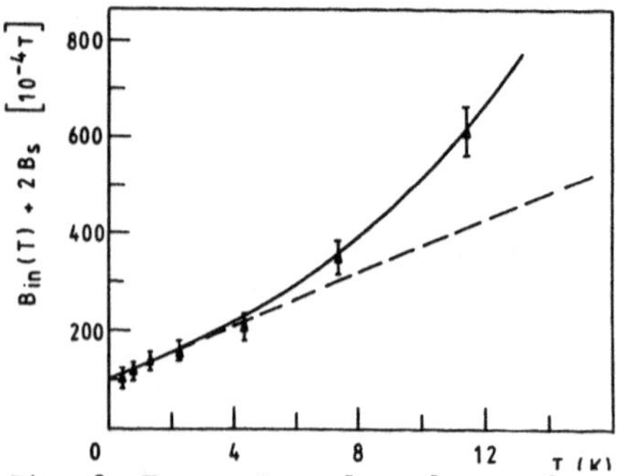

Fig. 3: Temperature dependence of the field $B_\phi(T) = B_{in}(T) + 2B$ for the Cu film shown in Fig. 2. The full curve is given by eq. 9; the dashed line assumes a linear temperature-dependence of $B_{in}(T)$

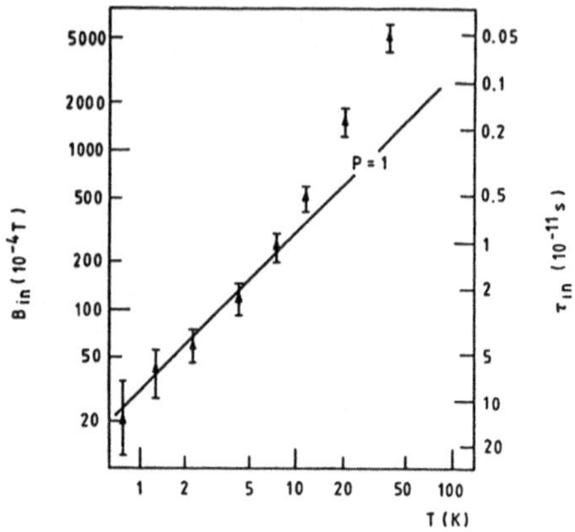

Fig. 4: Temperature dependence of the field $B_{in}(T)$ and the time τ_{in} for the Cu film shown in Fig. 2. The straight line is calculated with eq. 10.

contribution to the field B_ϕ. The influence of magnetic scattering upon the destruction of weak-localization will be discussed in more detail in Sect. 4.

From eq. 9 we may also conclude that below T = 4.2 K the inelastic scattering rate is a linear function of temperature. Such a behaviour is expected for two-dimensional electron-electron interaction in the presence of elastic impurity scattering [14]:

$$B_{in}(T) \propto \tau_{in}^{-1}(T) = AR_\square T \qquad (10)$$

where A is a constant which should be about $5 \times 10^6\ \Omega^{-1}K^{-1}$ for the Cu film. From the film thickness, the resistance ratio and the value for $R_\square(4.2\ K)$ we calculate that $l_{el}(d) = 6$ Å and $l_{el}(RR) = 13$ Å. With $v_F = 1{,}45 \times 10^6$ m/s and using eq. 4 we calculate that the linear term in eq. 9 should be in the region $20 \times 10^{-4}\ T < B_{in}(T) < 43 \times 10^{-4}\ T$. If the theoretical prediction (eq. 10) is indeed valid, our results suggest that the correct diffusion constant D is the microscopic value calculated from $l_{el}(RR)$. Substracting from eq. 9 the constant term due to the magnetic scattering, we obtain the temperature-dependence of the field B_{in} as shown in Fig. 4. Also shown is the temperature variation of τ_{in} assuming a diffusion constant $D = 5 \times 10^{-4}\ m^2/s$. Above $T \simeq 5$ K the classical three-dimensional electron-phonon scattering ($\tau_{in}^{-1} \propto T^3$) becomes evident (with respect to the phonons and due to the direct coupling between the Cu film and the glass substrate one is in the three-dimensional range even for a very thin film). It should be noted that the identification of the three scattering processes in eq. 9 is only possible if the MR is analyzed correctly and in a very broad temperature range (at least from 1K to 20 K). A similar analysis was performed on Al films and also showed the presence of both the linear and the cubic term [15]. When the analysis of the experimental MR is restricted to a smaller temperature range [5] or if the concentration of magnetic impurities is too large ($2B_s \gg B_{in}$) [16] the different contributions to $B_\phi(T)$ can not be obtained unambiguously.

To conclude this section, we present in Fig. 5 the $\tau_\phi(4.2)$-values obtained by different research groups using weak-localization experiments. Probably due to the uncertainty on the values of the diffusion constant,

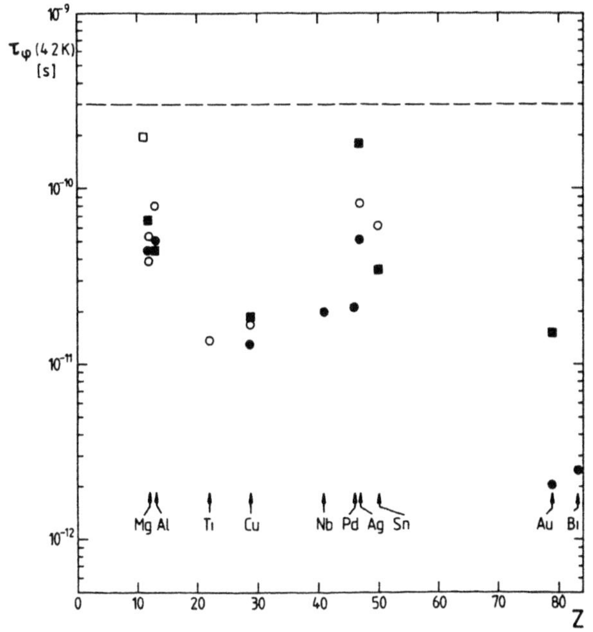

Fig. 5: Overview of the phase-breaking time τ_ϕ (4.2 K) reported by various research groups. The closed symbols refer to films with $R_\square \simeq 100\ \Omega/\square$; the open symbols to films with $R_\square \simeq 20\ \Omega/\square$. The distinction is made between films quench-condensed and measured in situ (squares) and films evaporated at room temperature (circles). The horizontal dashed line is related to the Aharonov-Bohm effect (see text). Reference numbers are coupled to the appropriate element as follows: Mg [6,8,17]; Al [18,19,20]; Ti [21,22]; Cu [5,7,23] ; Nb [24]; Pd [25]; Ag [5,23] ; Sn [7,20]; Au [5,26]; Bi [27].

the τ_ϕ-values for the films with $R_\square(4.2\ K) \simeq 20\ \Omega/\square$ (open symbols) are not always higher than for the films with $R_\square(4.2\ K) \simeq 100\ \Omega/\square$ (closed symbols) as would be expected from eq. 10. The fact that the values are comparable for films prepared at room temperature (circles) or quench-condensed at T = 4.2 K (squares), is an indication of the universality of two-dimensional weak-localization. The dashed horizontal line in Fig. 5 represents the minimum value for τ_ϕ(4.2 K) below which the famous Aharonov-Bohm effect induced by weak localization [28] will be difficult to observe in hollow metal cylinders with a radius of $\simeq 1\ \mu m$. The effect was observed in Mg [29, 30] and Li cylinders [31] for a film resistance $R_\square(4.2\ K) \simeq 1\ \Omega/\square$. The experimental data of Fig. 5 suggest that the Aharonov-Bohm effect should also be observable in cylindrical films of Ag. On the other hand, the effect will probably not be detectable in Cu or Ti cylinders, due to the important scattering by residual magnetic impurities.

4. Destruction of Localization by Magnetic Scattering

As indicated in the previous Section, an important part of the scattering rate τ_ϕ^{-1} below $T \simeq 5$ K may be due to the presence of residual magnetic impurities. In pure Cu films the magnetic scattering rate may even dominate when $R_\square < 10\ \Omega/\square$. Indeed, according to eq. 10 the inelastic scattering rate for such a low $R_\square$-film will be smaller than for the high $R_\square$-film shown in Fig. 2. Moreover, due to the increased film thickness, magnetic scattering centers will be less neutralized by oxygen, which diffuses into the Cu film. Figure 6 shows typical MR curves for a pure Cu film (starting material identical as for the $R_\square$ = 234 Ω/□ film) with $R_\square$(4.2 K) = 9.25 Ω/□ and d = 90 Å. At small fields (B $\lesssim$ 0.1 T) the MR is nearby temperature-independent, indicating that $\tau_{in}^{-1} \ll 2\tau_s^{-1}$ in eq. 1. The saturation around B = 1 T and the positive MR which appears at higher fields are due to the anisotropic scattering of the electrons on the Fermi surface. This normal positive MR is proportional to $(eB\tau_{el}/m)^2$ and therefore dominates at high fields the MR due to weak-localization. Moreover, it prevents the saturation of the MR which is predicted by eq. 2 for B ≃ B_{el}. The more pronounced temperature-dependence of the MR at B ≃ 1 T (see Fig. 6) is probably caused by the Kondo effect. Using the data of DAYBELL and STEYERT [32] we can explain the additional negative MR at lower temperatures if we assume the presence of about 1 ppm Cr in the Cu film. Doping the Cu films with about 1 at % Cr completely destroys the low field MR due to weak-localization. At higher fields the MR becomes isotropic, indicating that magnetic scattering dominates. Similar effects were also observed in CuMn alloys [33] and in Mg films covered with a Fe layer [34]. Using Kondo alloys (CuCr, AgMn, ...) with a small impurity content (about 100 ppm) it may be possible to obtain from weak-localization experiments valuable information about the anomalous Kondo scattering [35].

5. The Influence of Spin-Orbit Scattering

The low field positive MR in Cu-films (see Fig. 2) is a clear indication that spin-orbit scattering is important. According to eq. 2 the appearance of a positive MR is directly related to the ratio $B_{so}/B_\phi(T)$, and will be more pronounced when the temperature is lowered or when the magnetic impurity concentration is decreased. The intrinsic strength of the spin-orbit scattering rate is given by [36]:

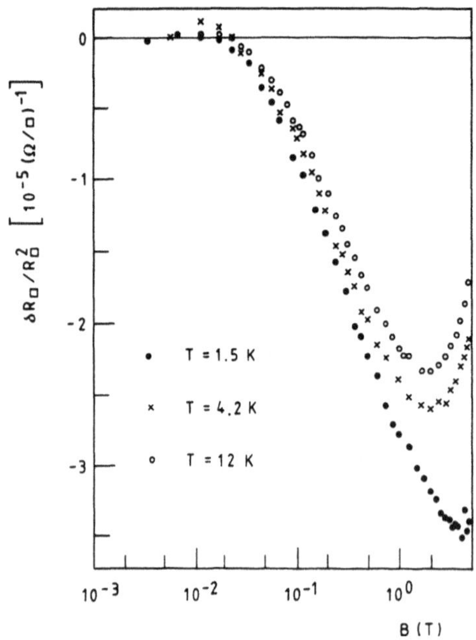

Fig. 6: Temperature-dependence of the normalized magnetoresistance for a Cu film with $R_\square$(4.2 K) = 9.25 Ω/□ and d = 90 Å. The scattering times at 4.2 K are:

τ_{in} = 6 x 10^{-11} s

τ_{so} = 0.5 x 10^{-11} s

τ_s = 3 x 10^{-11} s

$$\tau_{el}/\tau_{so} = (\alpha Z)^4 \qquad (11)$$

where $\alpha = 7.3 \times 10^{-3}$ is the fine structure constant and Z is the atomic number. Eq. 11 predicts an important spin-orbit interaction (pronounced low field positive MR) in heavy metal films. This was indeed experimentally observed in Au [5, 26] and Bi [37] film. On the other hand, the MR for quench-condensed and in situ-measured Mg films [17, 6] is only positive at $T \ll 4.2$ K. When the Mg films are, however, condensed at room temperature and exposed to air, the spin-orbit scattering is strongly enhanced (pronounced positive MR at $T \geq 4.2$ K). Fig. 7 a shows the temperature-dependence of the MR for a Mg film ($R_\square(4.2\ K) = 37.9\ \Omega/\square$; $d \simeq 160$ Å) evaporated in a vacuum of 7×10^{-5} Pa at room temperature [8]. The observed increase of the spin-orbit interaction supports the idea of MESERVEY and TEDROW [38] that in eq. 11 the diffuse elastic scattering at the oxidized Mg film surface is dominant. When the evaporation of Mg is performed in a reduced He atmosphere, the oxidation of the film surface is apparently reduced, resulting in a less pronounced positive MR as shown in Fig. 7b. These data were obtained for a Mg film with $R_\square(4.2\ K) = 26.1\ \Omega/\square$ and $d \simeq 150$ Å evaporated in a reduced He atmosphere of 2.1 Pa. If one tries to increase the spin-orbit scattering (and the $R_\square$) by evaporating the Mg film in a reduced oxygen atmosphere, the film becomes inhomogeneous (granular). Figure 7c shows the MR data for a Mg film with $R_\square(4.2\ K) = 271\ \Omega/\square$ and $d \simeq 140$ Å prepared in a partial oxygen pressure of 2.6×10^{-3} Pa. Since the characteristic scale of the inhomogeneities is comparable to the magnetic length L_B for 10^{-2} T < B < 1 T, the MR data can only be fitted when eq. 2 is multiplied by a correction factor $\alpha = 0{,}5$.

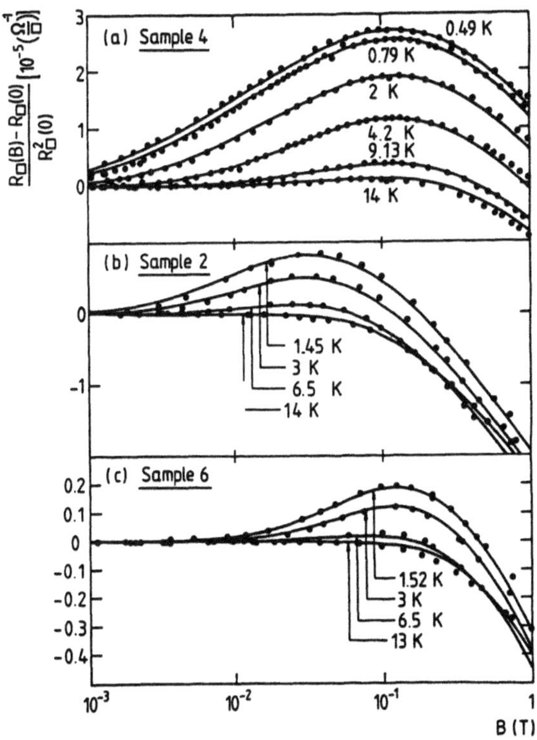

Fig. 7: Temperature-dependence of the normalized magneto-resistance for three Mg films condensed under different vacuum conditions (see text). The full curves are calculated using eq. 2. The scattering times at 2 K are:

(a) $\tau_{in} = 1.9 \times 10^{-10}$ s
$\tau_{so} = 4.2 \times 10^{-12}$ s

(b) $\tau_{in} = 1.1 \times 10^{-10}$ s
$\tau_{so} = 1.2 \times 10^{-11}$ s

(c) $\tau_{in} = 2.9 \times 10^{-11}$ s
$\tau_{so} = 2.9 \times 10^{-11}$ s

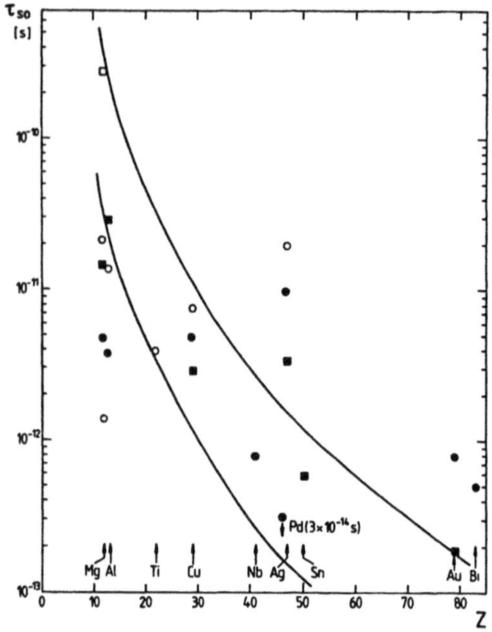

Fig. 8: Variation of the spin-orbit scattering time τ_{so} as a function of the atomic number Z. The full curves are calculated using eq. 11. The meaning of the different symbols, as well as the references coupled to the elements, are the same as in Fig. 5.

Finally, we present in Fig. 8 an overview of the published τ_{so}-values obtained from weak-localization experiments. Due to oxidation effects and the importance of surface scattering, the τ_{so} value for Mg and Al is longer in the films condensed at liquid He temperature (squares) than in the films evaporated at room temperature (circles). On the other hand, it is not clear that spin-orbit scattering is enhanced in the films with a higher $R_\square$ (see for instance Mg). The unknown contribution of the surface scattering and the uncertainty in the value of the diffusion constant D make a study of the proportionality between τ_{el} and τ_{so} (see eq. 11) very difficult. This may also explain why the dependence of τ_{so} on Z as shown by the full curves in Fig. 8 is not unambiguously confirmed by the experimental results. We also note the very small τ_{so}-value for Pd [25] which is probably related to the magnetic structure of this element.

6. Conclusions

A detailed analysis of the anomalous magnetoresistance in weakly localized metal films enables to obtain valuable information about the different scattering mechanisms in thin metal films. It is, however, well known that fitting of the magnetoresistance curves produces only values for the characteristic magnetic fields. Absolute values for the scattering times can therefore only be calculated in homogeneous films where the diffusion constant $D(l_{el})$ is well-defined.

An analysis of the different inelastic scattering mechanisms (electron-phonon, electron-electron) is possible when the magnetoresistance curves are carefully fitted in a very broad (0.5 K < T < 20 K) temperature region. In noble metal films the phase-breaking rate at low temperatures is dominated by magnetic scattering even for very low magnetic concentrations (≃ 1 ppm). Since the spin-orbit interaction is strongly modified by the surface properties, a systematic study of this interaction is only meaningful in well-defined and controlled film structures.

Acknowledgements

The authors would like to thank G. Deutscher for collaboration in various stages of the work. We acknowledge useful conversations with numerous colleagues, especially D.E. Prober, H. Fukuyama, and J.E. Mooij. This research is supported by the Belgian Interuniversitair Instituut voor Kernwetenschappen and one of us (M.G.) is a Research Fellow of the same Institute.

References

1. E. Abrahams, P.W. Anderson, D.C. Licciardello, T.V. Ramakrishnan: Phys. Rev. Lett. 42, 673 (1979)
2. G. Bergmann: Phys. Reports 107, 1 (1984)
3. B.L. Altshuler, A.G. Aronov and P.A. Lee: Phys. Rev. Lett. 44, 1288 (1980)
4. S. Hikami, A.I. Larkin and Y. Nagaoka: Prog. Theor. Phys. 63, 707 (1980)
5. G. Bergmann: Z. Phys. B 48, 5 (1982)
6. A.E.White, R.C. Dynes and J.P. Garno: Phys. Rev. B 29, 3684 (1984)
7. C. Van Haesendonck: Ph. D. Thesis, University of Leuven (1984)
8. M. Gijs, C. Van Haesendonck, Y. Bruynseraede and G. Deutscher: to be published
9. K.L. Chopra: Thin film phenomena (McGraw-Hill, New York, 1969)
10. J.E. Mooij: private communication
11. A.I. Larkin: Pis'ma Zh. Eksp. Teor. Fiz. 31, 239 (1980) (JETP Lett. 31, 219 (1980))
12. Y. Bruynseraede, M. Gijs, C. Van Haesendonck and G. Deutscher: Phys. Rev. Lett. 50, 277 (1983)
13. J.E. Mooij and T.M. Klapwijk: this volume
14. B.L. Altshuler, A.G. Aronov and D.E. Khmelnitzkii: J. Phys. G 15, 7367 (1982)
15. P. Santhanam, and D.E. Prober: Phys. Rev. B 29, 3733 (1984)
16. D. Abraham and R. Rosenbaum: Phys. Rev. B 27, 1413 (1983)
17. G. Bergmann: Phys. Rev. B 25, 2937 (1982)
18. J. Gordon, C.J. Lobb and M. Tinkham: Phys. Rev. B 28, 4046 (1983)
19. M.E. Gershenzon, V.N. Gubankov and Yu. E. Zhuravlev: Solid State Commun. 45, 87 (1983)
20. G. Bergmann: Phys. Rev. B 29, 6114 (1984)
21. A.C. Sacharoff and R.M. Westervelt: to be published
22. M.Gijs: unpublished

23. M.E. Gershenzon, V.N. Gubankov and Yu. E. Zhuravlev: Zh. Eksp. Teor. Fiz. 83, 2348 (1982) (Sov.Phys. JETP 56, 1362 (1982))

24. M.E. Gershenzon, V.N. Gubankov and Yu. E. Zhuravlev: Zh. Eksp. Teor. Fiz. 85, 287 (1983) (Sov. Phys. JETP 58, 167 (1983))

25. R.S. Markiewicz: private communication

26. T. Kawaguki and Y. Fujimori: J. Phys. Soc. Jpn. 52, 722 (1983)

27. F. Komori, S. Kobayashi and W.Sasaki: J. Phys. Soc. Jpn. 52, 4306 (1983)

28. B.L. Altshuler, A.G. Aronov and B.Z. Spivak: Pis'ma Zh. Eksp. Teor. Fiz. 33, 101 (1981) (JETP Lett. 33, 94 (1981))

29. D. Yu. Sharvin and Yu. V. Sharvin: Pis'ma Zh. Eksp. Teor. Fiz. 34, 285 (1981) (JETP Lett. 34, 272 (1981))

30. M. Gijs, C. Van Haesendonck and Y. Bruynseraede: Phys. Rev. Lett. 52, 2065 (1984)

31. B.L. Altshuler, A.G. Aronov, B.Z. Spivak, D. Yu Sharvin and Yu. V. Sharvin: Pis'ma Zh. Eksp. Teor. Fiz. 35, 476 (1982) (JETP Lett. 35, 588 (1982))

32. M.D. Daybell and W.A. Steyert: Phys. Rev. Lett. 20, 195 (1968)

33. F. Komori, S. Kobayashi and W. Sasaki: J. Phys. Soc. Jpn. 52, 4306 (1983)

34. G. Bergmann: Phys. Rev. Lett. 49, 162 (1982)

35. H.Suhl: Magnetism, vol. 5 (Academic, New York, 1973)

36. A.A. Abrikosov and L.P. Gorkov: Zh. Eksp. Teor. Fiz. 42, 1088 (1962) (Sov. Phys. JETP 15, 752 (1962))

37. P.M. Woerlee, G.C. Verkade and A.G.M. Jansen: J. Phys. C 16, 3011 (1983)

38. R. Meservey and P.M. Tedrow: Phys. Rev. Lett. 41, 805 (1978).

Inelastic Lifetime of Conduction Electrons as Determined from Non-Localization Methods

J.E. Mooij and T.M. Klapwijk

Department of Applied Physics, Delft University of Technology
Delft, The Netherlands

A review is given of methods to measure the inelastic scattering time of electrons in dirty metals, complementary to localization experiments. Most important are measurements of nonequilibrium superconductivity effects. In some cases the electron-phonon time can be determined separately. Values obtained for the scattering time in various superconducting materials are given.

1. Introduction

Inelastic scattering is of major importance in localization as it destroys the coherence of the wave function. Experimental results on localization provide information on the phase-breaking rate that is closely connected with inelastic scattering. As a cross-check on the interpretation and possibly also to explore the subtle relation between phase and energy scattering, it is interesting to compare the results with those obtained from different experimental methods. As we discuss non-localization methods we concentrate our attention on those that are relevant for the materials and the temperature-range of particular interest for this conference. Here, the elastic scattering rate strongly exceeds the inelastic one. Experiments on very pure single crystals, where inelastic scattering can be studied because it exceeds impurity scattering, will not be treated.

In metals, few properties depend in an essential way on the precise distribution of the electrons over the energy levels in a band. Superconductivity is exceptional in this respect, as all the main properties are directly coupled to the occupation of the quasiparticle states near the gap edge, and small deviations from the equilibrium distribution have important consequences. Nonequilibrium superconductivity is the common indication for a wide variety of effects that are all connected with a shifted quasiparticle distribution, and therefore limited by inelastic scattering. These effects have been studied for their own interest, and, with few exceptions, not as a means to obtain the scattering rate. We will discuss in particular those nonequilibrium phenomena that in our opinion are, by their experimental accessibility and their clear interpretation, suitable for a systematic study of energy relaxation rates. Unfortunately at low temperatures few quasiparticles are present in superconductors, and strong effects are seen mainly near the critical temperature. As also the theory has mostly been worked out unambiguously near T_c only, nonequilibrium superconductivity seems to be able to provide the scattering rate only at a single temperature, the critical one.

The main part of this paper is concerned with the nonequilibrium superconductivity methods. In addition, attention is paid to experiments in which, specifically, the electron-phonon scattering time is measured. A summary is given of results obtained with non-localization methods for inelastic times in dirty metals.

2. Inelastic Scattering Time from Nonequilibrium Superconductivity

2.1. Nonequilibrium Superconductivity

The properties of a superconductor depend very strongly on the occupation of its quasiparticle states. If $\xi_k = \hbar^2 k^2/2m - E_F$ is the free electron energy of a state with wave vector k, relative to the Fermi level, and Δ is the BCS gap, the quasiparticle energy in a superconductor is $E_k = (\Delta^2 + \xi_k^2)^{\frac{1}{2}}$. It will be assumed here, as is certainly most relevant for strong elastic scattering, that the occupation number f_k depends only on the magnitude and not the direction of $\vec{k}$. The value of Δ follows from the BCS gap equation

$$2/V = \sum_k (1-2f_k)/E_k \tag{1}$$

in which V is the attractive potential. In equilibrium $f_k = f^o(E_k) \equiv \{\exp(E_k/k_BT)+1\}^{-1}$, the Fermi function. As f_k depends on E_k and E_k on Δ, (1) has to be solved self-consistently. The direct relation between the quasiparticle distribution function and the superconducting properties is retained in a nonequilibrium state.

The charge of a quasiparticle in a superconductor is in general intermediate between that of an electron and a hole in the normal metal. It is given by $q_k = \xi_k/E_k$ in units of the electron charge. Unlike the normal state, the superconducting state allows for the possibility that the total quasiparticle charge is not zero, because overall electric neutrality can be maintained by a change in the charge of the condensate. As in equilibrium the occupation of the levels $+\xi_k$ and $-\xi_k$ is the same, the total quasiparticle charge per unit volume, in units of e

$$Q^* = \sum_k f_k q_k = \sum_k f_k \, \xi_k/E_k \tag{2}$$

is zero as long as $f_k = f^o$.

Nonequilibrium implies that the distribution function f_k is not equal to the Fermi function. We define $f^1 \equiv f_k - f^o$ where f^o depends on E_k only. The deviation f^1 can be decomposed into two modes: the energy mode or even mode for which $f^1(+\xi_k) = f^1(-\xi_k) = f^1(E_k)$ and the charge imbalance or odd mode where $f^1(+\xi_k) = -f^1(-\xi_k)$.

For the energy mode no charge effects occur. However, the gap changes in agreement with (1) as do other properties such as critical current and critical temperature. This mode can be induced by tunnel injection, by electromagnetic radiation (of various frequencies) or by phonons.

In the charge imbalance mode the gap is unchanged, but Q* is no longer equal to zero. A description can be given in terms of a two-fluid model with Cooper pairs and quasiparticles as the components. The electrochemical potential of the pairs μ_p is shifted with respect to the quasiparticle potential μ_n. A current from a normal metal into a superconductor will excite this mode near the N-S interface. Tunnel injection is another possibility. Charge imbalance occurs whenever there is a divergence of the supercurrent.

Inelastic scattering leads to relaxation of the induced nonequilibrium distribution. The effective relaxation time is usually considerably longer than the inelastic time, because not only the distribution over the levels but also the levels themselves change in the process. Only the states with energies between Δ and roughly 2Δ are determining the superconducting properties. Near T_c, this energy range is small with respect to the thermal energy kT. Here, the collective relaxation rates are considerably longer than, but roughly proportional to, the inelastic scattering rate in the normal state. A well-established theoretical framework is available near T_c for the calculation of these collective rates.

The inelastic scattering time for electrons at the Fermi surface at the critical temperature is called τ_E. Theory provides the relation between the collective relaxation rate and τ_E. It is usual to assume that τ_E is connected with electron-phonon scattering only. The consequences when inelastic scattering is dominated by electron-electron processes have not yet been considered. The same is true for recently studied departures from the usual electronic properties, like weak localization and dimensional effects.

General discussions of nonequilibrium superconductivity can be found in the proceedings of the 1980 Maratea summer school on Nonequilibrium superconductivity, phonons and Kapitza boundaries [1] and in the review article by TINKHAM [2].

2.2. Charge Imbalance

The effective relaxation time for charge imbalance near T_c is [3]

$$\tau_{Q*} = \frac{4k_BT_c}{\pi\Delta}\left(\frac{\hbar\tau_E}{2\Gamma}\right)^{\frac{1}{2}} \tag{3}$$

in which the pair breaking parameter Γ is given by:

$$\Gamma = \frac{\hbar}{\tau_s} + \frac{\hbar}{2\tau_E} + \frac{D}{2}\left\{4\hbar^{-1}m^2v_s^2 - \hbar\Delta^{-1}\frac{\partial^2\Delta}{\partial\vec{r}^2}\right\} \tag{4}$$

The last expression contains a term due to spin-flip scattering with a time τ_s, a term due to inelastic scattering and two contributions associated with elastic scattering: one from a superfluid velocity v_s and the other from anisotropy of the gap. When pair-breaking is dominated by inelastic scattering, $\Gamma \approx \hbar/2\tau_E$, (3) and (4) yield:

$$\tau_{Q*} = \frac{4k_BT_c}{\pi\Delta}\tau_E = 0.42\,\tau_E(1 - T/T_c)^{-\frac{1}{2}} \tag{5}$$

A thorough discussion of charge imbalance phenomena has been given by CLARKE [4]. The most important method to determine the inelastic scattering time with these effects makes use of a <u>tunneling injection</u> technique, as illustrated in Fig.1. Three overlapping films, separated by oxide barriers form two junctions. The middle film that serves as an electrode for both junctions, is the superconducting sample. Charge is injected by the injection junctions, of which the other electrode is either normal or superconducting, in which case the injection voltage must exceed the gap. The injection current I_{inj} generates charge density at a known rate, proportional to I_{inj}. Relaxation occurs at a rate Q^*/τ_{Q*} which leads to a stationary value of Q^*. In the sample volume, the difference between the electrochemical potentials of pairs and quasiparticles is proportional to $\tau_{Q*}I_{inj}$.

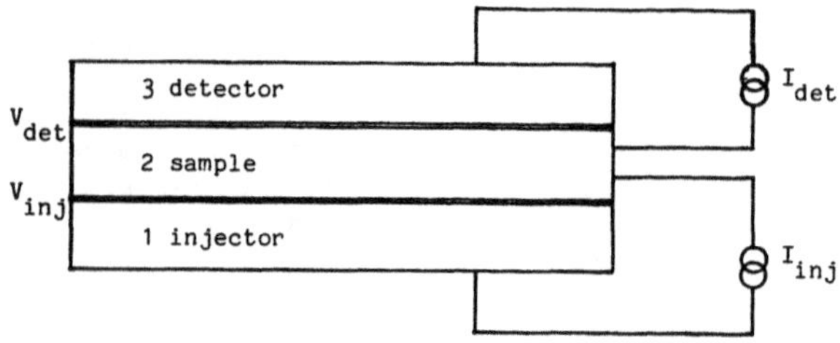

Figure 1: Geometry for tunneling injection. Films 1 and 2 are separated by a relatively low impedance tunneling barrier. They form the injection junction. The detection junction is formed by films 2 and 3, separated by a high impedance barrier. The area of the detection junction is smaller than that of the injector.

The second junction, with relatively high impedance, is used to measure $\mu_p - \mu_n$. The sample film extends beyond the volume in which injection is performed, μ_p is the same in all of the film and can be sensed with a contact away from the junction region. As long as no current passes in the detector junction, the detector electrode of normal metal will be at the quasiparticle potential μ_n. The voltage difference V_d across the detector junction at zero current is therefore equal to $(\mu_p - \mu_n)/e$. As a result it is found that:

$$\tau_{Q*}^{-1} = \frac{F*}{g_{NS}} \frac{1}{e^2 \Omega\ N(0)} \frac{I_{inj}}{V_d} \qquad (6)$$

where F* and g_{NS} are known correction functions, equal to 1 at T_c, Ω the sample volume and N(0) the single spin density of states at E_F. When I_{inj} and V_d are measured and N(0) is known, τ_{Q*} follows. This procedure can be followed at all temperatures. The detector voltage is typically in the (sub)nanovolt range. A SQUID is used as null instrument for the detector junction current.

The inelastic scattering time is obtained from the experimental results by fitting the temperature-dependent measured values of τ_{Q*} to a theoretical expression such as (5). Near T_c excellent agreement has been obtained, using one single value for τ_E. Theoretical predictions for temperatures significantly below T_c have been worked out numerically from the Boltzmann equation. The agreement between theory and experiment is considerably less satisfying in this temperature region.

As a method of determining the inelastic scattering time, the charge imbalance injection method is very reliable. The theory is well established near T_c. Experiments call for controlled fabrication and accurate measurements at a very low signal level. A disadvantage common to all methods based on charge imbalance effects is the possibility that other pair-breaking mechanisms such as spin-flip scattering or gap variation coupled with elastic scattering are significant. The experimental methods cannot distinguish between the terms. In particular at low temperatures, where the inelastic time is relatively large, these other terms may interfere.

A different experiment that yields τ_{Q*} studies phase slip centers in strips that are one-dimensional with respect to superconducting properties. A relatively high current is passed through the samples and, under certain conditions, it is found that the I-V characteristic exhibits a step-wise increase of the differential resistance. It has been shown [5] that each step corresponds with the formation of a phase-slip center, a localized dissipative region. A fast dynamic process in its middle generates charge imbalance that spreads by diffusion over a length $\Lambda_{Q*}=(D\tau_{Q*})^{\frac{1}{2}}$. The exponential decay of μ_n can be detected with the aid of small tunnel junction probes along the strip [6,7].

This method renders reliable results. One needs to know the diffusion constant to convert Λ_{Q*}. The uncertainty in the $\rho\ell$ product, used to determine the mean free path ℓ, is equivalent to the uncertainty in N(0), as used for determination of τ_{Q*} from tunnel injection. The fabrication of samples with probes is very complicated. For materials with high resistivity and for a short inelastic time, the diffusion length Λ_{Q*} gets too small to determine with probes of sizes around 1 μm.

The increase in differential resistance at the onset of a phase-slip center is equal to $2\Lambda_{Q*}\ R_n/L$ where R_n and L are the normal state resistance and the length of the whole strip. It would seem, therefore, that Λ_{Q*} can be determined much easier from the I-V characteristic. However, experience has shown that values of τ_{Q*} obtained in this way are significantly higher, due to heating effects. The voltage increases more with increasing current due to the strong temperature-dependence of the critical current, which leads to an increase of the differential resistance. At low temperatures heating becomes more important. The temperature in the phase-slip region may easily be several Kelvins higher than the bath temperature. CHAUDHARI et al.[8] have used I-V characteristics of phase-slip centers to determine the inelastic time in nanostrips of W-Re. In our opinion such measurements are certainly able to provide an order of magnitude for τ_{in}, but are not suited for a more precise study.

The diffusion length $\Lambda_{Q*} = (D\tau_{Q*})^{\frac{1}{2}}$ can also be measured as an additional resistance of a superconductor near an interface with a normal metal. Superconductor-normal metal-superconductor sandwiches without tunneling barriers have been used for this purpose [4]. Disadvantages are the large thickness of the sample film that is needed and the preparative complications that may occur at the interface.

2.3. Energy Mode

First we consider real time relaxation experiments. A short laser pulse is directed to one electrode of a tunnel junction. The photons break pairs, leading to an excess population of the quasiparticle states and a depression of the gap. The effective relaxation time for this effect is near T_c [3]:

$$\tau_\Delta = 1.2\ \tau_E(1 - T/T_c)^{\frac{1}{2}} \qquad (7)$$

Schuller and Gray [9] have measured the recovery of Δ in real time and derived values of τ_Δ at various temperatures. To determine the change of Δ, the tunnel junction is current-biased in such a way that the junction voltage is near the sum or the difference of the gaps of both electrodes. This voltage is detected. Fig. 2 gives an example. The results for the relaxation time show the temperature-dependence predicted by (7).

Translation of the measured τ_Δ to the inelastic time τ_E is more direct in the energy mode than in the charge imbalance mode. As long as the pair-breaking parameter Γ is small with respect to Δ, spin-flip

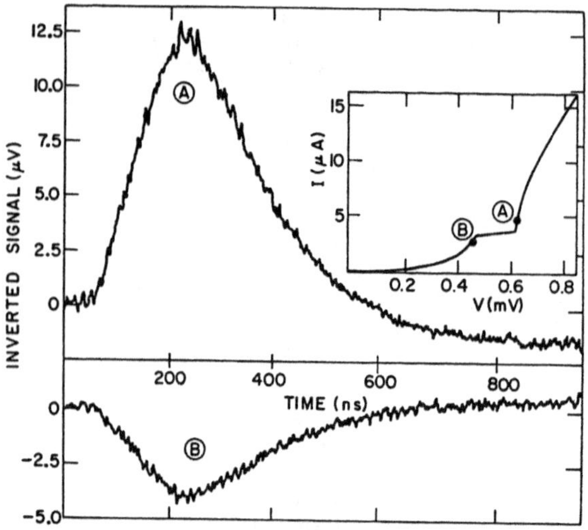

Figure 2: Real time relaxation [9]. Variation of voltage after applying the laser pulse, the junction being current-biased near the sum (A) or the difference (B) of the electrode gaps.

scattering and other factors are not important. As all effects are strongly temperature-dependent near T_c, the possible heating of electrons and phonons to an effective temperature higher than the bath temperature must always be kept in mind.

A different real time experiment studies the delay time between a short current pulse that exceeds the critical current and the onset of the resultant voltage. Under certain conditions this delay time yields τ_Δ. WOLTER et al.[10] have derived a value of τ_E from such measurements. However, the experiment is difficult to perform and it is also difficult to ensure that one is operating in the correct regime.

Stationary excitation of the energy mode can be realized in various ways. Tunneling injection is a very practical method. The geometry is the same as shown in Fig.1 for charge injection. Here, all three films are superconducting, usually the T_c of the detector electrode and sometimes that of the injector electrode is higher than the T_c of the sample film. Instead of the electrochemical potential, the gap in the sample film is measured with the detection junction. By injection or 'extraction' of quasiparticles by the injection junction, a nonequilibrium distribution of the quasiparticles over the energy levels is obtained. As follows from applying (1), the gap is either depressed or enhanced. Inelastic scattering tends to restore the equilibrium distribution and the stationary level of nonequilibrium is a measure of this inelastic rate.

A review of work with the double junction technique has been given by GRAY [11]. Of the experiments, those that aimed at obtaining enhancement of the gap are most relevant. Two extremes can be distinguished, depending on the T_c of the injection electrode. When the gap of the injector is higher than that of the sample and the injection junction is biased near the difference of the gaps, the number of quasiparticles passing from sample to injector is higher than in the other direction. The net result is an extraction of quasiparticles and a consequent enhancement of the gap in the sample. The effect has been predicted by PARMENTER [12] and studied experimentally by CHI and CLARKE [13]. When, on the other hand, the gaps of sample and injector are approximately equal, the number of quasiparticles remains the same, but a shift occurs to higher energies. As higher levels are less effective in blocking pair formation, the gap is increased. This type of enhancement is equivalent to stimulation by microwaves and phonons [14]. Experiments have been performed by GRAY [15] and VAN SON et al.[16].

For the theoretical interpretation of these injection experiments, Schön and TREMBLAY [17] have provided useful expressions of general validity in the temperature region near T_c. Their results indicate that for a bias voltage exceeding the sum of both gaps, terms cancel in such a way that no gap change is expected. In the practical experiment, the effect of heating (not considered in this theory) can therefore be separately determined by measurements at high voltages. In practice, only one fitting parameter remains in the comparison between theory and experiment, the inelastic scattering time.

Energy mode injection has received much less attention than charge imbalance injection as a means of measuring the inelastic scattering rate. Nevertheless, it has advantages in the weak dependence on spin-flip scattering and other pair breakers and in the easier measurement. Sample fabrication is of similar complexity.

A new method to obtain the inelastic time is based on the <u>minimum frequency for microwave enhancement</u>. As tunnel injection, a microwave field can lead to enhancement of the gap or the critical current [11]. Ignoring heating, one can write the gap equation for $\Delta \ll k_B T_c$ as:

$$1.06\left(\frac{\Delta}{k_B T_c}\right)^2 = \frac{T_c - T}{T_c} + \frac{\alpha_\omega}{4kT_c}\left\{\omega\tau_E G(\Delta/\hbar\omega) - 2\pi\right\} \qquad (8)$$

Here α_ω is a measure of the microwave power, G a weak function of $\Delta/\hbar\omega$ of order 1. Enhancement occurs if the term within brackets on the right hand side is positive. Equation (8) shows there is a minimum frequency for gap enhancement which depends directly on τ_E. The minimum frequency method has been developed by VAN SON et al.[18]. Their samples consist of a narrow strip, of which the critical current is measured. The theoretical expression for critical current enhancement is more complicated than for gap enhancement, but the principle is the same. Pair breaking by the supercurrent has to be taken into account. The limits in which theoretical predictions have been worked out so far are not valid for all materials.

Experimentally, the I_c depression at low frequencies and enhancement at higher frequencies are clearly observed and the minimum frequency can be measured accurately. In Fig. 3 the minimum frequency is shown for several samples as a function of temperature.

3. <u>Electron-Phonon Scattering Time</u>

Under certain conditions, the electron-phonon scattering rate can be measured directly or it can be derived from experimental results in a more indirect way. Obviously, when electron-electron scattering is strong this electron-phonon rate is different from the inelastic scattering rate.

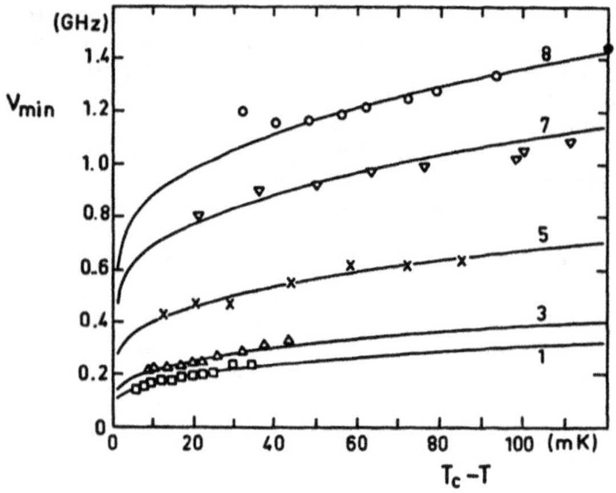

Figure 3: Minimum frequency for critical current enhancement in five Al strips [18]. The temperature-dependence follows the theory (drawn lines) with τ_E as the fitting parameter. τ_E varies from 13 ns for sample 1 to 0.8 ns for sample 8.

Simultaneous measurement of both inelastic and electron-phonon rates provides the interesting possibility of establishing the importance of electron-electron scattering.

Energy dissipated in the electron gas is transferred first to the metal phonons by electron-phonon scattering and next to the thermal bath by phonon transfer to substrate or liquid helium. In those cases where the first process is limiting, i.e.

$$\frac{c_v d}{\tau_{e-p}} \ll Y_k \qquad (9)$$

where c_v is the electronic specific heat, d the film thickness, τ_{e-p} the electron-phonon scattering time and Y_k the effective boundary conductance from the metal film to the surroundings, the effective temperature of the electrons increases with an amount

$$\delta T_{eff} = \frac{\tau_{e-p}}{c_v} P \qquad (10)$$

where P is the dissipation per unit volume. Certainly in the presence of strong electron-electron scattering such an effective temperature is a relevant quantity. The temperature rise is determined from a temperature-dependent property of the electron gas. The possibility of determining τ_{e-p} in this way has been pointed out by BERGMANN [19] in relation to the temperature-dependence of the resistance due to localization.

DESAILLY et al.[20] have measured τ_{e-p} by applying a short laser pulse to thin films. The real time behaviour provides the relaxation time τ_{e-p}. They studied thin Nb films in various regimes. In the dirtiest samples the temperature-dependence of the resistance was due to localization, in the cleaner samples to superconducting fluctuations. Taken together, ignoring their different resistivity levels, these samples exhibit a surprising T^5 or T^6 temperature-dependence of τ_{e-p}.

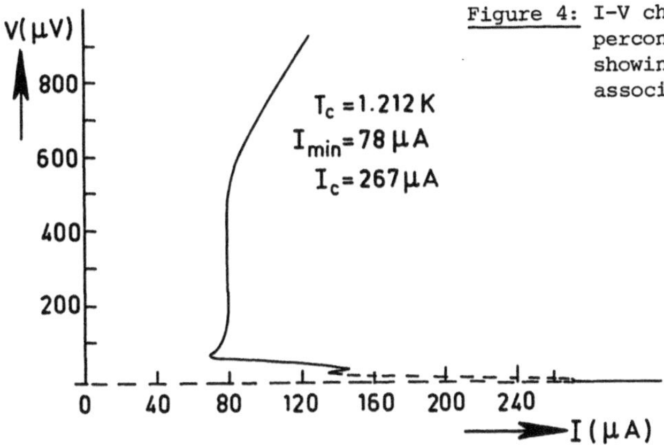

Figure 4: I-V characteristic of a superconducting Al strip, showing the current plateau associated with a 'hot spot'.

In narrow superconducting strips dissipation can lead to formation of 'hot spots', normal regions which increase in length with increasing voltage. The I-V characteristics show a plateau current, as illustrated in fig. 4. This plateau current is proportional to the square root of the effective heat transfer coefficient. If condition (9) holds this coefficient is equal to $c_v\tau_{e-p}^{-1}$. Discussions of these heating effects can be found in references 21 and 22. As with the methods based on nonequilibrium superconductivity, τ_{e-p} is only obtained in a narrow temperature region near T_c.

The interaction between electrons and phonons manifests itself not only in inelastic scattering of electrons, but also in scattering of phonons. The phonon mean free path is a measure of τ_{e-p}. CHI and CLARKE [24] give a relationship for this conversion.

4. Results Obtained

A large number of experimental studies of all various manifestations of nonequilibrium superconductivity have yielded a value of τ_E near T_c as a fitting parameter. Few have been performed with the explicit purpose of measuring τ_E. CHI and CLARKE [23] have used the charge imbalance injection technique to determine τ_{Q*} in a series of aluminum samples. By evaporation in an oxygen atmosphere, films were fabricated with different T_c and different resistivity. A clear correlation was observed between the increase of T_c, the decrease of ℓ and the decrease of τ_E. The thickness of Chi and Clarke's films was of the order of $(\hbar D/kT)^{\frac{1}{2}}$. VAN SON et al.[18] used the minimum frequency for critical current enhancement to measure τ_E in clean aluminum samples with decreasing thickness. A strong decrease of τ_E was found for the thinnest films. In these samples d was much smaller than $(\hbar D/kT)^{\frac{1}{2}}$. VAN DER PLAS et al.[24] extended the latter measurements to three-dimensional films. In the same samples, the electron-phonon scattering time was determined with the hot-spot method described in section 3. The difference between τ_E and τ_{e-p} increases with increasing resistivity.

From these results for aluminum films, it has become clear that the inelastic scattering rate is strongly dependent on resistivity and film thickness. Unless samples are much cleaner and thicker than is usual for films, it is not appropriate to think of one characteristic time for this material. The results are in line with the data obtained from localization. It seems that two categories can be distinguished: films where the thickness d is smaller then $(\hbar D/kT)^{\frac{1}{2}}$ in which the sheet resistance R_{sq}

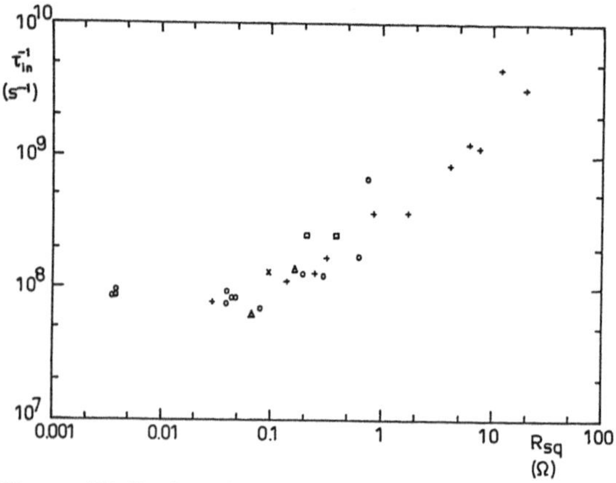

Figure 5: Inelastic scattering rates for Al films with thickness below $(\hbar D/kT)^{\frac{1}{2}}$, plotted versus sheet resistance. Data from:
- ✖ WOLTER et al.[10] current relaxation
- ◻ STUIVINGA et al.[7] spatial dependence of μ_n in phase-slip centers
- ○ CHI and CLARKE [23] tunnel injection of charge imbalance
- ▲ SCHULLER and GRAY [9] gap relaxation
- ✚ VAN SON et al.[18] minimum frequency for critical current enhancement

determines τ_E and thicker films where the resistivity is determining. For the latter category few data are available as yet.

Fig.5 plots the values of the inelastic scattering rate versus sheet resistance for samples where $d < (\hbar D/kT)^{\frac{1}{2}}$. We have included only data for films of which resistivity and thickness values are known. As the figure shows, a very satisfactory agreement exists on a factor of two level between the data, obtained with widely varying methods [25]. One should

Table 1. Inelastic scattering time τ_E of several superconducting elements at their critical temperature. Methods: PSC-IV phase-slip center I-V characteristics, N/S normal metal-superconductor interface, Q* tunneling injection of charge imbalance, PSC-μ_n spatial dependence of μ_n in phase-slip centers, τ_Δ real time current relaxation. Theoretical values are those for electron-phonon scattering given by KAPLAN et al.[26]. Experimental values in brackets are less reliable.

Material	T_c (K)	Method	Ref.	τ_E exp. (ps)	τ_{e-p} theor. (ps)
Nb	9.2	PSC-IV	27	(8)	18
Pb	7.2	N/S	28	25	23
Ta	4.4	N/S	29	89	210
Sn	3.7	N/S	28	260	270
		Q*	30	140	
		PSC-μ_n	6	120	
In	3.4	N/S	28	110	95
		τ_Δ	31	140	
Zn	0.9	PSC-IV	32	(100)	9×10^4

realize that the data points of fig.5 refer to different temperatures. The T_c for high sheet resistances is about 50% higher than for low sheet resistances.

For other materials only few data are available. No systematic study has been performed of inelastic times in films with different parameters. In Table 1 we give typical values of τ_E, obtained with different methods, for a number of superconducting elements. In the table we also include predictions for τ_{e-p} at T_c, as given by KAPLAN et al.[26]. These are based on tunneling data for the frequency-dependent electron-phonon interaction and theoretical values, indicating that electron-phonon relaxation dominates in most materials. Only for Zn a very significant deviation is found. A similar deviation is present for Al. For the materials where the electron-phonon interaction is weaker and T_c lower, such as Al and Zn, electron-electron scattering is much more important, in particular in thin and dirty samples.

5. Final Remarks

Nonequilibrium superconductivity can be used for accurate determination of τ_E. In practice, this is possible only at the critical temperature. At low temperatures, experimental signals are small and masked by other effects, moreover the theory is not developed sufficiently. Therefore, the temperature-dependence of inelastic scattering cannot be studied directly with these methods.

The theoretical interpretation of nonequilibrium effects is based on relaxation by electron-phonon processes. It has been shown, using the methods described, that electron-electron scattering in some systems strongly exceeds electron-phonon scattering. The consequences for the theory have not been studied yet.

The double junction technique, in which a stationary nonequilibrium is created by injection and studied with a detection junction, calls for well-controlled fabrication techniques. The theoretical interpretation is most direct. One needs to know the density of states to calculate τ_E. The sample film is always sandwiched between two other metal films. Both charge imbalance and gap variation can be used.

A number of other methods with limited applicability can be of value for particular materials. The minimum frequency for critical current enhancement is a good measure of the inelastic rate when τ_{in} is relatively large. The theory needs to be developed more fully. Real time measurements of the decay and recovery of the gap or the response to a supercritical current pulse require fast sampling techniques. From the I-V characteristics of phase-slip centers, only order of magnitude information can be derived unless heating is carefully considered.

From the available experimental data as yielded by nonequilibrium superconductivity, it follows that for most materials electron-phonon scattering is dominant near T_c. In Al and Zn electron-electron processes are much more prominent.

The phase-breaking time important for localization and the inelastic scattering time are not necessarily equal. An experimental contribution to the discussion on this point could be provided by a direct comparison between the times obtained in localization and nonequilibrium superconductivity in the same or very similar samples.

References

1. Nonequilibrium superconductivity, phonons and Kapitza boundaries, Proceedings NATO Advanced Study Institute, Maratea 1980, Ed. K.E. Gray, Plenum Press, New York 1981
2. M. Tinkham: Festkörperprobleme 19, 363(1979)
3. A Schmid: Chapter 14 of ref.1, page 423
4. J. Clarke: Chapter 13 of ref.1, page 353
5. W.J. Skocpol, M.R. Beasley, M. Tinkham: J. Low Temp.Phys. 16, 145 (1974)
6. G.J. Dolan and L.D. Jackel: Phys.Rev.Letters 39, 1628(1977)
7. M. Stuivinga, C.L.G. Ham, T.M. Klapwijk and J.E. Mooij: J. Low Temp.Phys.53, 627(1984)
8. P. Chaudhari, A.N. Broers, C.C. Chi, R. Laibowitz, E. Spiller and J. Viggiano: Phys.Rev.Letters 45, 930(1980)
9. I. Schuller and K.E. Gray: Phys.Rev.Letters 36, 429(1976) and Solid State Comm.23, 337(1977)
10. J. Wolter, P.M.T.M. van Attekum, R.E. Horstman and M.C.H.M. Wouters: Solid State Comm.40, 433(1981)
11. K.E. Gray: Chapter 5 of ref.1, page 131
12. R.H. Parmenter: Phys.Rev.Letters 7, 274(1961)
13. C.C. Chi and J. Clarke: Phys.Rev.B20, 4465(1979)
14. J.E. Mooij: Chapter 7 of ref.1, page 191
15. K.E. Gray: Solid State Comm.26, 633(1978)
16. P.C. van Son, J.E. Mooij and T.M. Klapwijk: Proceedings LT17 and to be published
17. G. Schön and A.M. Tremblay: Phys.Rev.Letters 42, 1086(1979)
18. P.C. van Son, J. Romijn, T.M. Klapwijk and J.E. Mooij: Phys.Rev.B29, 1503(1984)
19. G. Bergmann: Solid State Comm.46, 347(1983)
20. J. Desailly, B. Pannetier and J.-P. Maneval: Phys.Letters 99A, 125(1983) and J. Desailly, thesis Ecole Normale, Paris 1984 (unpublished)
21. W.J. Skocpol, M.R. Beasley and M. Tinkham: J.Appl.Phys.45, 4054(1974)
22. M. Stuivinga, T.M. Klapwijk, J.E. Mooij and A. Bezuijen: J.Low Temp.Phys.53, 673(1983)
23. C.C. Chi and J. Clarke: Phys.Rev.B19, 4495(1979)
24. P.A. van der Plas, T.M. Klapwijk, J.E. Mooij, M. Faas and J. Romijn: These Proceedings
25. H.J. Mamin, J. Clarke and D.J. Van Harlingen: Phys.Rev.Letters 51, 1480(1983) have found much higher scattering rates derived from spatial dependence of μ_n near a small injection junction. The discrepancy is not understood.
26. S.B. Kaplan, C.C. Chi, D.N. Langenberg, J.J. Chang and D.J. Scalapino: Phys.Rev.B14, 4854(1976)
27. R.B. Laibowitz, A.N. Broers, J.T.C. Yeh, and J.M. Viggiano: Appl.Phys. Letters 35, 891(1979)
28. T.Y. Hsiang and J. Clarke, Phys.Rev.B21, 945(1980)
29. M.L. Yu and J.E. Mercereau, Phys.Rev.B12, 4909(1975)
30. J. Clarke and J.L. Paterson, J.Low Temp.Phys.15, 491(1974)
31. D.J. Frank, M. Tinkham, A. Davidson and S.M. Faris, Phys.Rev.Letters 50, 1611(1983)
32. P.C. van Son, internal report Delft University of Technology 1981 (unpublished)

Comments of Some Aspects of the Localization and Interaction Problem

Elihu Abrahams

Serin Physics Laboratory, Rutgers University, Piscataway, NJ 08854, USA

This contribution has four sections:

1. Review of current status of localization with interactions
2. Conjectures about Phosphorous-doped Silicon
3. Comments about the inelastic scattering effect in localization
4. Magnetoresistance in disordered 2D superconductors

1. Localization with Interactions

In the absence of any main contribution to LITPIM on the subject of the scaling theory of localization with interactions included, it is worthwhile to review the current status.

We recall that when treating the electron-electron interaction perturbatively, logarithmic corrections to various quantities occur in the two-dimensional case and power-law corrections at d = 3. In particular, one finds a correction to the conductivity which is similar to that caused by localization. For a review of this aspect of the problem, see Fukuyama's LITPIM contribution.

An attempt at going beyond a perturbative treatment of interactions and achieving a scaling theory was made by FINKELSTEIN [1] who, neglecting the coherent backscattering terms which lead to ordinary Anderson localization [2], generalized the field theory formulation [3] of localization to include electron-electron interactions. His results were derived in an alternative formulation by CASTELLANI, DICASTRO, LEE and MA [4]. The surprising result obtained was that the conductivity scaled to large values as temperature T decreased to zero and there was no mobility edge in $2+\varepsilon$ dimensions.

An error was discovered in Refs. [1,4] by FINKELSTEIN [5] and independently by CASTELLANI et al [6]. With the correction, another surprising result was obtained. Now the scaling was to strong coupling so that the procedure broke down at a finite length scale. However, rather than scaling to a perfect conductor as previously, it appeared that the conductivity scaled to a finite value. Again, a mobility edge was not found.

In the new work of Refs. [5,6] a mobility edge is found in several special cases: 1) Strong impurity spin scattering or strong impurity spin-orbit scattering. 2) Strong magnetic field. In each of these cases, a universal behavior is found which is similar to that of ordinary localization [2].

2. Remarks about Si:P

One of the reasons for the great interest in the interactions problem has been the experimental results on Si:P, reviewed by Thomas at LITPIM. Here, the conductivity exponent t which enters

$$\sigma \propto (n-n_c)^t \qquad (1)$$

is close to 1/2 whereas in a variety of other three-dimensional systems it is close to 1. In (1) n is the impurity concentration, n_c its value at the mobility edge. Ordinary localization theory with coherent backscattering [2] predicts an exponent 1 ("orthogonal" case). If the backscattering is suppressed (by magnetic scattering, say) the exponent becomes 1/2 ("unitary" case). The orthogonal to unitary crossover is discussed at LITPIM by both Oppermann and Belitz.

For some time it was hoped that interaction effects would somehow explain the deviation of the measured Si:P exponent from the expected theoretical result for the orthogonal case, t = 1. Most recently, the discussion has been about how interactions might lead to the presence of localized slow electron spin fluctuations,which could then provide the spin-flip scattering mechanism which suppresses the coherent backscattering and leads to the unitary exponent t = 1/2. This is the mechanism of "slow spins" referred to by Anderson and Thomas in their LITPIM contributions. One argues that as the mobility edge is approached from the metallic side, the electron states are beginning to look localized. In that case,the Coulomb correlations are enhanced and (as in the Hubbard model) a short-range opposite spin interaction U plays the leading role. The situation is reminiscent of paramagnon theories [7] where the Stoner effect suppresses spin diffusion so that the spin diffusion constant becomes

$$D_s = D_0[1 - N_0U] \ , \qquad (2)$$

where the factor multiplying D_0, the zero-interaction diffusion constant, is just the Stoner factor. Some support for this scenario comes from the results of FINKELSTEIN [5] and CASTELLANI et al [6] which have D_s scaling to zero while charge diffusion D remains finite. This indicates the development of strong localized static spin fluctuations. In the "slow spins" picture,it is still not clear how the leading interaction contribution to the conductivity (described in the beginning oc Sec. 1), which gives exponent t = 1, is removed.

Part of the discussion about interaction effects and Si:P concerns the Brinkman-Rice mechanism [8] which has been referred to earlier at LITPIM (Anderson, Thomas). We may call it Gutzwiller-Brinkman-Rice (G-B-R). This refers to a variational treatment of the Hubbard model by G [9] which treats correlation more completely than the simplified Mott-Hubbard treatments. B-R [8] analyzed the solution of G [9] and found that as the Hubbard repulsion U increased,a metal-insulator (MI) transition occurs as $U > U_c$ where $m^*(E_F) \to \infty$. This heavy mass behaviour is reflected in the density of states, specific heat and spin susceptibility. Furthermore, states at the Fermi level become localized and slow localized spin fluctuations develop due to the density of states enhancement in the triplet particle-hole channel. This discussion is of course relevant [10] to the low-temperature behavior of liquid He^3 as well as to the properties of "heavy Fermion" metallic systems. In this context, we may ask if Si:P is a strongly correlated Mott-Hubbard or G-B-R system. If so, a new approach to the Si:P MI transition might be undertaken.

The possibility of strong Hubbard U being a decisive factor near the MI transition suggests that low-lying magnetic excitations may be important. Thus, future studies on strongly localized materials may well emphasize various magnetic and strong correlation effects. Already at LITPIM we have interesting contributions in the areas of spin fluctuation materials (Pd), Mott transition compounds (V_2O_3), Kondo systems (CuCr), ferromagnets (Fe).

3. Inelastic Scattering

The inelastic cutoff of the Cooperon propagator (maximally crossed graphs) is related to the electron inelastic lifetime; the former is measured in magnetoresistance experiments in the weak localization regime, especially in dimension two. Already in 1974, A. SCHMID [11] developed the formalism for treating the inelastic interactions of diffusing (rather than ballistic) electrons. In particular, his methods lead to the inelastic rate due to electron-electron interaction

$$1/\tau_{ee} \propto T^{d/2} \tag{3}$$

which is to be contrasted with the result for the pure case which leads to the replacement of d/2 by the dimension (d)-independent power 2.

The situation at d = 2 was examined in a diagrammatic analysis by ABRAHAMS et al [12]. They found, for the inelastic rate at the Fermi surface, a result similar to (3):

$$1/\tau_{ee}(d = 2) \propto T \ \ell n \ T_1/T \tag{4}$$

The scale T_1 was rather large so that the existence of the logarithmic correction was in principle experimentally verifiable. FUKUYAMA and ABRAHAMS [13] (FA) later pointed out that it was the Cooperon cutoff which is determined experimentally rather than the inelastic electron lifetime. They calculated this quantity for the zero wave-number Cooperon and found an expression identical to that of Ref. [12], namely (4).

Somewhat earlier, ALTSHULER, ARONOV and KHMEL'NITSKII [14](AAK) had calculated the effect of inelastic processes on the conductivity and found a temperature-independent logarithm for the phase-breaking rate, or effective Cooperon inelastic cutoff:

$$1/\tau_{\phi} \propto T \ \ell n \ E_F\tau \tag{5}$$

where τ is the elastic scattering time. Recently, FUKUYAMA [15] at LITPIM has pointed out that if the calculation of FA [13] is extended to the higher wave-number Cooperons which enter the conductivity, then the AAK [14] result of (5) is obtained. Although FA and AAK proceed by completely different methods, the physical processes involved are identical.

In recent years many experiments have been performed which are designed to distinguish between (4) and (5). The theorists appear now to agree on (5) for the purely 2D case. In metallic films [16] there are geometrical factors to be treated carefully and even in MOSFET structures any residual three-dimensional screening [16] will cooperate with finite wave-number effects to remove the $\ell n \ T_1/T$ of (4).

At dimension 1, the AAK prediction (as well as the revised FA one) is

$$1/\tau_{\phi} \propto T^{2/3} \ , \tag{6}$$

which should be measurable in an appropriate experimental setup.

4. Superconducting Fluctuations and Magnetoresistance

In the late 1960's, the effect of superconducting fluctuations on the conductivity of superconductors above T_c was discussed in dimensions 2 and 3 [17]. The two important contributions to the socalled paraconductivity were the Aslamazov-Larkin (AL) term (Cooper-pair fluctuations carry some current) and the Maki-Thompson (MT) term (superconducting fluctuations enhance the coupling of the external field to the current density). The magnetoresistance (MR) of the former was computed at that time [18]. Recently, Larkin considered the MR for the MT term [19]. For simplicity, he restricted his analysis to the regions

$$\ln(T/T_c) \gg 1/T\tau_\phi \tag{7}$$

$$4DeH/c < T \ln(T/T_c) \tag{8}$$

where τ_ϕ is the phase-breaking or Cooperon inelastic time as discussed in Sec. 3, D is the diffusion constant and H the magnetic field.

The restrictions (7), (8) on Larkin's MR formula have led to some confusion in the experimental literature. There is nothing fundamental about the limits, and they are easily removed by a somewhat more complicated calculation [20]. Limitation (7) arises from the approximation of neglecting a term of order $[T\tau_\phi \ln(T/T_c)]^{-1}$. Eq. (8) results from neglecting the influence of the magnetic field on the superconducting fluctuation. LOPES DOS SANTOS and ABRAHAMS [20] find by removing the limitations that the zero field MT paraconductivity of a thin film is given <u>for all T</u> by

$$\sigma_{MT}(T,H = 0) = (e^2/2\pi^2 d)\ \beta(T,\tau_\phi)\ \ln\left[\frac{\ln(T/T_c)}{\delta}\right] \tag{9}$$

Here, d is the film thickness and δ is the MT pair-breaking parameter $\delta = \pi/8T\tau_\phi$. The function $\beta(T, \tau_\phi)$ is a generalization of the $\beta(T)$ introduced by LARKIN [19]. Here we give only its form when $T \approx T_c$:

$$\beta(T, \tau_\phi) \approx (\pi^2/4)\left[\ln\frac{T}{T_c} - \delta\right]^{-1} \tag{10}$$

The second important result of Ref. [20] gives the MR of the MT term in regions of T and H not previously discussed. In particular, the restriction of (8) is replaced, <u>for all $T > T_c$</u> by

$$4DeH/c < \min\,[T \ln(T/T_c),\, T] \tag{11}$$

and the form of Larkin's result is recovered with the more general $\beta(T, \tau_\phi)$ discussed above. Finally, close to T_c, the MR has been found for fields exceeding (11) [20]. Of course, at high fields the superconducting fluctuations are quenched and the MR saturates:

$$[\sigma_{MT}(T,0) - \sigma_{MT}(T,H)]_{max} = \sigma_{MT}(T,0) \quad .$$

I would like to conclude by expressing the thanks of all the LITPIM participants to Bernhard Kramer and his colleagues G. Bergmann, Y. Bruynseraede, L. van Gerven and W. Gey for the wonderful organization of this conference which made possible many stimulating interchanges, both formal and informal, among theorists and experimentalists in the lively field of LITPIM. Thanks are due also to the charming and efficient hospitality of the Physikalisch-Technische Bundesanstalt.

References

1. A.M. Finkelstein: Zh. Eksp. Teor. Fiz. 84, 168 (1983) [Sov. Phys. JETP 57, 97 (1983)]
2. See for example, E. Abrahams, P.W. Anderson and T.V. Ramakrishnan: Phil. Mag. B 42, 827 (1980)
3. See for example, K.B. Efetov, A.I. Larkin and D.E. Khmel'nitskii: Zh. Eksp. Teor. Fiz. 79, 1120 (1980) [Sov.Phys.JETP 52, 568 (1980)]
4. C. Castellani, C. DiCastro, P.A. Lee and M. Ma: to be published
5. A.M. Finkelstein: Z. Phys. B 56, 189 (1984)
6. C. Castellani, C. DiCastro, P.A. Lee, M. Ma, S. Sorella and E. Tabet: to be published
7. W.F.Brinkman and S. Engelsberg: Phys. Rev. Lett. 21, 1187 (1968); P. Fulde and A. Luther: Phys.Rev. 170, 570 (1968); 175, 337 (1968)
8. W.F. Brinkman and T.M. Rice: Phys. Rev. B 2, 4302 (1970)
9. M.C. Gutzwiller: Phys. Rev. 137A, 1726 (1965)
10. D. Vollhardt: Rev. Mod. Phys. 56, 99 (1984)
11. A. Schmid: Z. Phys. 271, 251 (1974)
12. E. Abrahams, P.W. Anderson, P.A. Lee and T.V. Ramakrishnan: Phys. Rev. B 24, 6783 (1981)
13. H. Fukuyama and E. Abrahams: Phys. Rev. B 27, 5976 (1983)
14. B.L. Altshuler, A.G. Aronov and D.E. Khmel'nitskii: J.Phys.C 15, 7367 (1982)
15. H. Fukuyama: submitted to J. Phys. Soc. Japan
16. J.M.B. Lopes dos Santos and E. Abrahams: in preparation
17. For a review see W.J. Skocpol and M. Tinkham: Rep. Prog. Phys. 38, 1049 (1975)
18. K. Usadel: Z. Phys. 227, 268 (1969); E. Abrahams, R.E. Prange and M.J. Stephen: Physica 55, 230 (1971); M. Redi: Phys. Rev. B 16, 2027 (1977)
19. A.I. Larkin: Pis'ma Zh. Eksp. Teor. Fiz. 31, 239 (1980) [JETP Lett. 32, 219 (1980)]
20. J.M.B. Lopes dos Santos and E. Abrahams: submitted to Phys. Rev. B

Part II

Titles of Contributed Papers

As the contributed papers are published in full length in a supplementary volume (Proc. LITPIM Suppl., PTB-PG-1, PTB Braunschweig, 1984), they are given here by title only. The supplementary volume may be ordered from: E. Meyer, Physikalisch-Technische Bundesanstalt, Postfach 3345, D-3300 Braunschweig, F.R. Germany.

Localization in One Dimension

Search for Flux Periodicity in Quasi-One-Dimensional AuPd, AuGe, and Si MOSFET Rings
By W.J. Skocpol, L.D. Jackel, R.E. Howard, P.M. Mankiewich, R.E. Behringer, L.A. Fetter, D.M. Tennant

Strong Localization Effects in Ultra-Small Metallic Wires
By J.H. Claassen, S.A. Wolf, C. Leemann, J.H. Elliot, and R. Orbach

Chains of Random Impedances
By E. Akkermans and R. Maynard

Localization and Electron-Electron Interaction Effects in Thin Bi Wires
By D.E. Beutler and N. Giordano

The Statistics of One-Dimensional Resistances
By P.D. Kirkman and J.B. Pendry

Phase Resonances in the Transport Properties of a Disordered Solid
By C.J. Lambert

Theory of DC Conductance in a One-Dimensional System with Static Disorder
By D. Lenstra and W. van Haeringen

Numerical Simulation of the Level Repulsion and the Dynamical Conductivity of One-Dimensional Disordered Systems
By Tetsuro Saso

Low Temperature Electrical Conduction in Quasi One-Dimensional Inversion Layers
By P.H. Woerlee and G.A.M. Hurkx

On Transmission and Localization for Finite One-Dimensional Disordered Systems
By G. García-Calderón

Classical Waves in Random Media: Localization Effects
By Robert Johnston, Hervé Kunz, Cristina Flesia

Statistical Properties of Stark-Ladder Resonances in Disordered Chains
By Jorge V. Jose and G. Monsivais

Localization in One-Dimensional Correlated Potentials
By Robert Johnston and Bernhard Kramer

Exactly Soluble Quantities in Random Chains: Relation with Anomalies
By Th. M. Nieuwenhuizen

Weak Localization in Three Dimensions

Resistivity and Magnetoresistivity in Non-Superconducting Non-Magnetic Y-Al Metallic Glasses
By J.O. Strom-Olsen, M. Olivier, Z. Altounian, and R.W. Cochrane

Electrical Transport in Three-Dimensional Bismuth Films at Low Temperatures
By Yoji Koike, Masami Okamura, and Tetsuo Fukase

Weak Antilocalization, Orbital and Zeeman Interaction Effects in Two-and Three-Dimensional Bismuth Films
By D.S. McLachlan and H. White

Anisotropic Temperature-Dependent Resistivity of Sn-In Alloys
By K. Mori and Y. Saito

Quantum Corrections to the Boltzmann Conductivity and Electron Transport in Metallic Glasses
By Davor Pavuna

A Transfer Matrix Approach to Localization in 3D
By J.B. Pendry

A Monte-Carlo Study of Localization and Coherence in Two-Dimensional Systems with Weak Gaussian Disorder
By Michael Schreiber

Inelastic and Spin-orbit Scattering in Two- and Three-Dimensional Aluminum
By K.C. Mui, P. Lindenfeld, and W.L. McLean

Influence of Eléctron-Electron Collisions on Weak Localization
By Walter Eiler

Anomalous Transport Properties in Nb_3Ge Samples Irradiated by Electrons or Heavy Ions
By F. Rullier-Albenque, L. Zuppiroli, and F. Weiss

Localization and Interaction in Metallic Glasses
By D. Greig and M.A. Howson

Localization and Interaction Effects in Metallic Glasses
By J.B. Bieri, A. Fert, and G. Creuzet

Flux Quantization in Normal Rings

Magnetic Flux Quantization in a 2D Normal Metal Network
By B. Pannetier, J. Chaussy, and P. Gandit

Quantum Oscillations in Small Normal Metal Rings
By M. Büttiker

Quantum Oscillations and Electron Localization in Normal Metal Cylinders
By M. Gijs, C. Van Haesendonck, and Y. Bruynseraede

Quantum Coherence Effects in Submicron Al Cylinders
By James M. Gordon, C.J. Lobb, and M. Tinkham

Two-Dimensional Systems

Weak Localization and Fluctuation Electron-Electron Interaction in Disordered Indium Films
By B.I. Belevtsev, Yu. F. Komnik, and A.V. Fomin

2D-Conductivity of Thin Pd-Films with Different Fe-Impurity Concentrations
By G. Dumpich, H. Kristen, and E.F. Wassermann

Resistive Minimum in Thin Fe Films
By W.W. Fuller, M. Rubinstein, G.A. Prinz, and S.A. Wolf

Electron Localization and Interaction in Thin Mg Films
By M. Gijs, C. Van Haesendonck, Y. Bruynseraede, and G. Deutscher

Localization and Size Effects in Single Crystal Gold Films
By H.-U. Habermeier and P. Chaudhari

Dimensional Crossover of Localization and Interaction Effects in Thick Films
By Fumio Komori, Satoshi Okuma, and Shun-ichi Kobayashi

Localization in Ultrathin Niobium Films
By C. Leemann, S.A. Wolf, J. Elliott, and R.L. Orbach

2D Localization and Interaction Effects in Pd, PdH_x and Pd-Ni Thin Films
By H. Raffy, P. Nêdellec, L. Dumoulin, D.S. MacLachlan, and J.P. Burger

Parallel Magnetoconductance Measurements
By Ralph Rosenbaum

Effect of Finite Range Impurity Potential on Localisation
By Kurt Scharnberg

The Observation and Interpretation of Large Localisation Corrections to the Conductivity of 2D Silicon MOSFETs
By R.P. Upstone and M. Pepper

Dependence of the Phase-Coherence Time in Weak Localization on Electronic Mean Free Path and Film Thickness
By R.P. Peters and Gerd Bergmann

Localization, Magnetism and Superconductivity

Localized Magnetic Excitations in Metallic V_2O_3:
Electron Specific-Heat Enhancement and T^2-Resistance
By P. Hertel and J. Appel

Weak Localization and Superconductivity in Quench Condensed MgSi Thin Films
By J. Kästner and R. Mölleken

Localization and Interaction Effects in Amorphous Nb_3Ge Films
By P.H. Kes, C.C. Chi, and C.C. Tsuei

Percolation, Localization and Superconductivity in Si-Al Films
By S.P. McAlister and A.D. Inglis

Correlation of Superconducting and Localization Parameters in Metal Films: Pd/Si
By R.S. Markiewicz, C.J. Rollins, and J.S. Brooks

Magnetic Susceptibility and Specific Heat of $1T\text{-}Ta_{1-x}Ti_xS_2$ in the Anderson Localized States
By Y. Nishio, N. Kobayashi, and Y. Muto

Localization and Interaction Effects in Titanium Films
By A.C. Sacharoff and R.M. Westervelt

Influence of Giant Spin-Splitting on Magnetoresistance in Weakly Localized Regime in $n\text{-}Cd_{1-x}Mn_xSe$
By M. Sawicki, T. Dietl, and J. Kossut

Microwave Absorption in Granular Superconducting Al Films
By E. Stocker, J. Buttet

Magnetic Scattering and Weak Localization in CuCr Films
By C. Van Haesendonck and Y. Bruynseraede

Low-Temperature Properties of Mo/Ni Superlattices
By Ctirad Uher, Roy Clarke, Guo-Guang Zheng, and Ivan Schuller

Quantum-to-Classical Crossover in Granular Superconductor
By T.K. Kopec

Spikes in the Orbital Magnetic Susceptibility of a 2D Electron Gas with Disorder and Interaction
By I.D. Vagner and Tsofar Maniv

N Orbital Models
By R. Oppermann

Metal-Insulator Transition

Electron Diffusion and Localization in a Narrow Band
By D. Belitz and W. Götze

Magnetic Field-Induced Metal-Insulator Transition in InP: New Results at Very Low Temperatures
By A. Briggs, G. Remenyi, G. Biskupski, H. Dubois, J.L. Wojkewicz

Metallic Conductivity in Compensated InP near the MNM Transition
By D.M. Finlayson and I.F. Mohammad

Quantum Effects in Two-Dimensional Films of Superconductors at $T > T_c$
By M.E. Gershenson, V.N. Gubankov, Yu. E. Zhuravlev

Quasitransition Phenomena in the Effective Medium Framework
By W. Götze and J. Metzger

Screening and Localization in Disordered Electron Systems
By A. Gold and W. Götze

Dynamical Approach to the Anderson-Mott Transition
By Reiner Kree

The Percolation Transition of the Classical Lorentz Model
By E. Leutheusser

A Model for the Hybridization Effect in Disordered Systems
By D. Mayou, A. Pasturel, D. Manh Nguyen, and F. Cyrot-Lackmann

Carrier Concentration Dependence of Physical Parameters of the Anderson Localization in Metallic n-GaAs and n-InSb.
By Seizo Morita, Nobuo Mikoshiba, Yoji Koike, Tetsuo Fukase, Michiharu Kitagawa, and Shuichi Ishida

Phonon-Induced Delocalization in Amorphous Semiconductors
By P. Fenz, H. Müller, P. Thomas

The Thouless Formula as a Test for Anderson Localization
By J. Canisiús, J.L. van Hemmen, and Th.M. Nieuwenhuizen

Theory of Phonon-Controlled Conductivity in High-Resistivity Conductors: The Influence of Coulomb Interactions
By Walter Schirmacher

Self-Consistent Solution of Localization Problem in Real Lattices
By Vipin Srivastava

Interplay Between Superconductivity and the Metal-Insulator Transition in Sn_xO_y
By C. Van Haesendonck and Y. Bruynseraede

Electrical Resistivity of Organic Metals and A-15 Intermetallic Compounds with a Small Mean Free Path
By M. Weger, O. Entin Wohlman, H. Gutfreund, and M. Kaveh

Localization and Interaction in the Magnetic Field

Influence of a Weak Magnetic Field on Electron Localization
By D. Belitz

Anderson Localization in a Magnetic Field
By S. Hikami

Quantised Hall Effect of Electrons on a Disordered Lattice
By Hideo Aoki

Simple Disorder Models for Two-Dimensional Electronic Transport in Strong Magnetic Fields
By Gerd Czycholl

N-Orbital Model for a Two-Dimensional Disordered System in a Strong Magnetic Field
By Thomas S.J. Streit

Frequency Produced Fractional Quantisation in Si and GaAs
By A.P. Long, C. McFadden, H.W. Myron, M. Pepper, D. Andrews, and G.J. Davies

Localization and Scaling in the Magnetic Field
By A. MacKinnon, B. Kramer, and L. Schweitzer

A Model for the Quantum Hall Effect
By L. Schweitzer, B. Kramer, and A. MacKinnon

Inelastic Scattering Times

The Effects of Inelastic Processes on Quantum Interference in Small Systems
By Y. Gefen and G. Schön

Localization, Interaction and Superconductivity in 3D Crystalline Copper Alloys
By W. Eschner, W. Gey, and P. Warnecke

Study of the Electron-Electron Scattering Rate in the Noble and Transition Metals
By V.A. Gasparov, P.-A. Probst, R. Stubi, and R. Huguenin

Electron-Electron and Electron-Phonon Inelastic Times in GaAs in the Weak Localization Regime
By H. Godfrin, P. Averbuch, M. Laviron

Inelastic Scattering Time in Disordered Metals
By Yoshimasa Isawa

Inelastic Scattering Time in Metallic n-GaAs
By Seizo Morita, Nobuo Mikoshiba, Yoji Koike, Tetsuo Fukase, Michiharu Kitagawa, and Shuichi Ishida

Inelastic Scattering Times in Superconducting Aluminum Films
By P.A. van der Plas, T.M. Klapwijk, J.E. Mooij, M. Faas, and J. Romijn

Destruction of Phase-Coherence by Electron-Phonon Interaction in Weakly Disordered Conductors
By Jørgen Rammer and Albert Schmid

Localization Effects in Thin Potassium Wires
By Mary Eileen Farrell, Marilyn F. Bishop, and N. Kumar

Conference Photograph

Index of Contributors

Springer Series in Solid-State Sciences

Editors: **M. Cardona, P. Fulde, H.-J. Queisser**

A selection:

Volume 51
Phonon Scattering in Condensed Matter
Proceedings of the Fourth International Conference, University of Stuttgart, Federal Republic of Germany, August 22–26, 1983
Editors: **W. Eisenmenger, K. Laßmann, S. Döttinger**
1984. 321 figures. XIV, 472 pages. ISBN 3-540-12954-5

Volume 53
Two-Dimensional Systems, Heterostructures, and Superlattices
Proceedings of the International Winter School Mauterndorf, Austria, February 26–March 2, 1984
Editors: **G. Bauer, F. Kuchar, H. Heinrich**
1984. 231 figures. IX, 293 pages. ISBN 3-540-13584-7

Volume 45
B. I. Shklovskii, A. L. Efros
Electronic Properties of Doped Semiconductors
Translated from the Russian by S. Luryi
1984. 106 figures. XII, 388 pages. ISBN 3-540-12995-2
Originally published in Russian by Nauka Publishing House, Moscow 1979

Volume 40
K. Seeger
Semiconductor Physics
An Introduction
2nd corrected and updated edition. 1982. 288 figures. XII, 462 pages
ISBN 3-540-11421-1

Volume 39
Anderson Localization
Proceedings of the Fourth Taniguchi International Symposium, Kyoto, Japan, November 3–8, 1981
Editors: **Y. Nagaoka, H. Fukuyama**
1982. 98 figures. XII, 225 pages. ISBN 3-540-11518-8

Volume 28
The Structure and Properties of Matter
Editor: **T. Matsubara**
With contributions by T. Matsubara, H. Matsuda, T. Murao, T. Tsuneto, F. Yonezawa
1982. 229 figures. XI, 446 pages. ISBN 3-540-11098-4

Volume 11
B. R. Nag
Electron Transport in Compound Semiconductors
1980. 148 figures, 15 tables. XVI, 461 pages. ISBN 3-540-09845-3